CW01359970

Developmental Social Cognitive Neuroscience

The Jean Piaget Symposium Series
Series Editor:
Ellin Scholnick
University of Maryland
Available from Psychology Press / Taylor and Francis

Overton, W.F. (Ed.): The Relationship Between Social and Cognitive Development.

Liben, L.S. (Ed): Piaget and the Foundations of Knowledge.

Scholnick, E.K. (Ed): New Trends in Conceptual Representations: Challenges to Piaget's Theory?

Niemark, E.D., DeLisi, R. & Newman, J.L. (Eds.): Moderators of Competence.

Bearison, D.J. & Zimiles, H. (Eds.): Thought and Emotion: Developmental Perspectives.

Liben, L.S. (Ed.): Development and Learning: Conflict or Congruence?

Forman, G. & Pufall, P.B. (Eds.): Constructivism in the Computer Age.

Overton, W.F. (Ed.): Reasoning, Necessity, and Logic: Developmental Perspectives.

Keating, D.P. & Rosen, H. (Eds.): Constructivist Perspectives on Developmental Psychopathology and Atypical Development.

Carey, S. & Gelman, R. (Eds.): The Epigenesis of Mind: Essays on Biology and Cognition.

Beilin, H. & Pufall, P. (Eds.): Piaget's Theory: Prospects and Possibilities.

Wozniak, R.H. & Fisher, K.W. (Eds.): Development in Context: Acting and Thinking in Specific Environments.

Overton, W.F. & Palermo, D.S. (Eds.): The Nature and Ontogenesis of Meaning.

Noam, G.G. & Fischer, K.W. (Eds.): Development and Vulnerability in Close Relationships.

Reed, E.S., Turiel, E. & Brown, T. (Eds.): Values and Knowledge.

Amsel, E. & Renninger, K.A. (Eds.): Change and Development: Issues of Theory, Method, and Application.

Langer, J. & Killen, M. (Eds.): Piaget, Evolution, and Development.

Scholnick, E., Nelson, K., Gelman, S.A. & Miller, P.H. (Eds.): Conceptual Development: Piaget's Legacy.

Nucci, L.P., Saxe, G.B. & Turiel, E. (Eds.): Culture, Thought, and Development.

Amsel, E. & Byren, J.P. (Eds.): Language, Literacy, and Cognitive Development: The Development and Consequences of Symbolic Communication.

Brown, T. & Smith, L. (Eds.): Reductionism and the Development of Knowledge.

Lightfoot, C., LaLonde, C. & Chandler, M. (Eds.): Changing Conceptions of Psychological Life.

Parker, J., Langer, J. & Milbrath, C. (Eds.): Biology and Knowledge Revisited: From Neurogenesis to Psychogenesis.

Goncu, A. & Gaskins, S. (Eds.): Play and Development: Evolutionary, Sociocultural, and Functional Perspectives.

Overton, W., Mueller, U. & Newman, J. (Eds.): Developmental Perspectives on Embodiment and Consciousness.

Wainryb, C., Turiel, E. & Smetana, J. (Eds.): Social Development, Social Inequalities, and Social Justice.

Muller, U., Carpendale, J., Budwig, N. & Sokol, B. (Eds.): Social Life and Social Knowledge: Toward a Process Account of Development.

Zelazo, P. D., Chandler, M. & Crone, E. (Eds.): Developmental Social Cognitive Neuroscience.

Developmental Social Cognitive Neuroscience

Philip David Zelazo
University of Minnesota

Michael Chandler
University of British Columbia

Eveline Crone
University of Leiden

Psychology Press
Taylor & Francis Group

New York London

Psychology Press
Taylor & Francis Group
270 Madison Avenue
New York, NY 10016

Psychology Press
Taylor & Francis Group
27 Church Road
Hove, East Sussex BN3 2FA

© 2010 by Taylor and Francis Group, LLC
Psychology Press is an imprint of Taylor & Francis Group, an Informa business

Printed in the United States of America on acid-free paper
10 9 8 7 6 5 4 3 2 1

International Standard Book Number: 978-1-84169-767-3 (Hardback)

For permission to photocopy or use material electronically from this work, please access www.copyright.com (http://www.copyright.com/) or contact the Copyright Clearance Center, Inc. (CCC), 222 Rosewood Drive, Danvers, MA 01923, 978-750-8400. CCC is a not-for-profit organization that provides licenses and registration for a variety of users. For organizations that have been granted a photocopy license by the CCC, a separate system of payment has been arranged.

Trademark Notice: Product or corporate names may be trademarks or registered trademarks, and are used only for identification and explanation without intent to infringe.

Library of Congress Cataloging-in-Publication Data

Developmental social cognitive neuroscience / editors, Philip David Zelazo, Michael Chandler, Eveline Crone.
    p. cm. -- (The Jean Piaget Symposium series)
  Includes bibliographical references and index.
  ISBN 978-1-84169-767-3 (hardcover : alk. paper)
   1. Cognitive psychology--Congresses. 2. Cognitive neuroscience--Congresses. 3. Developmental psychology--Congresses. I. Zelazo, Philip David. II. Chandler, Michael J. III. Crone, Eveline.

BF201.D48 2010
155.4'13--dc22   2009027140

Visit the Taylor & Francis Web site at
http://www.taylorandfrancis.com

and the Psychology Press Web site at
http://www.psypress.com

CONTENTS

Contributors		ix
Preface		xiii

PART I Introduction

Chapter 1 The Birth and Early Development of a New Discipline: Developmental Social Cognitive Neuroscience   3
PHILIP DAVID ZELAZO, MICHAEL CHANDLER, AND EVELINE A. CRONE

PART II The Typical and Atypical Development of Social Cognition in Childhood

Chapter 2 Motor Cognition: Role of the Motor System in the Phylogeny and Ontogeny of Social Cognition and Its Relevance for the Understanding of Autism   13
VITTORIO GALLESE AND MAGALI ROCHAT

Chapter 3 The Construction of Commonsense Psychology in Infancy   43
CHRIS MOORE AND JOHN BARRESI

Chapter 4 Theory of Mind and Executive Functioning: A Developmental Neuropsychological Approach   63
JEANNETTE E. BENSON AND MARK A. SABBAGH

Chapter 5	The Development of Iterative Reprocessing: Implications for Affect and Its Regulation WILLIAM A. CUNNINGHAM AND PHILIP DAVID ZELAZO	81
Chapter 6	Brain Mechanisms in the Typical and Atypical Development of Social Cognition SUSAN B. PERLMAN, BRENT C. VANDER WYK, AND KEVIN A. PELPHREY	99
Chapter 7	Autism and the Empathizing–Systemizing (E-S) Theory SIMON BARON-COHEN	125

PART III Social Cognition in Adolescence

Chapter 8	The Neural Foundations of Evaluative Self-Knowledge in Middle Childhood, Early Adolescence, and Adulthood JENNIFER H. PFEIFER, MIRELLA DAPRETTO, AND MATTHEW D. LIEBERMAN	141
Chapter 9	Neurodevelopment Underlying Adolescent Behavior: A Neurobiological Model MONIQUE ERNST AND MICHAEL HARDIN	165
Chapter 10	The Terrible Twelves ABIGAIL A. BAIRD	191
Chapter 11	Paradoxes in Adolescent Risk Taking LINDA VAN LEIJENHORST AND EVELINE A. CRONE	209
Chapter 12	Between Neurons and Neighborhoods: Innovative Methods to Assess the Development and Depth of Adolescent Social Awareness ROBERT L. SELMAN AND LUBA FALK FEIGENBERG	227

PART IV The Developmental Social Cognitive Neuroscience of Moral Reasoning

Chapter 13	Crucial Developmental Role of Prefrontal Cortical Systems in Social Cognition and Moral Maturation: Evidence From Early Prefrontal Lesions and fMRI PAUL J. ESLINGER AND MELISSA ROBINSON-LONG	251

Chapter 14	Contributions of Neuroscience to the Understanding of Moral Reasoning and Its Development R. JAMES BLAIR	269
Chapter 15	Is a Neuroscience of Morality Possible? JEREMY CARPENDALE, BRYAN W. SOKOL, AND ULRICH MÜLLER	289
Chapter 16	The Relevance of Moral Epistemology and Psychology for Neuroscience ELLIOT TURIEL	313

Author Index	333
Subject Index	351

CONTRIBUTORS

Abigail A. Baird—Vassar College, Department of Psychology, Poughkeepsie, New York

Simon Baron-Cohen—Autism Research Centre, Cambridge University, Cambridge, United Kingdom

John Barresi—Department of Psychology, Dalhousie University, Halifax, Nova Scotia

Jeannette E. Benson—Department of Psychology, Queen's University, Ontario, Canada

R. James Blair—National Institutes of Health, Washington, DC

Jeremy Carpendale—Department of Psychology, Simon Fraser University, British Columbia, Canada

Michael Chandler—Department of Psychology, University of British Columbia, British Columbia, Canada

Eveline A. Crone—University of Leiden, Leiden, The Netherlands

Wil Cunningham—Department of Psychology, Ohio State University, Columbus, Ohio

Mirella Dapretto—FPR-UCLA Center for Culture, Brain, & Development, University of California–Los Angeles, and Ahmanson-Lovelace Brain Mapping Center, Jane & Terry Semel Institute for Neuroscience and Human Behavior, Los Angeles, California

Monique Ernst—Neurodevelopment of Reward Systems, Mood and Anxiety Programs, NIMH/NIH/DHHS, Washington, DC

Paul J. Eslinger—Departments of Neurology, Neural & Behavioral Sciences, Pediatrics, and Radiology, Penn State/Hershey Medical Center, Penn State University, Hershey, Pennsylvania

Luba Falk Feigenberg—Graduate School of Education, Harvard University, Cambridge, Massachusetts

Vittorio Gallese—Department of Neuroscience, Section of Physiology, University of Parma, Parma, Italy

Michael Hardin—Neurodevelopment of Reward Systems, Mood and Anxiety Programs, NIMH/NIH/DHHS, Washington, DC

Linda van Leijenhorst—University of Leiden, Leiden, The Netherlands

Matthew D. Lieberman—FPR-UCLA Center for Culture, Brain, & Development, and Department of Psychology, University of California–Los Angeles, Los Angeles, California

Chris Moore—Dalhousie University, Department of Psychology, Halifax, Nova Scotia, Canada

Ulrich Müller—University of Victoria, British Columbia, Canada

Kevin A. Pelphrey—Yale Child Study Center, Yale University, New Haven, Connecticut

Susan B. Perlman—Yale Child Study Center, Yale University, New Haven, Connecticut

Jennifer H. Pfeifer—Department of Psychology, University of Oregon, Eugene, Oregon

Melissa Robinson-Long—Departments of Neurology, Neural & Behavioral Sciences, Pediatrics, and Radiology, Penn State/Hershey Medical Center, Hershey, Pennsylvania

Magali Rochat—Department of Neuroscience, Section of Physiology, University of Parma, Parma, Italy

Mark Sabbagh—Department of Psychology, Queen's University, Ontario, Canada

Robert L. Selman—Graduate School of Education, Harvard University, Cambridge, Massachusetts

Bryan W. Sokol—Department of Psychology, Saint Louis University, St. Louis, Missouri

Elliot Turiel—Graduate School of Education, University of California–Berkeley, Berkeley, California

Brent C. Vander Wyk—Yale Child Study Center, Yale University, New Haven, Connecticut

Philip David Zelazo—Institute of Child Development, University of Minnesota, Minneapolis, Minnesota

PREFACE

This volume drew its inspiration from the 37th Annual Meeting of the Jean Piaget Society (JPS), which was held in Amsterdam, The Netherlands, from 31 May through 2 June 2007. The focus of this meeting was the confluence of social cognitive neuroscience and developmental psychology, a topic that generated enormous enthusiasm and resulted in the highest level of attendance at a JPS meeting to date. As planned, the meeting showcased a wide range of exciting research and provided a valuable forum for constructive dialogue among scholars working from diverse approaches and theoretical orientations.

This volume brings forward in printed form many of the ideas first aired at that meeting, and our hope is that it will help crystallize the emergence of a new field, developmental social cognitive neuroscience. The intended audience is researchers and students in neuroscience and developmental, cognitive, and social psychology, as well as the growing number of scientists whose work is inherently interdisciplinary. In addition to offering the interested reader a convenient and representative look at new directions in research on human experience, the book is appropriate for graduate seminars and upper-level undergraduate courses on Social Cognitive Neuroscience, Developmental Neuroscience, Social Development, and Cognitive Development, among other topics.

The meeting was made possible by generous financial support from several sources including the University of Amsterdam (Psychology), the University of Leiden (Developmental Psychology), and The Netherlands' Organisation for Scientific Research (NWO). We are very grateful for that support. We also thank the local organizers, Dr. Mariette Huizinga (University of Amsterdam) and Dr. Jan Boom (University of Utrecht), for their outstanding logistical support. Finally, we are grateful to Debra Riegert, Senior Editor at Taylor and Francis, for her unflagging interest in this timely volume.

Philip David Zelazo

Michael Chandler

Eveline A. Crone

I
Introduction

1

THE BIRTH AND EARLY DEVELOPMENT OF A NEW DISCIPLINE
Developmental Social Cognitive Neuroscience

Philip David Zelazo
University of Minnesota

Michael Chandler
University of British Columbia

Eveline A. Crone
University of Leiden

INTRODUCTION

Understanding is often accomplished through a sequential process of analysis followed by synthesis: we tear things apart, scrutinize the pieces in relative isolation, and only then begin to explore how the pieces interact to comprise a whole. Research on human development is now emerging from a protracted period of analysis during which various aspects of human function were studied one piece at a time. The worst excesses of this period, including our fascination with genetic determinism and extreme domain specificity, are now considered untenable, and interest has shifted to the complicated ways in which processes at cultural, social, cognitive, neural, and molecular levels of analysis operate together to yield human behavior and human development. The aim of this volume is to showcase outstanding examples of this new synthesis and, by doing so, to herald the birth of a new, more ecumenical discipline: "developmental social cognitive neuroscience."

The birth of this new developmental discipline followed quickly on the heels of its immediate predecessor—social cognitive neuroscience, which also aimed at identifying the neural bases of socioemotional behavior (see, for example, Ochsner & Liebermann, 2001), but did so without the signature ontogenetic emphasis that marks the newer work featured in this volume. Since 2001, nondevelopmental research in social cognitive neuroscience has appeared regularly in leading journals, such as *Science* and *Nature*, and has been collected in numerous volumes and special issues. Among the many well-publicized findings are reports concerning the neural correlates of sympathy, moral reasoning, theory of mind, and the evaluation of socially relevant stimuli (faces, persons vs. objects, biological motion, etc.). These demonstrated neural correlates, together with research on infant behavior and rapid advances in developmental evolutionary psychology (e.g., Bjorklund & Pellegrini, 2002), were seen by many to provide converging evidence that human beings are born with a predisposition to acquire considerable socio-emotional competence very early in life.

What was largely missing from this new social cognitive neuroscience literature, however, was an account of how all of these well-established relations between manifest behaviors and their neural underpinnings actually came into being. No one any longer doubts that young persons are born with something of a biologically based head start in accomplishing their important life tasks—genetic resources, if you will, that are exploited differently in different contexts. Nevertheless, it is also true that socially relevant neural functions develop slowly during childhood, and that this development is owed to various complex interactions among genes, social and cultural environments, and children's own behavior. A key challenge lies in finding appropriate ways of describing these complex interactions and the way in which they unfold in real developmental time. This is the challenge that motivates research in developmental social cognitive neuroscience.

The concatenation of three adjectives (i.e., "developmental," "social," and "cognitive") and one prefix ("neuro-") required to describe the scope of this new science, while admittedly a mouthful, does more than simply work to rally a crowd of historically distinct scholars together under a common banner. Rather, it signals a new, multidimensional characterization of a once-fractured subject matter—one in which human beings are newly viewed as dynamic organisms that must be described at multiple, interacting levels of analysis. On this view, it is arbitrary and potentially misleading simply to study in isolation some single aspect of a human being at one point during the lifespan. We would be mistaken, for example, to assume that cognition can be adequately characterized independently of the sociocultural context in which it occurs. Context may not be everything, but it certainly is a lot—a fact that seems to be true at all levels of analysis, from the protein-coding functions of specific genes to the characteristics of neural systems, to the subjective interpretation of arousal. At all levels of analysis, our subject matter is best understood, in the views of those who have contributed to this volume, as a dynamic event, conditioned by (but not determined by) history, and unfolding in time and space. There is little use nowadays for simple-and-sovereign accounts according to which human behavior is "explained" by exclusive reference to any one of a number of supposed causes, whether these be Piagetian stages, children's "theories," mental modules, neural activation in some region of cortex, or the presence of particular genetic polymorphisms. In most cases, these so-called "causes" are merely correlates or constituent parts of the phenomena in question, and, more often than not, they are simply alternative descriptions of the same phenomena at different levels of analysis. From the perspective of the contributors to this volume, the shared hope is that all such piecemeal descriptions will eventually find their proper place in a far more complex characterization of overall human growth and development.

THE PRESENT VOLUME

The contributors to this volume include both social cognitive neuroscientists who recognize the importance of human development and developmental psychologists who appreciate the promise and potential of a social cognitive neuroscience approach. Collectively, their contributions illustrate the latest, best work in this emerging field. At the same time, however, these multidisciplinary efforts and collaborations are not without their risks and possible pitfalls, and we have also included contributors who reflect critically on this new enterprise in an effort to foster its early development.

The first set of chapters concern the typical and atypical development of social cognition in childhood. One of the most exciting recent advances in social cognitive neuroscience has been the discovery of mirror neurons in monkeys by Rizzolatti, Gallese, and colleagues (e.g., see Rizzolatti, Fogassi, & Gallese, 2001). Gallese and Rochat (Chapter 2) review this work and suggest that the mirror neuron system plays a key, foundational role in the ontogeny of social cognition, providing us with a prereflexive understanding of others' actions as these actions are mapped onto our own action repertoires. Whereas Gallese and

Rochat emphasize the conservation of the mirror neuron system during primate evolution, Moore and Barresi (Chapter 3) focus on the way in which self-other matching develops in human infants. They review recent research suggesting that infants' representations of intentional action are initially quite limited (e.g., restricted to particular actions) and are only gradually, through social interactions, integrated into more complex characterizations of independent intentional agents.

One aspect of social cognition that develops substantially during the course of childhood is children's understanding of their own and others' mental states—their theory of mind. Benson and Sabbagh (Chapter 4) address the possible neural changes associated with the development of an understanding of the intentional or representational character of mental states. Their approach is to consider the possible roles of executive function—and by extension, prefrontal cortex—in children's developing theory of mind. As Benson and Sabbagh show, it is now well established that theory of mind and executive function are closely related, but the nature of this relation is still a matter of debate.

Cunningham and Zelazo (Chapter 5) also address neural changes associated with both theory of mind and executive function, in the context of their work on the iterative reprocessing of information via neural circuits involving prefrontal cortex. In this chapter, they consider the implications of this work for the early development of affective experience and its regulation.

The last two chapters in this section are concerned, in large part, with the atypical development of social cognition that is characteristic of the neurodevelopmental disorder of autism. Perlman, Vander Wyk, and Pelphrey (Chapter 6) present a new hypothesis regarding the neural changes involved in the development of two interrelated aspects of social cognition, social perception (i.e., the initial stages of evaluating intentions and dispositions of others), and action understanding (i.e., the appreciation of others' actions in terms of mental states). They then discuss how early insults to the brain mechanisms involved in these processes might influence the subsequent development of higher-order social cognition, including theory of mind. Baron-Cohen (Chapter 7) focuses more squarely on autism and, in particular, he reviews the evidence for his empathizing-systemizing theory, according to which autism can be characterized in terms of below average empathy and average or above-average systemizing—the drive to analyze or construct systems.

The second set of chapters addresses the developmental changes taking place in later childhood and adolescence, including changes in impulsivity, novelty seeking, risk taking, self-awareness, and executive function. Pfeifer, Dapretto, and Lieberman (Chapter 8) focus on evaluative self-knowledge, which undergoes considerable changes during adolescence, and they summarize the first set of imaging studies meant to examine the neural basis of self-referential processing in children and adolescents. On the basis of their findings, they describe a new developmental model of the neural systems supporting self-evaluation.

Ernst and Hardin (Chapter 9) also present a model of the neurodevelopmental changes occurring during adolescence, with a special focus on affective decision making. They show how their triadic model of decision making, which includes systems associated with approach, avoidance, and executive function, can account for an emerging corpus of data on reward-related behavior in adolescence, and they use the model to generate predictions for future research.

The potentially problematic behaviors associated with adolescence are also considered by Baird (Chapter 10), who refers to early adolescence as "the terrible twelves." By drawing a parallel between behavioral changes in early childhood and early adolescence, she provides a neurobiological account of noncompliant behavior that may shed common light on behavior during both developmental periods—and may help parents to view their teenagers' behavior with more compassion.

Leijenhorst and Crone (Chapter 11) address an apparent paradox in the literature on adolescent decision-making: there is evidence that risk taking increases during adolescence even though risk taking might be expected to decrease with age during this same period as a consequence of increased cognitive control. To resolve the paradox, they present a new neurodevelopmental model that describes how social and emotional context may contribute to both linear and nonlinear developmental changes, and how emotion-related and control-related neural networks compete in relatively hot, emotional contexts (such as when large rewards are at stake, or in the presence of peers).

Selman and Feigenberg (Chapter 12) present several innovative methods to assess the growing depth of self and social awareness during adolescence, and they particularly highlight the importance of considering social context when interpreting adolescent behavior. They also note the opportunity for synergy between innovative behavioral methods and the techniques of neuroscience, but they remind researchers not to lose sight of the meaningful subjective experience that is associated with neural activity, and that mediates between this neural activity and behavior.

The final section of the volume contains four chapters concerned with moral reasoning—two chapters that review exciting advances in the developmental neuroscience of morality and two that consider the potential limitations of a neuroscience perspective. Eslinger and

Robinson-Long (Chapter 13) review evidence not only from neuroimaging studies, but also from rare studies of patients who sustained lesions to prefrontal cortex during infancy or childhood. This evidence informs their characterization of the crucial role of prefrontal cortex in the development of moral reasoning and moral action. Blair (Chapter 14) also presents a characterization of the neurodevelopmental changes associated with moral development—in the context of a broad consideration of the contributions of neuroscience to the understanding of moral reasoning.

As a counterpoint to the optimism of these two earlier chapters, Carpendale, Sokol, and Müller (Chapter 15) argue that neural models alone are insufficient to understand what morality is and how it operates in peoples' lives. They discuss the essential roles of normativity, interpretation, and social context in moral judgments, and they caution researchers to avoid the various forms of reductionism that are too often associated with research in neuroscience. A note of caution is also sounded by Turiel (Chapter 16), who argues in the final chapter that the study of the neural correlates of morality and its development would benefit from the inclusion of both philosophical considerations and a careful look at the large literature on moral development. In his opinion, neuroscientific research on morality has emphasized unconscious and emotionally driven decisions to the exclusion of the important role of reasoning.

CONCLUSION

Human beings are complex creatures, continually in flux, and a confluence of perspectives is clearly needed to understand them. After decades of increasing professional specialization during which psychologists and neuroscientists studied relatively narrow aspects of human functioning from within a particular disciplinary approach, the time has come to work together to create a more unified view—to put the pieces back together again. The chapters in this volume all look beyond the limits of single disciplines or approaches, and collectively, they provide an excellent illustration of what we hope will be the future of research on the developing nature of human experience.

REFERENCES

Bjorklund, D. F., & Pellegrini, A. D. (2002). *The origins of human nature: Evolutionary developmental psychology.* Washington, DC: American Psychological Association.

Ochsner, K. N., & Lieberman, M. D. (2001). The emergence of social cognitive neuroscience. *American Psychologist, 56,* 717–734.

Rizzolatti, G., Fogassi, L., & Gallese, V. (2001). Neurophysiological mechanisms underlying the understanding and imitation of action. *Nature Reviews Neuroscience, 2,* 661–670.

II

The Typical and Atypical Development of Social Cognition in Childhood

2

MOTOR COGNITION

Role of the Motor System in the Phylogeny and Ontogeny of Social Cognition and Its Relevance for the Understanding of Autism

Vittorio Gallese
Magali Rochat
University of Parma

The traditional view in the cognitive sciences holds that humans understand the behavior of others in terms of their mental states—intentions, beliefs, and desires—by exploiting what is commonly designated as "folk psychology." According to a widely shared view, nonhuman primates are, instead, focused on the statistical recurrence of observable causal aspects of reality rather than attributing a causal role to opaque mental states. This view prefigures a sharp distinction between nonhuman species, confined to behavior reading, and our species, whose social cognition makes use of a more abstract level of explanation: mind reading. The claim according to which behavior reading and mind reading constitute two autonomous realms has nevertheless been radically questioned by recent findings in both animal and human infants social cognition.

From the ontogenetic perspective, a fundamental problem has been posed to the supposed correlation between the maturation of linguistic competence and mind-reading abilities (as posited by the mainstream representationalist account) by the demonstration that 15-month-old infants may already understand false beliefs (Onishi & Baillargeon, 2005). Those results suggest that typical constitutive aspects of what mentalizing is taken to be could be explained on the basis of low-level mechanisms, which develop well before full-blown linguistic competence, or even without requiring language. From the phylogenetic perspective, the claim of a unique human capacity for understanding the causal role of mental states has been smoothed by recent evidence showing that nonhuman primates do indeed infer the intentional states of others on the basis of visible observable cues such as behavior and context.

We posit that human social cognition cannot only rely on the capacity to explicitly think about the contents of someone else's mind by means of symbols or other representations in propositional format. Although those sophisticated mentalizing abilities can certainly be used to "explain" the behavior of others, most of our social transactions do require an immediate, automatic, and almost reflex-like understanding of the ongoing situation. As emphasized by Bruner (1990, p. 40), "When things are as they should be, the narratives of Folk Psychology are unnecessary."

Recent findings in cognitive neuroscience shed light on the existence of a common neural mechanism, the mirror neuron system (MNS) that could account for direct understanding of abilities of action and intention in both humans and nonhuman primates. The discovery of mirror neurons in the motor cortex, long-confined to a mere action programming and execution function, has changed our views on the relations among action, perception, and cognition, and strongly suggests adoption of a bottom-up approach, by thoroughly investigating the nonmetarepresentational aspects of human social cognition, so far unduly minimized or even neglected.

Our stance on the study of human social cognition embraces an evolutionary perspective, complemented by the neurophysiological and psychological investigation of the functional mechanisms implicated in nonhuman primates' social cognition. By doing so, it might be established to which extent the apparently different social cognitive abilities and strategies adopted by different species of primates may be produced by similar functional mechanisms, which in the course of evolution acquired increasing complexity.

In the present chapter, we introduce a new notion of the phylogenetic and ontogenetic origin of social cognition, which capitalizes upon the motor system's organization, the motor cognition hypothesis.* We posit that motor cognition provides both human and nonhuman primates with a direct, prereflexive understanding of biological actions of others, as they are mapped on the observers' own action repertoire. Motor cognition finds its neural substrate in the brain areas involved in mapping action perception on action execution (MNS). A consequence of our hypothesis is that action understanding is tightly related to the motor expertise individuals acquire during their development.

We will first consider the neuroscientific evidence relating the existence and functions of a neural mechanism underpinning action understanding's abilities in macaque monkeys and humans, the MNS. The possible contribution of this direct-mapping mechanism to the development of social cognitive abilities that are similar in both infants and nonhuman primates will be successively discussed from ontogenetic and phylogenetic perspectives. Finally, we will introduce a new approach to the origin of intentional understanding, the motor cognition hypothesis, and discuss its relevance for the understanding of autistic spectrum disorder.

THE MIRROR NEURON SYSTEM IN MONKEYS

About 16 years ago a new class of motor neurons was discovered in a sector of the macaque monkey's ventral premotor cortex, known as area F5. These neurons discharge not only when the monkey executes goal-related hand movements like grasping objects, but also when observing other individuals (monkeys or humans) executing similar acts. These neurons were called "mirror neurons" (Gallese, Fadiga, Fogassi, & Rizzolatti, 1996; Rizzolatti, Fadiga, Gallese, & Fogassi, 1996).

Neurons with similar properties were later discovered in a sector of the posterior parietal cortex reciprocally connected with area F5 (Gallese, Fogassi, Fadiga, & Rizzolatti, 2002; Fogassi et al., 2005; see Figure 2.1).

For most mirror neurons in F5, the effective motor actions coincide both in terms of goal (like grasping) and in terms of how the goal is achieved (such as, for example, a specific type of grip). However, for a good number of these visuomotor neurons, this congruence appears

---

* Our focus on the cognitive relevance of the motor system doesn't mean that we downplay the enormous importance played for the development of social cognitive skills by the proper functioning and regulatory role on action of affect and emotional systems. We are fully aware that motor cognition is only one—and a too-long neglected—component of a larger integrated picture.

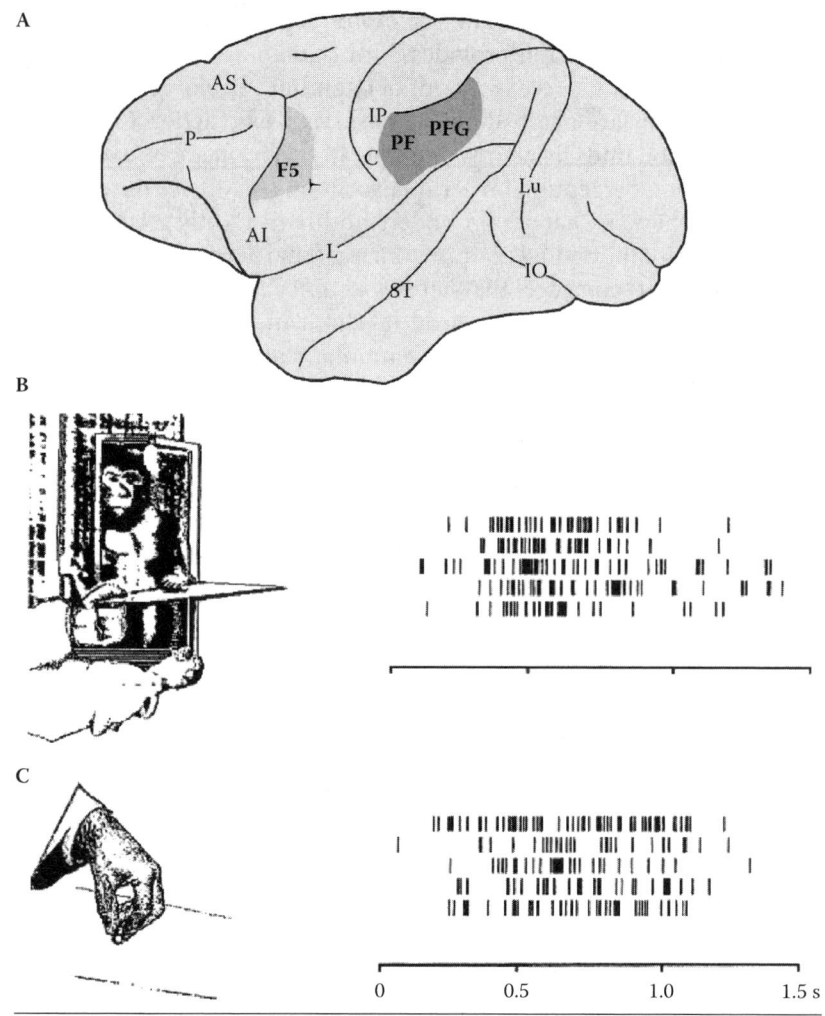

Figure 2.1 Mirror neuron system in macaque monkeys. (A) Lateral view of macaque monkey brain showing the frontal premotor (F5) and parietal (PF/PFG) areas where mirror neurons have been localized. As, superior arcuate sulcus; Ai, inferior arcuate sulcus; C, central sulcus; IO, inferior occipital sulcus; IP, intraparietal sulcus; L, lateral sulcus; Lu, lunate sulcus; P, principal sulcus; STS, superior temporal sulcus. (B) Visual and motor response of a hand-grasping mirror neuron. The behavioral responses are schematically represented on the left side of each panel. On the right side are shown the mirror neuron's discharge frequency (raster display) during a series of five consecutive trials. In a typical experimental session, the experimenter places a piece of food on a tray and presents it to the monkey that grasps it. The raster display shows an increased discharge during the precise phase of grasping execution, while no neural activation is elicited by the presentation of the tray towards the monkey or the motor act succeeding the grasping (food holding). (C) The monkey observes the experimenter grasping a piece of food placed on a tray. The neuronal discharge is driven by the observation of the grasping motor act only. Abscissae: time. (B and C modified from Rizzolatti et al. (1996). *Cognitive Brain Research*, 3, 131–141.)

to be broader, linking observed and executed motor acts on the basis of the achievement of a similar goal rather than on the similarity of movement kinematics. These broadly congruent mirror neurons seem to be endowed with a more abstract level of action coding that allows them to generalize the goal of the observed action across its many instances (Rizzolatti, Fogassi, & Gallese, 2001). Moreover, the interaction between object and effectors appears to be the principal triggering element to elicit a neural discharge, during both action execution and observation. Typically, mirror neurons in monkeys do not respond to the observation of an object alone, or to the sight of a hand mimicking an action without a target.

Action observation causes in the observer the automatic activation of the same neural mechanism triggered by action execution. The novelty of these findings resides in the fact that, for the first time, a neural mechanism allowing a direct mapping between the sensory description of a motor act and its execution has been identified. The idea of perception and action as two intrinsically intertwined and interdependent processes is not new. The common-coding theory (Prinz, 1987; Hommel, Müsseler, Aschersleben, & Prinz, 2001) proposed that actions are coded in terms of their perceivable effects. Association between a perceived effect and the action that produced it would linearly strengthen with experience. The shared coding of action and its effect within the same motor format successively allows automatically retrieving a motor act by anticipating its effects and vice versa. This mechanism can also be extended to observed actions to the extent that the perceived action and the one motorically coded in the observer's brain are similar (Knoblich & Flach, 2003).

The discovery of the existence of a shared neural substrate coding perceived and executed actions provides a parsimonious solution to the problem of translating the results of the visual analysis of observed motor behavior—in principle, devoid of meaning for the observer—into something that the observer is able to understand to the extent that the observer "experientially owns" it already. It was proposed that this direct-mapping mechanism could be at the basis of a direct form of action understanding, where the observer's motor knowledge is used to understand the other's goal-directed behavior (Gallese et al., 1996; Rizzolatti et al., 1996; Rizzolatti & Gallese, 1997).

The proposal that mirror neurons' activity reflects an internal description of the perceived action's meaning rather than a mere a visual description of its features has been demonstrated in two seminal experiments.

In the first study, Umiltà et al. (2001) found a subset of F5 mirror neurons that discharged also during the observation of partially hidden actions, coding the action outcome even in the absence of the complete visual information about it. Macaque monkey's mirror neurons therefore code observed acts not exclusively on the basis of their visual description, but on the basis of the anticipation of their final goal-state, simulated through the activation of its motor neural "representation" in the observer's premotor cortex.

Those data do not exclude the coexistence of a system that visually analyzes and describes the acts of others, most likely through the activation of extra-striate visual neurons sensitive to biological motion. However, such "pictorial" analysis per se is most likely insufficient to provide an understanding of the observed act. Without reference to the observer's internal "motor knowledge," this description is devoid of factual meaning for the observing individual.

A second study (Kohler et al., 2002) demonstrated that mirror neurons also code the actions' meaning on the basis of their related sound. A particular class of F5 mirror neurons ("audio-visual mirror neurons") responds not only when the monkey executes and observes a given hand action, but also when it just hears the sound typically produced by the action. These neurons respond to the sound of actions and discriminate between the sounds of different actions, but do not respond to other similarly interesting sounds such as arousing noises, or monkeys' and other animals' vocalizations. The recent discovery of audio-motor mirror neurons in singing birds (Prather, Peters, Nowicki, & Mooney, 2008) demonstrates that the mirroring mapping mechanism is evolutionarily old and widespread among living species.

Mirror neurons' activity reveals the existence of a mechanism through which perceived events as different as sounds, images, or voluntary acts of the body are nevertheless coded as similar to the extent that they represent the assorted sensory aspects of the motor act's goal. It has been proposed that the MNS, by matching observed, implied, or heard goal-directed motor acts on their motor neural substrate in the observer's motor system, allows a direct form of action understanding, through a mechanism of embodied simulation (Gallese, 2005).

A major step forward in the research on the MNS consisted in the discovery that parietal mirror neurons not only code the goal of an executed/observed motor act such as grasping an object, but they also discriminate identical motor acts (like grasping) according to the final goal of the action in which the act is embedded (e.g., grasping an object to bring it to the mouth or into a container, Fogassi et al., 2005). The MNS maps integrated sequences of goal-related motor acts (grasping, holding, bringing, placing) so as to obtain different and parallel chained sequences of motor acts properly assembled to accomplish a more distal goal-state. Each embedded motor acts appears to be facilitated by the previously executed one, reorganizing itself so as to map the fulfillment of the overarching goal. These results suggest—at least at the level of basic actions—that the "prior intention" of eating or placing the food is also coded by parietal mirror neurons. Of course, this doesn't imply that monkeys explicitly represent prior intentions as such. Preliminary results show that similar properties are instantiated by F5 mirror neurons (Ferrari et al., 2006a).

MIRRORING MECHANISMS IN HUMANS

Several studies using different experimental methodologies and techniques have documented the existence of a common neural activation during action observation and execution also in the human brain (for review, see Rizzolatti et al., 2001; Gallese 2003a,b, 2006; Gallese, Keysers, & Rizzolatti, 2004; Rizzolatti & Craighero, 2004).

Neuroimaging studies have demonstrated that the observation of actions activated the likely human homologue of the monkey areas in which mirror neurons were originally described. In humans, the lower part of the precentral gyrus, the posterior part of the inferior frontal gyrus (IFG), the rostral part of the inferior parietal lobule, and regions within the intraparietal sulcus are described as "forming the core of human mirror system" (Rizzolatti & Craighero, 2004; see Figure 2.2).

Moreover, during the observation/execution of mouth-, hand-, and foot-related acts, the activation of distinct cortical regions within the premotor and posterior parietal cortices reflect the presence of a coarse somatotopic organization, similar to the one found in monkeys' homologue areas (Buccino et al., 2001; Aziz-Zadeh, Wilson, Rizzolatti, & Iacoboni, 2006). Similar results have been found using transcranial magnetic stimulation (TMS, see Rizzolatti & Craighero, 2004). Furthermore, Fadiga, Craighero, Buccino, and Rizzolatti (2002) showed that listening to speech is associated with an increase of motor-evoked potentials recorded from the listener's tongue muscles when the presented words strongly involved tongue movements.

Experimental evidence seems to suggest that the involvement of the MNS during action observation is strictly correlated to species and individuals' motor history. Mirror areas are significantly more activated when observing goal-directed actions executed by conspecifics (Buccino et al., 2004a). Moreover, several neuroimaging studies underlined the formative role played by motor experience in modeling action comprehension (Järveläinen et al., 2004; Calvo-Merino, Glaser, Grezes, Passingham, & Haggard, 2005; Calvo-Merino, Grèzes, Glaser,

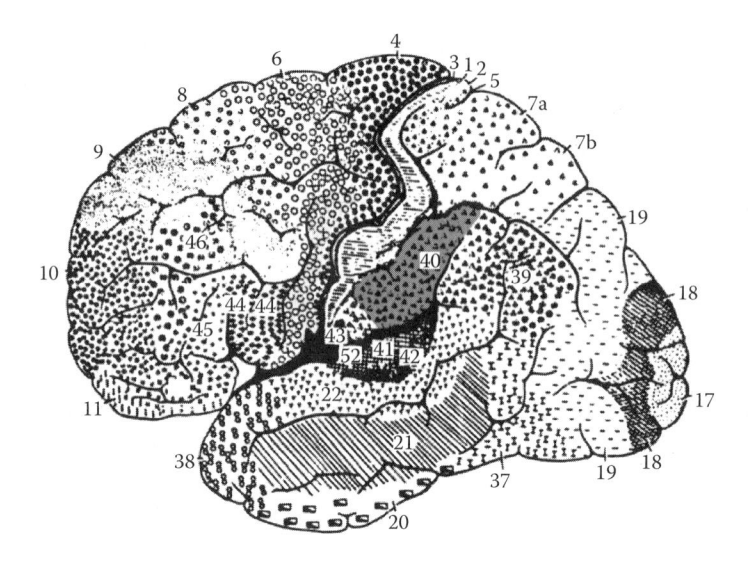

Figure 2.2 Mirror neuron system in humans. Lateral view of a human brain illustrating the areas activated during the execution of goal-related motor acts and their observation when executed by others. The posterior part of inferior frontal gyrus corresponds to BA 44. The inferior part of the precentral gyrus, corresponding to the ventral premotor cortex, is indicated with BA 6. The rostral part of the inferior parietal lobule corresponds to BA 40.

Passingham, & Haggard, 2006). Those results corroborate the hypothesis that actions may be differently perceived—and understood—on the basis of the individual's motor capabilities and experience.

The existence of shared neural and cognitive representations of one's own and other's action could at least partially account for the human ability to imitate actions. fMRI evidence shows that mirror areas in humans are selectively activated during simple movements imitation (Iacoboni et al., 1999), and during imitation learning of complex skills (Buccino et al., 2004b; Vogt et al., 2007). Buccino and colleagues (2004b) proposed that during learning of new motor patterns by imitation, the observed actions are decomposed into elementary motor acts that automatically activate the corresponding motor maps. The prefrontal cortex would then recombine the activated motor maps according to the observed model. A recent study by Buxbaum, Johnson, and Bartlett-Williams (2005), on posterior parietal neurological patients with ideomotor apraxia, has shown that they were not only disproportionately impaired in the imitation of transitive as compared to intransitive gestures, but they also showed a strong correlation between imitation deficits and the incapacity of recognizing observed goal-related meaningful hand actions.

The MNS seems to be also involved in the detection of action intentions (Iacoboni et al., 2005). In this study participants witnessed three kinds of videos portraying grasping hand actions without a context, context only (a scene containing objects), and grasping hand actions embedded in contexts. In the latter condition, the context suggested the intention associated with the grasping action (either drinking or cleaning up). Actions embedded in contexts, compared with the other two conditions, yielded a significant signal increase in the posterior part of the inferior frontal gyrus and the adjacent sector of the ventral premotor cortex where hand actions are represented. Thus, premotor mirror areas—areas active during the execution and the observation of motor acts—previously thought to be involved only in action recognition are actually also involved in understanding the intention by relying on the context to infer a forthcoming new goal.

These results seem to suggest that most of the time even humans do not explicitly represent intentions as such when understanding them in others. Action intentions are embedded within the intrinsic intentionality of action; that is, the intrinsic relatedness to an end-state, a goal. Most of the time we do not ascribe intentions to others, we simply detect them. By means of embodied simulation, when witnessing others' behaviors, their motor intentional contents can be directly grasped without the need of representing them in propositional format.

THE ONTOGENETIC DEVELOPMENT OF MIRRORING MECHANISMS

One crucial issue still not clarified is how the MNS develops in ontogeny. To date, no existing study clarified to which extent the described mirroring mechanisms are innate, and how development might shape and model them.

We know that motor skills mature much earlier on than previously thought. Chiron et al. (1992) demonstrated that at birth, the primary somatosensory and motor cortex show an advanced maturation compared with other brain areas. But even before birth, a recent study (Zoia et al., 2007) showed that fetal hand movements display a specific spatial and temporal coordination pattern. By 22 weeks of gestation, fetal hand movements show kinematics patterns that apparently depend on the end-state of the different motor acts fetuses perform. This result led the authors of this study to argue that 22-week-old fetuses show a surprisingly advanced level of motor planning, already compatible with the execution of "intentional actions."

One might speculate that during prenatal development specific connections may develop between the motor centers controlling mouth and hand goal-directed behaviors and the brain regions that will become recipient of visual inputs after birth. Such connectivity could provide functional templates (e.g., specific spatio-temporal patterns of neural firing) to areas of the brain that, once reached by visual inputs after birth, would be ready to specifically respond to the observation of biological motion, such as hand or facial gestures. In other words, neonates and infants, by means of a specific connectivity developed during the late phase of gestation between motor and "to-become-visual" regions of the brain, would be ready to mirror and imitate the gestures performed by adult caregivers in front of them, and would therefore be endowed with the neural resources enabling the reciprocal behaviors that characterize our postnatal life since its very beginning (see Braten, 1988, 1992, 2007; Meltzoff & Moore, 1977, 1998; Meltzoff & Brooks, 2001; Stern, 1985; Trevarthen, 1979, 1993; Tronick, 1989).

The newborns' ability to reproduce a model facial expression (Meltzoff & Moore, 1977) can be viewed as an automatic process resulting from a lack of inhibitory control on the motor resonance mechanism. However, evidence for the newborns' ability to produce a delayed imitation (Meltzoff & Moore, 1994) and their real efforts to accurately imitate the model (Kugiumutzakis, 1993) indicates that this process is something more than a simple reflex. Interestingly, 5- to 8-week-old infants imitate the tongue protrusion behavior of a human model only, and not the one of a nonbiological agent (Legerstee, 1991), showing that early imitation behavior is selective for conspecifics.

Recently, some experiments have confirmed the early existence of a resonance mechanism in the infants' premotor and posterior parietal cortices. In a one-case electroencephalographic study with intracranial recordings, Fecteau et al. (2004) showed that corresponding areas of the sensorimotor cortex were activated when the 36-month-old child watched another person drawing with his right hand and when the child drew with her own right hand. The earliest indirect evidence available to date of a MNS in infants comes from a study by Shimada and Hiraki (2006) who demonstrated by means of near-infrared spectroscopy (NIRS) the presence of an action execution/observation mapping system in 6-month-old human infants. Interestingly, this study showed that the sensory-motor cortex of infants (but not that of adult participants) was also activated during the observation of a moving lifeless object when presented on a TV screen. These findings suggest that during the early developmental stages, even nonbiological moving objects are "anthropomorphized" by means of their mapping onto motor representations pertinent to the observers' motor skills. As recently proposed by Lepage and Theoret (2007), the development of the MNS can be conceptualized as a process whereby the child learns to refrain from acting out the automatic mapping mechanism linking action perception and execution. We propose that such development can be viewed as a process leading from mandatory reenaction to mandatory embodied simulation.

It can be hypothesized that an innate rudimental MNS is already present at birth and can be henceforth flexibly modulated by motor/affective experience and gradually enriched by visuomotor learning.

THE ONTOGENESIS OF ACTION AND INTENTION UNDERSTANDING

The ability to assign mental states to others has been explained by some authors by introducing the notion of a "theory of mind" model (ToM; Premack & Woodruff, 1978). This ability has been thought to develop gradually during the first 3 to 5 years of life, grounded on some ToM precursor abilities such as understanding gaze direction and shared attention. ToM is considered as full-blown when infants pass the false-belief task (Baron-Cohen, Leslie, & Frith, 1985), that is, when they understand that others' behavior is driven by their own representation of the world and that such representation might not accurately reflect reality (Wimmer & Perner, 1983; Baron-Cohen et al., 1985).

A relevant amount of experimental evidence shows that during the first year of life, infants' action understanding is already fairly well developed. According to a common consensus, those first forms of intentional understanding do not imply any metarepresentational capacity, nor can they be interpreted in terms of mind reading—even though the suggested underlying mechanisms are very different.

According to some theorists, specialized developmental mechanisms that are in place at birth (Premack, 1990, Leslie, 1994, Baron-Cohen, 1994), allow infants to interpret actions as goal-directed very early in life. Innate sensitivity to ostensive behavioral cues like animacy, self-propelledness, temporal contingency, and equifinal variations of action would enable infants to ascribe goal-relatedness to the action of a wide range of entities, largely encompassing their experience-related knowledge.

In a similar vein, Gergely and Csibra's Teleological stance hypothesis (Gergely, Nàdasdy, Csibra, & Bìrò, 1995; Csibra, Gergely, Birò, Koòs, & Brockbank, 1999; Csibra, Birò, Koòs, and Gergely, 2003), posits that by 9 months of age, infants are equipped with an inferential system applied to factual reality (action, goal-state and current situational constraints) for generating nonmentalistic, goal-directed action representations. According to these authors, an action is represented as teleological only if it satisfies a principle of rational action, stating that an action can be explained by its goal-state if the agent reaches its goal through the most efficient mean, given the contextual constraints.

A different theoretical view on the emergence of infants' goal-directed action interpretation, stresses the intrinsic link between action understanding and experience. Jean Piaget (1970) stated that "... human knowledge is essentially active ... knowing an object does not mean (passively) copying it—it means acting upon it." Following Piaget (1952), several scholars emphasize the constructional effect of observational and self-agentive experience on infants' understanding of actions' goal-relatedness (see Sommerville & Woodward, 2005). In particular, infant research employing habituation/dishabituation paradigms showed that previous motor experience facilitates 3-month-old infants' perception of goal-directed actions performed by others (Sommerville, Woodward, & Needham, 2005). Moreover, 10-month-old infants' ability to construe an action representation as hierarchically organized towards a distal goal strictly depends on their ability to perform similarly structured action sequences (Sommerville & Woodward, 2005).

Interestingly, congruency between the observed action and the observer's motor repertoire seems to be crucial for goal prediction. In a recent study, it has been shown that just like adults using their own action plans to anticipate the actions of others (Flanagan & Johansson, 2003), infants produce proactive goal-directed eye movements when observing a placing action, only to the extent they can perform it (Falck-Ytter, Gredeback, & von Hofsten, 2006). In addition, seminal studies demonstrated that infants' early goal discrimination is initially restricted to actions executed by conspecifics. Six-month-old infants are sensitive to the action goals of others only when performed by a human agent (Woodward, 1998), and Meltzoff (1995) showed that older infants imitate the unseen motor goal of a human model but not of an inanimate object.

Goal detection is thought to form the core ability of action understanding and social learning through imitation. Both adults (Baird & Baldwin, 2001) and children represent actions as constituted by units hierarchically organized with respect to an overarching goal. Ten-month-old children share with adults the ability to parse actions in units whose boundaries correspond to the completion of a goal (Baldwin, Baird, Saylor, & Clark, 2001). Imitation tasks clearly reflect children's ability to represent actions' units as organized towards a distal goal. When asked to imitate the action of another person, preschoolers reproduce the higher order goal of the action (Bekkering, Wohlschläger, & Gattis, 2000). Eighteen-month-old infants reproduce the goal they inferred from the failed attempts of a human demonstrator (Meltzoff, 1995). Carpenter, Call, and Tomasello (2005) showed that infants could flexibly interpret the goal of an observed sequence of movements according to the context and therefore reenact either the goal of an observed action or the means by which it had been produced. Similarly, Gergely, Bekkering, and Kiràly (2002) found that 14-month-old infants reproduce both observed means and goals only when the reason according to which the agent chose a specific means appeared to surpass children's knowledge. Underlying this cognitive flexibility is the fundamental ability to discriminate between means and ends.

The neurophysiological discovery that goal-relatedness is the functional organizing principle of primates' motor systems provides a possible phylogenetic explanatory framework to these empirical findings, lending support to a deflationary, motor account of the development of intentional understanding.

THE MNS AND ITS RELEVANCE IN THE EVOLUTION OF SOCIAL COGNITION

The presence of mirror neurons in different species of primates such as macaques and humans seem to favor a continuist view of the evolution of social cognition. However, it is also true that the very same evidence must be reconciled with the uniqueness of human social cognition.

Just like humans, nonhuman primates are social beings living in highly cohesive groups. The basic requirement of primate group living is associated with a particularly sharp sensitivity to subtle social cues (Humphrey, 1976) such as visual (Dasser, 1988; Humphrey, 1974) and acoustic (Snowdon, 1986) discrimination of others. During their social interactions, primates exhibit several complex behaviors such as gaze-following, deception, or reconciliation that are apparently very similar to human behavior. The main difference between human and nonhuman primates' social cognition consists in the degree of intentional understanding—that is, of the understanding of what causes a goal-directed action. According to a mainstream view, humans would have the unique ability to go below behavior's surface to infer mental states such as intentions, beliefs, and desires, that might drive the current or future agents' behavior (Povinelli & Eddy, 1996; Tomasello & Call, 1997), while nonhuman primates' knowledge is thought to rely on the extraction of procedural rules from observable environments' regularities (Köhler, 1927;

Visalberghi & Tomasello 1998; Povinelli 2000). Tomasello & Call (1997), initially echoing the position of Povinelli & Eddy (1996), proposed that nonhuman primates basically learn to understand and predict an imminent action by associating a certain antecedent-consequent sequence of behavior in a recurrent contextual situation.

However, evidence in comparative psychology has recently challenged the traditional conception of nonhuman primates as mere behavior-readers, essentially unable to interpret movements in terms of intentional actions.

The chimps' inability to understand others as intentional agents turned out to be only apparent in particular cooperative contexts. Recent evidence showed that when engaged in a competitive setting, chimps deduce what others know on the basis of what they are looking at (Hare, Call, & Tomasello, 2001; Tomasello, Carpenter, Call, Behne, & Moll, 2005). Furthermore, Call and Tomasello (1998) reported that chimpanzees succeed in discriminating between an intentionally teasing action and an unwillingly clumsy one performed by a human. Even more important, it has been recently shown that rhesus monkeys can establish a cognitive link between seeing and knowing, by systematically choosing to steal food from the human competitor that could not see the food, while refraining from doing it when the human competitor could see it (Flombaum & Santos, 2005). Similarly, it has been shown that rhesus monkeys choose to obtain food silently only in situations in which silence is crucial to remain undetected by a human competitor (Santos, Nissen, & Ferrugia, 2006).

One of the characteristics distinguishing human societies from other social groups is humans' ability to transfer their cultural knowledge. Traditionally, social learning strategy through imitation is considered a human beings' prerogative, as imitation requires the ability to inhibit one's own motor strategies to reach a goal in order to reproduce the action of a model with the highest fidelity (Tomasello, Kruger, & Ratner, 1993).

Although, for a long time, nonhuman primates have been considered better emulators than imitators, achieving the same goal of a model by following their own strategy (Tomasello, Davis-Dasilva, Camak, & Bard, 1987), recent experiments showed that apes and macaque monkeys actually display unexpected abilities in imitation tasks.

Horner and Whiten (2005) demonstrated that, similarly to 14-month-old children (Gergely et al., 2002), wild-born chimpanzees can solve a tool-using task by flexibly switching from an emulation to an imitation strategy when causal information was insufficient. In an observational learning task, rhesus macaques learned to respond in a particular sequence to items that were simultaneously displayed on a touch screen. Results showed that monkeys didn't acquire new sequences by simple motor imitation but by detecting and copying a cognitive rule underlying the demonstrator motor's behavior (Subiaul et al., 2004).

The presence of a mirroring mechanism in primates could account for the neural underpinning of neonatal facial imitation, originally described in humans by Meltzoff and Moore (1977), and subsequently described in chimpanzees (Myowa-Yamakoshi, Tomonaga, Tanaka, & Matsuzawa, 2004) and macaque monkeys (Ferrari et al. 2006b). Further experimental evidences suggests that the MNS can be exploited to make sense of the conspecifics' behavior for various social purposes such as social facilitation (Ferrari, Maiolini, Addessi, Fogassi, & Visalberghi, 2005a), or, like 9-month-old babies (Agnetta & Rochat, 2004), the recognition of being imitated (Paukner, Anderson, Borelli, Visalberghi, & Ferrari, 2005).

In a recent experiment, Rochat and colleagues (2008) further explored the issue of whether the MNS could account for the phylogenetic continuity of the ability to evaluate and predict the goal-directed action of others. Nonhuman primates' ability to discriminate between means and end, and to use contextual cues to evaluate the ecological validity of a chosen means, has been tested by adapting a looking-time paradigm previously used with human babies (Gergely & Csibra, 1995). Results showed that macaque monkeys, similarly to 9- to 12-month-old human infants, detect the goal of an observed motor act and, according to the physical characteristics of the context, construe expectancies about the most likely action the agent will execute in a given context. This, however, is true only to the extent that observed motor acts are consonant to the observer's motor repertoire, whereas inadequate motor acts, nongoal-related movements, or unfamiliar goal-related motor acts do not allow any simulation and prediction. Although this study does not provide direct evidence about the neural mechanisms underpinning the results, it is reasonable to hypothesize that monkeys evaluate the observed acts by mapping them on their own motor neural substrate, through the activation of the MNS.

These results reveal that nonhuman primates are endowed with the ability to understand the intentional meaning of others' behavior by relying upon visible behavioral cues; hence, they seriously argue against the traditional dichotomous account of primate social cognition based on a sharp evolutionary discontinuity between behavior- and mind-readers (Gallese & Umiltà, 2006; Gallese 2007). These results further corroborate the notion that motor behavior contains elements that can be detected and used to understand it and construe predictions about it, without necessarily relying on mental representations in propositional format, certainly precluded to nonhuman primates.

THE MOTOR COGNITION HYPOTHESIS

Challenging a traditional purely mentalistic view of intersubjectivity, we posit that the capacity to understand others' intentional behavior—both from a phylogenetic and ontogenetic point of view—relies on a more basic functional mechanism, which exploits the intrinsic functional organization of the primates motor system. Abilities like goal detection, action anticipation, and hierarchical representation of action with respect to a distal goal can be considered as the direct consequence of the peculiar functional architecture of the motor system, organized in terms of goal-directed motor acts (Rizzolatti et al. 1988, Rizzolatti, Fogassi, & Gallese, 2000, Gallese & Rizzolatti, 1997). We also posit that the correct development of such a mechanism is required to scaffold more cognitively sophisticated social mentalistic abilities.

Primates constantly use their distal effectors, hands and mouth, to explore and exploit the external world. Neuroscientific investigation during the last decades has shed light on the neurophysiological mechanisms underpinning such abilities. In primates a central role in grasping actions is played by the inferior part of the ventral premotor cortex (area F5), which sends its output to the hand field of the primary motor cortex (F1/M1). Some neurons in F5 code a specific type of grasping, others the specific temporal phase of grasping, while one-third of them discharge when the monkey grasps an object using different effectors—for example, the right hand, the left hand, or the mouth. On the basis of this finding, it was suggested that these neurons code the goal of grasping rather than a specific motor behavior (Rizzolatti et al., 1988; Rizzolatti et al., 2000).

A formal quantitative testing of this hypothesis was recently carried out by Umiltà et al. (2008). In this study hand-related neurons were recorded from premotor area F5 and the primary motor cortex (area F1) in monkeys trained to grasp objects using two different tools: "normal pliers" and "reverse pliers." These tools require opposite movements to grasp an object: With normal pliers the hand has to be first opened and then closed, as when grasping is executed with the bare hand, while with reverse pliers the hand has to be first closed and then opened. The use of the two tools enabled to dissociate the neural activity related to hand movement from that related to the goal of the motor act.

All tested neurons in area F5 and half of neurons recorded from the primary motor cortex discharged in relation to the accomplishment of the goal of grasping when the tool closed on the object, regardless of whether during this phase the hand opened or closed—that is, regardless of the movements employed to accomplish the goal. The data of Umiltà et al. (2008) indicate that goal-coding structures the way action is mapped in area F5 and, although to a minor extent, even in the primary motor cortex. Furthermore, in the same study, the recording of mirror neurons during tool use confirmed that also this set of neurons has the same goal-relatedness as the other F5 neurons (Umiltà, personal communication). When F5 mirror neurons discharge, the goal of the motor act is specified, both when they are internally activated and when they discharge in response to an observed action.

Goal coding as a distinctive functional feature of the nonhuman primates cortical motor system organization can also shed light on the debate on the relative importance either of motor or perceptual experience to grasp the meaning of an observed action. Several studies have documented the impact of the action's visual familiarity on its motor representations (Porro, Facchin, Fusi, Dri, & Fadiga, 2007; Ertelt et al., 2007). Moreover, a recent neurophysiological study has reported that a particular class of ventral premotor mirror neurons starts to respond to the observation of unfamiliar actions after extensive visual exposure to them (Ferrari, Rozzi, & Fogassi, 2005b). Along with the results of Umiltà et al. (2008), showing the impact of a motor training on action coding, the results of both experiments seem to suggest that when an action performed by others becomes familiar, independently from the perceptual or motor source of its familiarization, it is nevertheless always mapped onto the motor representation of a similar goal (to take possession of an object) belonging to the observing individual. Furthermore, several brain-imaging studies conducted on human beings have shown that the intensity of the MNS activation during action observation depends on the similarity between the observed actions and the participants' action repertoire (Buccino et al., 2004a; Calvo-Merino et al., 2005). In particular, one fMRI study (Calvo-Merino et al., 2006) focused on the distinction between the relative contribution of visual and motor experience in processing an observed action. The results revealed greater activation of the MNS when the observed actions were frequently performed with respect to those that were only perceptually familiar but never practiced.

These discoveries emphasize the crucial role played by the motor system in providing the building blocks upon which more sophisticated social cognitive abilities can be built. What is more important, these neurophysiological findings provide a neuro-functional basis to interpret the ever-growing evidence coming from developmental psychology research on the role played by experience-based motor knowledge in shaping the ontogenetic development of social cognitive skills. The motor cognition hypothesis also provides a new approach to the study of the developmental breakdown of social cognition, as exemplified by autistic spectrum disorder (ASD).

MOTOR COGNITION AND ITS BREAKDOWN IN THE CASE OF AUTISM

The most striking features of autistic spectrum disorder (ASD) are the impairments (displayed with various degrees of severity) of the capacity to establish meaningful social communications and bonds, to establish visual contact with the world of others, to share attention with others, and to understand others' intentions, emotions, and sensations (Dawson et al., 2002).

Not surprisingly, a dominant paradigm in the study of autism, consonant with the cognitive science mainstream view on social cognition, portrayed the aloneness of these children as the consequence of a defective theory of mind, as a sort of "mind-blindness" (Baron-Cohen, 1995). Recent advances in neurophysiology, however, challenge this view and shed new light on the relationship between ASD and impaired motor cognition.

Diamond (2000) has suggested that motor and cognitive developments are more interrelated than previously thought. Children with ASD regularly show motor coordination problems that might be associated with cerebellar dysfunctions (Mostofsky, Reiss, Lockhart, & Denckla, 1998). Unlike typically developing children, autistic children use motor strategies basically relying on feed-back information, rather than on feed-forward modes of control. Such motor disturbance prevents autistic children from adopting anticipatory postural adjustments (Schmitz, Martineau, Barthélemy, & Assaiante, 2003).

The theoretical relevance of these findings has became fully clear by a very recent breakthrough series of experiments (Cattaneo et al., 2008) documenting that autistic children fail to activate a specific action chain from its very outset, thus being deprived of an internal copy of the whole action before the execution thereof. These EMG experiments have documented that high-functioning autistic children are unable to organize their own motor acts in the intentional motor chains shown by typically developing children. Participants in this study were typically developed (TD) and high-functioning autistic children who were required both to execute and observe two different actions: grasping with the right hand a food item placed on a plate, bringing it into the mouth and eating it, or grasping a piece of paper placed on the same plate and putting it into a box. During the execution and observation conditions of both actions the activity of participants' mouth-opening mylohyoid muscle (MH) was recorded using EMG surface electrodes.

The results showed that during the execution and observation of the eating action, a sharp increase of MH activity was recorded in TD children, starting well before the food was grasped. No increase of MH activity was present during the observation of the placing action. This means that one of the muscles responsible for the action final goal (opening the mouth to eat a piece of food) is already activated during the initial phases of the action. The motor system anticipates the motor consequences of the action final goal (to eat), thus directly mapping the action intention, both when the action is executed and observed when done by others. In contrast with TD children, high-functioning autistic children showed a much later activation of the MH muscle during eating action execution and no activation at all during eating action observation.

The results of Cattaneo et al. (2008) reveal that autistic children are impaired in smoothly chaining sequential motor acts within a reaching-to-grasp-to-eat intentional action. This impairment is mirrored in the action observation condition, and most likely accounts for their difficulty in directly understanding the intention of the observed action when executed by others.

Additional sides of the motor domain, such as action simulation, mimicry, and imitation, have been recently explored by a number of studies, all confirming a deep impairment of the core mechanisms of motor cognition in autistic children.

Two recent studies, for instance, employing different techniques such as EEG (Oberman et al., 2005) and transcranial magnetic stimulation (TMS; Théoret et al., 2005) show that individuals with ASD might be suffering an action simulation deficit induced by a dysfunction of their MNS. The study by Oberman et al. (2005) showed that ASD individuals, at difference with healthy controls, did not show *mu* frequency suppression over the sensory-motor cortex during action observation. The study by Théoret et al. (2005) showed that, again at difference with healthy controls, ASD individuals did not show TMS-induced hand-muscle facilitation during hand action observation.

A further indication of motor cognitive deficits in ASD is exemplified by imitation deficits. Autistic children have problems in both symbolic and nonsymbolic imitative behaviors, in imitating the use of objects, in imitating facial gestures, and in vocal imitation (Rogers, 1999; for review, see Rogers & Williams, 2006). These deficits characterize both high- and low-functioning forms of autism. Furthermore, imitation deficits are apparent not only in comparison with the performances of healthy subjects, but also with those of mentally retarded nonautistic subjects. As previously proposed (Williams, Whiten, Suddendorf, & Perrett, 2001, Williams, Whiten, & Singh, 2004), imitation deficits in autism could stem from the incapacity to establish a motor equivalence between demonstrator and imitator, most likely due to a malfunctioning of the mirror neuron system, and/or because of a disrupted emotional/affective regulation of the same system (Gallese, 2003b, 2006; Oberman & Ramachandran, 2007).

Dapretto et al. (2006) investigated the neural correlates of the capacity of imitating the facial expressions of basic emotions in high-functioning ASD individuals. The results of this fMRI study showed that during observation and imitation autistic children did not show activation of the MNS in the pars opercularis of the inferior frontal gyrus. Activity in this area inversely correlated with symptoms severity in the social domain. McIntosh, Reichman–Decker, Winkielman, and Wilbarger (2006) recently showed that individuals with ASD, at difference with healthy controls, do not show automatic mimicry of the facial expression of basic emotions, as revealed by EMG recordings, although they can voluntary do it upon request.

Finally, a recent study (Cossu et al., submitted) has simultaneously investigated the imitation of meaningful and meaningless actions with and without objects, the production of meaningful actions from visual and oral cues, respectively, and the comprehension of pantomimes by choosing the corresponding visual (from alternative pictures) and oral target (from the name spoken by the examiner). Fifteen high-functioning autistic children (mean age 11.4 years) were selected, along with two much younger control groups (4 to 5 years and 5 to 6 years, respectively). The results showed that each autistic child failed significantly in all of the 10 tasks presented; neither imitation or production, nor comprehension of the presented actions reached the degree of accuracy observed in both groups of much younger TD children. This study indicates that the whole architecture of motor cognition is developmentally impaired in autistic children.

Hence, converging evidence from a variety of studies suggest that a dysfunction of motor cognition might be at the basis of many of the social cognitive impairments shown by ASD individuals. In contrast with what a long-standing mainstream account of autism would hold (see Baron-Cohen et al., 1985; Baron-Cohen, 1995), the lack of social understanding of autistic individuals is not due to their incapacity to theorize about the mind of others. Theorizing is most likely the only compensatory strategy available to them (see Gallese, 2001). Many of the social cognitive impairments manifested by ASD individuals appear to be rooted in their incapacity to organize and directly grasp the intrinsic goal-related organization of motor behavior.

CONCLUSION

The existence of a direct-mapping mechanism for action understanding—the MNS—offers a deflationary, motor account of the development of important aspects of social cognition, both from phylogenetic and ontogenetic perspectives.

The recent discovery that in primates' motor cortex, actions are coded and structured in terms of their goal has further shed light on the neural processes underlying the ability to execute but also to understand an action, thanks to the direct recognition of its goal that has become familiar through motor experience. This mechanism, common among primates, could therefore also account for the development of the ability to discriminate between means and goals and, therefore, for the capacity to map as similar different actions leading to the same goal, and, more generally, to produce a bedrock for social identification. The specific social cognitive flexibility of our species, as reflected in our pervasive mimetic abilities, in our propensity for pedagogy, and in the sophisticated quality of our social understanding, likely exceeds the functional properties of the MNS. However, we posit that a proper development of the MNS is a necessary prerequisite for scaffolding the development of proper human social cognitive skills.

In contrast with the mainstream view in the cognitive sciences, social cognition cannot be merely identified with "social metacognition," that is, explicitly thinking about the contents of someone else's mind by means of symbols or other representations in propositional format. We can certainly "explain" the behavior of others by using our most sophisticated mentalizing abilities. It must be added, though, that the neural mechanisms underpinning such abilities are far from being understood, unless willing to acritically commit oneself to a mere correlational use of brain imaging techniques, a very dangerous enterprise, which seriously faces the risk of resurrecting a high-tech version of phrenology. A blind faith in folk psychology as the sole characterization of social cognition should be also carefully scrutinized. Chances that we will find "boxes" in our brain containing the neural correlates of beliefs, desires, and intentions as such, and independently from their specific content, probably amount to next to zero. Such a search looks to us like an ill-suited form of reductionism leading us nowhere.

In conclusion, we believe that a thorough understanding of social cognition and of the human condition must start from the multilevel comparative study of the premotor system of primate brains.

ACKNOWLEDGMENTS

This work was supported by MIUR (Ministero Italiano dell'Università e della Ricerca) and by the EU grants DISCOS and NESTCOM.

REFERENCES

Agnetta, B., & Rochat, P. (2004). Imitative games by 9-, 14-, and 18-month-old infants. *Infancy*, 6, 1–36.

Aziz-Zadeh, L., Wilson, S. M., Rizzolatti, G., & Iacoboni M. (2006) Congruent Embodied Representations for Visually Presented Actions and Linguistic Phrases Describing Actions. *Current Biology*, 16, 1818–1823.

Baird, J. A., & Baldwin, D. A. (2001). Making sense of human behavior: Action parsing and intentional inferences. In B. F. Malle, L. J. Moses, & D. A. Baldwin (Eds.), *Intentions and intentionality* (pp. 193–206). Cambridge, MA: MIT Press.

Baldwin, D. A., Baird, J. A, Saylor, M. M., & Clark, M. A. (2001). Infants parse dynamic action. *Child Development*, 72, 708–717.

Baron-Cohen, S., Leslie, A. M., & Frith, U. (1985). Does the autistic child have a "theory of mind"? *Cognition*, 21, 37–46.

Baron-Cohen, S. (1994). How to build a baby that can read minds: Cognitive mechanisms in mindreading. *Cahier de Psychologie Cognitive/Current Psychology of Cognition*, 13, 1–40.

Baron-Cohen, S. (1995). *Mindblindness*. Cambridge, MA: MIT Press.

Bekkering, H., Wohlschläger, A., & Gattis, M. (2000). Imitation of gestures in children is goal-directed. *Quarterly Journal of Experimental Psychology*, 53A, 153–164.

Buxbaum, L. J., Johnson, S. H., & Bartlett–Williams, M. (2005). Deficient internal models for planning hand-object interactions in ideomotor apraxia. *Neuropsychologia*, 43, 917–929.

Braten S. (1988). Dialogic Mind: The infant and the adult in protoconversation. In M. Carvallo (Ed.), *Nature, cognition, and system*, vol. I (pp. 187–205). Dordrecht: Kluwer Academic Publishers.

Braten, S. (1992). The virtual other in infants' minds and social feelings. In H. Wold (Ed.), *The dialogical alternative* (pp. 77–97). Oslo: Scandinavian University Press.

Braten, S. (2007). *On being moved: From mirror neurons to empathy* (p. 333). Philadelphia: John Benjamins Publishing Company.

Bruner, J. (1990). *Acts of meaning*. Cambridge, MA: Harvard University Press.

Buccino, G., Binkofski, F., Fink, G. R., Fadiga, L., Fogassi, L., Gallese, V. et al. (2001). Action observation activates premotor and parietal areas in a somatotopic manner: An fMRI study. *European Journal of Neuroscience, 13*, 400–404.

Buccino, G., Lui, F., Canessa, N., Patteri, I., Lagravinese, G., Benuzzi, F., Porro, C. A., & Rizzolatti, G. (2004a). Neural circuits involved in the recognition of actions performed by nonconspecifics: An fMRI study. *Journal of Cognitive. Neuroscience, 16*, 114–126.

Buccino, G., Vogt, S., Ritzl, A., Fink, G. R., Zilles, K., Freund, H.-J., & Rizzolatti, G. (2004b). Neural circuits underlying imitation learning of hand actions: An event-related fMRI study. *Neuron, 42*, 323–334.

Call, J., & Tomasello, M. (1998). Distinguishing intentional from accidental actions in orangutans, chimpanzees, and human children. *Journal of Comparative Psychology, 112*, 196–206.

Calvo-Merino, B., Glaser, D. E, Grezes, J., Passingham, R. E., & Haggard, P. (2005). Action observation and acquired motor skills: An fMRI study with expert dancers. *Cerebral Cortex, 15*, 1243–1249.

Calvo-Merino, B., Grèzes, J., Glaser, D. E., Passingham, R. E., & Haggard, P. (2006). Seeing or doing? Influence of visual and motor familiarity in action observation. *Current Biology, 16*(19), 1905–1910.

Carpenter, M., Call, J., & Tomasello, M. (2005). Twelve- and 18-month-olds copy actions in terms of goals. *Developmental Science, 8*, F13–F20.

Cattaneo, L., Fabbi-Destro, M., Boria, S., Pieraccini, C., Monti, A., Cossu, G., & Rizzolatti, G. (2008). Impairment of actions chains in autism and its possible role in intention understanding. *Proceedings of the National Academy of Sciences, 104*, 17825–17830.

Chiron, C., Raynaud, C., Mazière, B., Zilbovicius, M., Laflamme, L., Masure, M. C. et al. (1992). Changes in regional cerebral blood flow during brain maturation in children and adolescents. *Journal of Nuclear Medicine, 33*, 696–703.

Csibra, G., Gergely, G., Biró, S., Koòs, O., & Brockbank, M. (1999). Goal attribution without agency cues: The perception of "pure reason" in infancy. *Cognition, 72*, 237–267.

Csibra, G., Biró, S., Koòs, O. and Gergely, G. (2003). One-year-old infants use teleological representations of actions productively. *Cognitive Science, 27*, 111–133.

Cossu, G., Boria, S., Copioli, C. Bracceschi, R., Dalla Vecchia, A., Santelli, E., & Gallese, V. (submitted). Action representation in autistic children.

Dapretto, L., Davies, M. S., Pfeifer, J. H., Scott, A. A., Sigman, M., Bookheimer, S. Y., & Iacoboni, M. (2006). Understanding emotions in others: Mirror neuron dysfunction in children with autism spectrum disorders. *Nature Neuroscience, 9*, 28–30.

Dasser, V. (1988). Mapping social concepts in monkeys. In R.W. Byrne & A. Whiten (Eds.), *Machiavellian intelligence. Social expertise and the evolution of intellect in monkeys, apes and humans* (pp. 85–93). New York: Oxford University Press.

Dawson, G., Webb, S., Schellenberg, G.D., Dager, S., Friedman, S., Aylward, E., and Richards, T. (2002). Defining the broader phenotype of autism: Genetic, brain, and behavioral perspectives. *Development and Psychopathology, 14,* 581–611.

Diamond, A. (2000). Close interrelation of motor development and cognitive development and of the cerebellum and prefrontal cortex. *Child Development, 71,* 44–56.

Ertelt, D., Small, S., Solodkin, A., Dettmers, C., McNamara, A., Binkofski, F., and Buccino, G. (2007). Action observation has a positive impact on rehabilitation of motor deficits after stroke. *NeuroImage, 36,* 164–173.

Fadiga, L., Craighero, L., Buccino, G., and Rizzolatti, G. (2002). Speech listening specifically modulates the excitability of tongue muscles: A TMS study. *European Journal of Neuroscience, 15,* 399–402.

Falck-Ytter, T., Gredeback, G., & von Hofsten, C. (2006) Infants predict other people's action goals. *Nature Neuroscience, 9,* 878–879.

Fecteau, S., Carmant, L., Trambley, C., Robert, M., Bouthillier, A., & Théoret, H. (2004). A motor resonance mechanism in children? Evidence from subdural electrodes in a 36-months-old child. *NeuroReport, 15,* 2625–2627.

Ferrari, P. F., Maiolini, C., Addessi, E., Fogassi, L., & Visalberghi, E. (2005a). The observation and hearing of eating actions activates motor programs related to eating in macaque monkeys. *Behavioral Brain Research, 161,* 95–101.

Ferrari, P. F., Rozzi, S., & Fogassi, L. (2005b). Mirror neurons responding to observation of actions made with tools in monkey ventral premotor cortex. *Journal of Cognitive Neuroscience, 17*(2), 212–226.

Ferrari, P. F., Bonini, L., Ugolotti, F., Simone, L., Rozzi, S., Rizzolatti, G., & Fogassi, L. (2006a). Coding of motor intention in monkey parietal and premotor grasping neurons. *Social Neuroscience Abstracts,* 255.1/X13.

Ferrari, P. F., Visalberghi, E., Paukner, A., Fogassi, L., Ruggiero, A. et al. (2006b). Neonatal imitation in rhesus macaques. *PLoS Biology, 4*(9): e302.

Flanagan, J. R., & Johansson, R. S. (2003). Action plans used in action observation. *Nature, 424,* 769–770.

Flombaum, J. L., & Santos, L. R. (2005). Rhesus monkeys attribute perceptions to others. *Current Biology, 15,* 447–452.

Fogassi, L., Ferrari, P. F., Gesierich, B., Rozzi, S., Chersi, F., & Rizzolatti, G. (2005). Parietal lobe: From action organization to intention understanding. *Science, 302,* 662–667.

Gallese, V., Fadiga, L., Fogassi, L., & Rizzolatti, G. (1996). Action recognition in the premotor cortex. *Brain, 119,* 593–609.

Gallese, V., Fogassi, L., Fadiga, L., & Rizzolatti, G. (2002). Action Representation and the inferior parietal lobule. In W. Prinz, & B. Hommel, (Eds.), *Attention and Performance XIX* (pp. 247–266). Oxford, Oxford University Press.

Gallese, V. (2001). The "Shared Manifold" Hypothesis: From mirror neurons to empathy. *Journal of Consciousness Studies: 8,* N° 5–7; 33–50.

Gallese, V. (2003a). The manifold nature of interpersonal relations: The quest for a common mechanism. *Philosophical Transactions of the Royal Society of London, 358*, 517–528.
Gallese, V. (2003b). The roots of empathy: The shared manifold hypothesis and the neural basis of intersubjectivity. *Psychopathology*, 36(4), 171–180.
Gallese, V., Keysers, C., & Rizzolatti, G. (2004). A unifying view of the basis of social cognition. *Trends in Cognitive Sciences*, 8: 396–403.
Gallese, V. (2005). Embodied simulation: From neurons to phenomenal experience. *Phenomenology and the Cognitive Sciences, 4*, 23–48.
Gallese, V. (2006). Intentional attunement: A neurophysiological perspective on social cognition and its disruption in autism. *Experimental Brain Research: Cognitive Brain Research, 1079*, 15–24.
Gallese, V., & Umiltà, M. A. (2006). Cognitive continuity in primate social cognition. *Biological Theory*, 1(1), 25–30.
Gallese, V. (2007). Before and below theory of mind: Embodied simulation and the neural correlates of social cognition. *Proceedings of Royal Society of Biology, 362*, 659–669.
Gergely, G., Nàdasdy, Z., Csibra, G., & Bìrò, S. (1995). Taking the intentional stance at 12 months of age. *Cognition, 56*, 165–193.
Gergely, G., Bekkering, H., & Kiràly, I. (2002). Rational imitation in preverbal infants. *Nature, 415*, 755.
Hare, B., Call, J., & Tomasello, M. (2001). Do chimpanzees know what conspecifics know? *Animal Behaviour*, 61(1), 139–151.
Hommel, B., Müsseler, J., Aschersleben, G., & Prinz, W. (2001). The theory of event coding (TEC): A framework for perception and action planning. *Behavioral and Brain Sciences, 24*, 849–78.
Horner, V., & Whiten, A. (2005). Causal knowledge and imitation/emulation switching in chimpanzees (Pan troglodytes) and children (Homo sapiens). *Animal Cognition, 8*, 164–181.
Humphrey, N. K. (1974). Species and individuals in the perceptual world of monkeys. *Perception, 3*, 105–114.
Humphrey, N. K. (1976). The social function of intellect. In: P. Bateson, & R. A. Hinde, R.A. (Eds.), *Growing points in ethology* (pp. 303–321). Cambridge: Cambridge University Press.
Iacoboni, M., Woods, R. P., Brass, M., Bekkering, H., Mazziotta, J. C., & Rizzolatti, G. (1999). Cortical mechanisms of human imitation. *Science*, 286(5449), 2526–2528.
Iacoboni, M., Molnar-Szakacs, I., Gallese, V., Buccino, G., Mazziotta, J., & Rizzolatti, G. (2005). Grasping the intentions of others with one's owns mirror neuron system. *PLOS Biology, 3*, 529–535.
Järveläinen, J., Schurmann, M., & Hari, R. (2004). Activation of the human primary motor cortex during observation of tool use, *NeuroImage, 23*, 187–192.
Knoblich, G., & Flach, R. (2003). Action identity: Evidence from self-recognition, prediction, and coordination. *Consciousness & Cognition, 12*, 620–632.
Köhler, W. (1927). *The mentality of apes*. Vintage: New York.

Kohler, E., Keysers, C., Umiltà, M. A., Fogassi, L., Gallese, V., & Rizzolatti, G. (2002). Hearing sounds, understanding actions: Action representation in mirror neurons. *Science, 297*, 846–848.

Kugiumutzakis, G. (1993). Intersubjective vocal imitation in early mother–infant interaction. In J. Nadel & L. Camaioni (Eds.), *New perspectives in early communicative development* (pp. 23–47). London: Routledge.

Legerstee, M. (1991). The role of person and object in eliciting early in imitation. *Journal of Experimental Child Psychology, 51*, 423–433.

Lepage, J. F., & Théoret, H. (2007). The mirror neuron system: Grasping other's actions from birth? *Developmental Science, 10*, 513–529.

Leslie, A. M. (1994). *ToMM, ToBy, and agency: Core architecture and domain specificity*. In L. A. Hirschfield & S. A. Gelman (Eds.), *Mapping the mind: Domain specificity in cognition and culture* (pp. 119–148). New York: Cambridge University Press.

McIntosh, D. N., Reichman-Decker, A., Winkielman, P., & Wilbarger, J. (2006). When the social mirror breaks: Deficits in automatic, but not voluntary mimicry of emotional facial expressions in autism. *Developmental Science, 9*, 295–302.

Meltzoff, A. N., & Moore, M. K. (1977). Imitation of facial and manual gestures by human neonates. *Science, 198* (4312), 74–78.

Meltzoff, A., & Moore, M. K. (1994). Imitation, memory, and the representation of persons, *Infant Behavior and Development, 17*, 83–99.

Meltzoff, A. N. (1995). Understanding the intentions of others: Re-enactment of intended acts by 18-month-old children. *Developmental Psychology, 31*, 838–850.

Meltzoff, A. N., & Moore, M. K. (1998). Infant inter-subjectivity: Broadening the dialogue to include imitation, identity and intention. In S. Braten (Ed.) *Intersubjective communication and emotion in early ontogeny* (pp. 47–62). Paris: Cambridge University Press.

Meltzoff, A. N., & Brooks, R. (2001). "Like Me" as a building block for understanding other minds: Bodily acts, attention, and intention. In B. F. Malle, L. J. Moses, & D. A. Baldwin (Eds.), *Intentions and intentionality: Foundations of social cognition* (pp. 171–191). Cambridge, MA: MIT Press.

Mostofsky, S. H., Reiss, A. L, Lockhart, P., & Denckla, M. B. (1998). Evaluation of cerebellar size in attention-deficit hyperactivity disorder. *Journal of Child Neurology, 17*, 83–99.

Myowa-Yamakoshi, M., Tomonaga, M., Tanaka, M., & Matsuzawa, T. (2004). Imitation in neonatal chimpanzees (Pan troglodytes). *Developmental Science, 7*, 437–442.

Oberman, L. M., Hubbard, E. H., McCleery, J. P., Altschuler, E., Ramachandran, V. S., & Pineda, J. A. (2005). EEG evidence for mirror neuron dysfunction in autism spectrum disorders. *Cognitive Brain Research, 24*, 190–198.

Oberman, L. M., & Ramachandran, V. S. (2007). The simulating social mind: the role of the mirror neuron system and simulation in the social and communicative deficits of autism spectrum disorders. *Psychological Bulletin, 133,* 310–327.

Onishi, K. H., & Baillargeon, R. (2005). Do 15-month-olds understand false beliefs? *Science, 308,* 255–258.

Paukner, A., Anderson, J. R., Borelli, E., Visalberghi, E., & Ferrari, P. F. (2005). Macaques (Macaca nemestrina) recognize when they are being imitated. *Biology Letters, 1,* 219–222.

Piaget, J. (1952). *The origins of intelligence in children.* New York: International University Press.

Piaget, J. (1970). *Genetic epistemology.* New York: W.W. Norton & Co.

Povinelli, D. J., & Eddy T. J. (1996). What young chimpanzees know about seeing. *Monographs of the Society for Child Development, 61:* 1–152.

Povinelli, D. J. (2000). *Folk physics for apes: A chimpanzee's theory of how the world works.* Oxford: Oxford University Press.

Porro, C.A., Facchin, P., Fusi, S., Dri, G., & Fadiga, L. (2007). Enhancement of force after action observation: Behavioural and neurophysiological studies. *Neuropsychicologia, 45,* 3114–3121.

Prather, J. F., Peters, S., Nowicki, S., & Mooney, R. (2008). Precise auditory-vocal mirroring in neurons for learned vocal communication. *Nature, 451,* 305–310.

Premack, D., & Woodruff, G. (1978). Does the chimpanzee have a theory of mind? *Behavioral and Brain Sciences, 1,* 515–526.

Premack, D. (1990). The infant's theory of self-propelled objects. *Cognition, 36,* 1–16.

Prinz, W. (1987). Ideo-motor action. In H. Heuer & A. F. Sanders (Eds.), *Perspectives on perception and action.* Hillsdale, NJ: Erlbaum.

Rizzolatti, G., Camarda, R., Fogassi M., Gentilucci M., Luppino G., & Matelli M. (1988). Functional organization of inferior area 6 in the macaque monkey: II. Area F5 and the control of distal movements. *Experimental Brain Research, 71,* 491–507.

Rizzolatti, G., Fadiga, L., Gallese, V., & Fogassi, L. (1996). Premotor cortex and the recognition of motor actions. *Cognitive Brain Research, 3,* 131–141.

Rizzolatti, G., & Gallese, V. (1997). From action to meaning. In J.-L. Petit (Ed.), *Les neurosciences et la philosophie de l'action* (pp. 217–229). Paris: Librairie Philosophique J. Vrin.

Rizzolatti, G., Fogassi, L., & Gallese, V. (2000). Mirror neurons: Intentionality detectors? *International Journal of Psychology, 35,* 205–205.

Rizzolatti, G., Fogassi, L., & Gallese, V. (2001) Neurophysiological mechanisms underlying the understanding and imitation of action. *Nature Neuroscience Reviews, 2,* 661–670.

Rizzolatti G., & Craighero, L. (2004). The mirror neuron system. *Annual Review of Neuroscience, 27,* 169–192.

Rochat, M., Serra, E., Fadiga, L., & Gallese, V. (2008). The evolution of social cognition: Goal familiarity shapes monkeys' action understanding. *Current Biology, 18*, 227–232.

Rogers, S. (1999). An examination of the imitation deficit in autism. In J. Nadel & G. Butterworth (Eds.), *Imitation in infancy* (pp. 254–279). Cambridge: Cambridge University Press.

Rogers, S., & Williams, J. (Eds.) (2006). *Imitation and the social mind: Autism and typical development*. New York: Guilford Press.

Santos, L. R., Nissen, A. G., & Ferrugia, J. A. (2006). Rhesus monkeys, Macaca mulatta, know what others can and cannot hear. *Animal Behavior, 71*, 1175–1181.

Shimada, S., & Hiraki, K. (2006). Infant's brain responses to live and televised action. *NeuroImage, 32*, 930–939.

Schmitz, C., Martineau, J., Barthélemy, C., & Assaiante, C. (2003). Motor control and children with autism: Deficit of anticipatory function? *Neuroscience Letters, 348*, 17–20.

Snowdon, C. (1986). Vocal communication. In G. Mitchell & J. Erwin (Eds.), *Comparative primate biology, vol. 2A: Behaviour, conservation, and ecology* (pp. 3–38). New York: Alan R. Liss.

Sommerville, J. A., & Woodward A. (2005). Pulling out the intentional structure of action: The relation between action processing and action production in infancy. *Cognition, 95*, 1–30.

Sommerville, J. A., Woodward, A., & Needham, A. (2005). Action experience alters 3-month-old's perception of other's actions. *Cognition, 96*, 1–11.

Stern, D.N. (1985). *The interpersonal world of the infant*. London: Karnac Books.

Subiaul, F., Cantlon, J. F., Holloway, R. L., & Terrace H. S. (2004). Cognitive imitation in rhesus macaques. *Science, 305*, 407–410.

Theoret, H., Halligan, E., Kobayashi, M., Fregni, F., Tager-Flusberg, H., & Pascual-Leone, A. (2005). Impaired motor facilitation during action observation in individuals with autism spectrum disorder. *Current Biology, 15*, 84–85.

Tomasello, M., Davis-Dasilva, M., Camak, L., & Bard, K. (1987). Observational learning of tool use by young chimpanzees. *Journal of Human Evolution, 2*, 175–83.

Tomasello, M., Kruger, A., & Ratner, H. (1993). Cultural learning. *Behavioral and Brain Sciences, 16*, 450–488.

Tomasello, M., & Call, J. (1997). *Primate cognition*. Oxford: Oxford University Press.

Tomasello, M., Carpenter, M., Call, J., Behne, T., & Moll, H. (2005). Understanding and sharing intentions: The origins of cultural cognition. *Behavioral and Brain Sciences, 28*, 675–691.

Trevarthen, C. (1979). Communication and cooperation in early infancy: A description of primary intersubjectivity. In M. Bullowa (Ed.), *Before speech: The beginning of interpersonal communication* (pp. 321–347). New York: Cambridge University Press.

Trevarthen, C. (1993). The self born in intersubjectivity: An infant communicating. In U. Neisser (Ed.) *The perceived self* (pp. 121–173). New York: Cambridge University Press.

Tronick, E. (1989). Emotion and emotional communication in infants. *American Psychologist, 44*, 112–119.

Umiltà, M. A., Kohler, E., Gallese, V., Fogassi, L., Fadiga, L., Keysers, C., & Rizzolatti, G. (2001). "I know what you are doing": A neurophysiological study. *Neuron, 32*, 91–101.

Umiltà, M. A., Escola, L., Intskirveli, I., Grammont, F., Rochat, M., Caruana, F., Jezzini, A., Gallese, V., & Rizzolatti, G. (2008). How pliers become fingers in the monkey motor system. *Proceedings of the National Academy of Science, 105*, 2209–2213.

Visalberghi, E., & Tomasello, M. (1998). Primates causal understanding in the physical and psychological domains. *Behavioural Process, 42*, 189–203.

Vogt, S., Buccino, G., Wohlschläger, A. M., Canessa, N., Shah, N. J., Zilles, K., Eickhoff, S. B., Freund, H. J., Rizzolatti, G., & Fink, G. R. (2007). Prefrontal involvement in imitation learning of hand actions: Effects of practice and expertise. *NeuroImage, 37*, 1371–1383.

Williams, J. H., Whiten, A., Suddendorf, T., & Perrett, D. I. (2001). Imitation, mirror neurons and autism. *Neuroscience & Biobehavioral Reviews, 25*, 287–95.

Williams, J. H., Whiten, A., & Singh, T. (2004). A systematic review of action imitation in autistic spectrum disorder. *Journal of Autism Developmental Disorders, 34*, 285–299.

Wimmer, H., & Perner, J. (1983). Beliefs about beliefs: Representation and constraining function of wrong beliefs in young children's understanding of deception. *Cognition, 13*, 103–128.

Woodward, A. L. (1998). Infants selectively encode the goal object of an actor's reach. *Cognition, 69*, 1–34.

Zoia, S., Blason, L., D'Ottavio, G., Bulgheroni, M., Pezzetta, E., Scabar, A., & Castiello, U. (2007). Evidence of early development of action planning in the human foetus: A kinematic study. *Experimental Brain Research, 176*, 217–226.

3

THE CONSTRUCTION OF COMMONSENSE PSYCHOLOGY IN INFANCY

Chris Moore
John Barresi
Dalhousie University

CONSTRUCTION OF COMMONSENSE PSYCHOLOGY IN INFANCY

The discovery of "mirror neurons" in the mid-1990s (Gallese, Fadiga, Fogassi, & Rizzolatti, 1996; Rizzolatti, Fadiga, Gallese, & Fogassi, 1996) was greeted with considerable excitement by many in the field of the psychology of social understanding because these specialized cells appeared to provide a neural substrate for how humans, and their close phylogenetic relatives are able to make sense of the goal-directed actions of others. The existence of mirror neurons appealed particularly to a subset of developmentalists who had been arguing that the development of specifically human forms of social understanding, including the understanding of intentionality and children's "theory of mind," must depend upon the capacity to integrate information derived from the observation of others' actions and the execution of the self's actions. This is because human social understanding recognizes the similarity of both self and others as intentional and mental agents even though the information available about self and others is radically different and in some ways incommensurable. In a paper reviewing theories of the origins of social understanding in infancy, Moore (1996) termed such accounts "matching" theories. Matching theories are those theories that assign a key developmental role to infants' participation in social interactive situations in which the actions of self and other are matched. In such situations, infants will be provided with information about others' actions and information about their own actions so that information processing mechanisms such as intermodal integration can serve to generate amodal representations of those actions, representations that can be applied equivalently to the observation of both others' and one's own action.

In the mid-1990s, there were two main variants of matching theories. One theory, expressed most comprehensively by Meltzoff and Gopnik (Gopnik & Meltzoff, 1994; Meltzoff & Gopnik, 1993; see also Rogers & Pennington, 1991) was that infants are born with a fundamental ability to recognize the similarity between their own actions and those of others. This ability is expressed through bodily imitation, such that the observation of others' action leads to an attempt by the infant to reproduce the observed action. This starting state provides the foundation of matched action of self and other upon which more sophisticated representations of intentionality are built. In recent years, Meltzoff (e.g., 2007) has elaborated this account and linked it explicitly with the existence of the mirror neuron system (MNS). In short the MNS is supposed to be present at birth and the basis upon which early imitation is made possible.

The second theory (e.g., Barresi & Moore, 1993; 1996; Moore & Corkum, 1994) argued similarly for the importance of building an understanding of intentionality on the cooccurrence of matched actions of self and other. However, in contrast to Meltzoff's theory, the claim was that matching of actions was not provided through infants imitating others but through a variety of social activities in which infants and others would come to share an intentional orientation. Moore (2006a) provided a more expanded treatment of this account. In brief, the core idea is that social interactive structures present in infancy, including dyadic interactions, and then triadic or joint attentional interactions, provide the appropriate information conditions within which mirrored representations of the actions of self and other are generated. In this account, therefore, it is not necessary for the MNS to be present at birth; its important properties could be acquired through experience (see also Barresi & Moore, 2008).

In this chapter, we are not going to attempt to resolve the issue of whether the MNS is present at birth or depends on postnatal experience of matched actions of self and others. This question is clearly an important one, and it has been perhaps too often implicitly assumed that the existence of the MNS in the brain implies that it is innate. On reflection, this assumption is obviously false; an enormous amount of neural development depends on experience. So there is certainly no reason to believe that the mere existence of mirror neurons in mature brains means that they must have been present all along. Some have interpreted evidence on neonates' apparent ability to match observed action with action execution as support for the existence of a MNS at birth (e.g., LePage and Theoret, 2007). However, because there is as yet no strong consensus on the existence or meaning of neonatal matching abilities, it is premature to conclude that the evidence comes down strongly in favor of a MNS as a starting state for social understanding.

Our goal in this chapter is to address a different, albeit related issue: To what extent do infants understand intentional action as a characteristic of individual persons as opposed to a characteristic of action systems? As adults, we have an individualistic form of social understanding whereby we assign intentionality to persons, so it is natural to assume that the personal level of understanding intentionality is the default one and that it also characterizes the understanding of intentionality all the way down, developmentally. However, we (Barresi & Moore, 1996; Moore, 2006a,b) have long argued that intentional understanding only becomes elaborated at the level of persons at the end of infancy. It is at this point that both self and others are understood to be independent intentional agents of the same kind; in particular, agents that have an existence as independent objective entities with separate subjective perspectives. Prior to this point, we argue that the understanding of intentionality is piecemeal and subpersonal or, in some interactive contexts, interpersonal. We argue that the individualistic conception of persons as centers of intentionality is the culmination of the development of social understanding during infancy. In this chapter we will review some of the evidence that supports this interpretation of infant social understanding and elaborate on how the development of social understanding proceeds through essentially constructive information processing mechanisms.

In this review, we will draw on two main types of evidence—experiments that use looking time approaches as a window onto infants' representation of intentional action, and experiments that have manipulated the interactive context in which infants can respond to the intentionality of others. The first line of research is consistent with the idea that young infants' representations of intentional action are acquired in a relatively piecemeal way, initially limited to particular action systems, and only gradually integrated into more complex coordinations of intentional relations. The second line of research is consistent with the view that infants first understand the intentionality of others to the extent that they can participate with them in shared object-centered activity. Only during the second year of life do children develop an understanding of both self and other as independent intentional agents (Moore, 2007).

LOOKING-TIME APPROACHES TO ASSESSING INFANT UNDERSTANDING OF INTENTIONALITY

In the last ten years, it has become increasingly popular to explore infants' representation of intentional action by using looking-time methods. This approach was pioneered by Woodward (1998), and her group has produced the most systematic body of work to date (Woodward, 2005). Woodward (1998) introduced the switch methodology whereby the infant is initially habituated to an event involving an action directed at one of two objects. The logic of the method is based on the idea that intentional action can potentially be decomposed into the spatiotemporal form of the action and its object- or goal-directness. After habituation is achieved, the position of the two objects is switched, and the infant shown two test events—one involving the same action (same spatiotemporal form) now directed at the other object and the other a different action directed at the same object as in the habituation trials. In this way the action is separated into its spatiotemporal form and its object-directedness, and the particular feature that infants selectively attend to can be ascertained. To represent an action as intentional, it is essential that the object-directedness be selectively attended. This basic design logic is consistent with that used in many habituation studies of infant concepts (Cohen, Chaput, & Cashon, 2002; Moore & Corkum, 1994) and can be used to examine infants' representations of the intentionality of a range of actions. It can also be extended to examine the extent to which infants attribute intentionality to particular action systems, to action complexes, and to individual agents.

Horizontal Decalage

In her initial work, Woodward (1998) showed that infants as young as 5 to 6 months of age represent manual reaching and grasping actions as intentional. Follow-up studies showed that under certain condition, even infants as young as 3 months will represent such actions in an object-directed way (Sommerville, Woodward, & Needham, 2005). In subsequent work, Woodward and her colleagues (Woodward, 2003; Woodward & Guajardo, 2002) applied the switch methodology to epistemic actions, such as gaze and pointing. These studies revealed that it was not until later in the first year that such actions were represented by infants as object-directed. Woodward (2003) found that 12-month-olds, but not younger, infants appear to represent gaze as object-directed. Similarly, Woodward and Guajardo (2002) found that 12-month-olds but not 9-month-olds represented pointing gestures that directly contacted an object as object-directed. Notably, Woodward (2003) directly compared infants' performance on reaching and gaze versions of the switch procedure and found that whereas 7- and 9-month-olds clearly represented reaching as object-directed, they did not represent gaze in this way. In general studies from other labs have been consistent, with reaching understood as object-directed relatively early (e.g., Király, Jovanovic, Prinz, Aschersleben, & Gergely, 2003), whereas "epistemic" actions, such as gaze, and communicative actions, such as pointing, are understood relatively late in the first year (e.g., Johnson, Ok, & Luo, 2007; Moore, 1999; Sodian & Thoermer, 2004).

These studies show obvious lags in the onset of intentional representations for different kinds of action—lags that Piaget (e.g., Piaget, 1976) would have referred to as "horizontal décalage." The reason for these lags is unclear at present but a few points should be noted. First, there appears to be independent developmental onset of the intentional representations of different actions. Clearly, some actions (particularly manual contact actions such as reaching and grasping) are represented as object-directed rather earlier than others (particularly epistemic actions at a distance). In addition, correlational work reveals that the onsets of representations of two actions quite similar in the timing of their occurrence as components of triadic interactions, as well as their referential function—pointing and gaze—are at best only weakly associated. Brune and Woodward (2007) report no significant correlation between infants' understanding of gaze as object-directed and their understanding of pointing as object-directed, although there was some indication from factor analysis of a common underlying factor. This finding suggests that these representations of quite similar or cooccurring intentional actions may have somewhat independent developmental histories and are not initially related manifestations of the same underlying understanding of the intentionality of human attention.

Second, the developmental onset of intentional representation of different actions appears to be in part associated with infants' own abilities to perform such actions in intentional ways. Again, Woodward and her colleagues have provided the clearest evidence on this issue. In one study, Woodward and Guajardo (2002; see also Brune & Woodward, 2007) found that at a transitional point in the development of pointing production around 9- to 10-months, it was the infants who themselves had started to point that were more likely to understand pointing as object-directed. Although understanding gaze as object-directed is obviously unrelated to infants' ability to gaze at objects themselves, Brune and Woodward (2007) report that it is related to infants' tendency to engage in shared attention to objects with others. Even more compelling evidence comes from training research. Sommerville, and colleagues (Sommerville et al., 2005) trained 3-month-old infants who had not yet started to show visually guided reaching to catch objects by "batting" at them with Velcro-covered mittens. They then participated in a habituation experiment in which the switch method was used to show them mitten-covered hands reaching for one of two objects. A second group of 3-month-olds were tested in the habituation phase and then given the training experience. Only the infants who had first received the action experience responded to the new-object events in the test phase of the habituation experiment.

Together these lines of evidence suggest that infants do not have a single concept of intentional action that guides their understanding of all object-directed activity (Woodward, Sommerville, & Guajardo, 2001). Rather, representations of object-directed action come online at different points in development during the first year and are relatively independent of each other. One explanation for these lags appears to be linked to developmental lags in infants' own production of intentional action. Of particular interest are the results that for those actions where there is intramodal perceptual overlap between own and others' action, such as manual actions like reaching and pointing, action production and action understanding go hand-in-hand.

Whereas the understanding of the actions of others is correlated with infants' production of those actions, the capacity for mere production of a movement is not sufficient. The intention involved in the movement must also be appreciated. Thus, the capacity to point involves more than mere movement of one's hand forward with outstretched finger, which occurs early in infant development (Masataka, 2003; Tomasello, Carpenter & Liskowski, 2007). The difference between reaching and pointing has a social basis. While one can reach and perhaps thereby also understand the reach of another from the experience of reaching of self and other in noninteractive contexts, pointing and understanding pointing requires an appreciation of the meaning of the act in an interactive context.

The intentional activity of gazing at an object requires a further differentiation, since gaze requires more than an intramodal matching of an action that one can produce and a similar action that one can observe. Unlike reaching and pointing, gaze activity is not perceptually available through the common visual modality. Therefore, another means of matching self and other is required. In this case it is matching through the context of action; infants' understanding of gaze is linked to their participation in shared activity with others (Brune & Woodward, 2007). Thus, shared or matched activity appears particularly important for the understanding of the object-directedness of those intentional relations for which information is radically different across observation and execution.

Personal Versus Subpersonal Representations of Intentional Action

Although it has sometimes been assumed that response patterns using the kinds of stimuli reviewed so far may reveal an understanding of the intentionality of agents, many of these studies are fundamentally indeterminate on whether young infants represent object-directedness as a property of individual persons or of particular actions. Many of the looking-time studies using manual actions have used stimuli that include only an arm and hand (Király et al., 2003; Woodward, 1998; Sommerville et al., 2005; but see Woodward, 2003). Thus, the stimuli themselves have been presented at a "subpersonal" level, and it is impossible to conclude anything about the extent to which infants' understanding of object-directedness is at a personal or subpersonal level. Those studies examining epistemic actions such as attention and communicative actions like pointing have typically used stimuli involving a whole person as the actor (e.g., Moore, 1999; Sodian & Thoermer, 2005; Woodward, 2003; Woodward & Guajardo, 2002). Although infants have been shown to represent object-directedness of such actions, it is not possible to know whether the object-directedness is linked to the whole person or just their particular action.

One approach to this issue is to determine whether infants limit their intentional representations to particular agents or whether they will transfer such representations across individuals. Empirically, the approach is to use the switch paradigm but change the adult across habituation and test trials. If infants understand the object-directedness as a property of actions rather than people, then they should not be disturbed if the identity of the actor changes from habituation to test, so long as the same action is involved. In contrast, an infant who understands object-directedness as a property of the actor rather than just the action should not show the usual pattern if the identity of the actor changes from habituation to test.

Only two studies have directly addressed this issue—one using reaching (Buresh & Woodward, 2007) and the other using distal pointing (Moore, 1999)—so the data so far are very limited. Buresh and Woodward (2007) showed infants of 9 and 12 months of age reaching and grasping actions similar to those used by Woodward (1998). After habituation to trials showing one adult reaching for one of two objects, the positions of the two objects were switched and then infants saw test trials involving a different adult reaching for the same object in the new location or the other object in the same location. They reported that infants at both ages failed to discriminate between the two test trial types. In contrast, control infants shown the same adult in both habituation and test trials looked more to the new object test events as Woodward (1998) had previously reported. Moore used a similar approach for pointing actions with 13-month-olds. He reported that the infants did discriminate the test trial types even when different adults were present in habituation and test events.

These two studies therefore report different results with Buresh and Woodward (2007) finding infants as young as 9 months appearing to understand reaching as a property of individual persons and Moore (1999) finding that 13-month-olds appearing to understand pointing as property of the action system, not the person. Although this difference may appear to be an inconsistency, it should be remembered that these two types of action are first understood as object-directed at different points in the infancy. So the time at which infants understand object-directedness as a property of individual agents may well depend upon the particular action that is involved. Evidently, further systematic research is needed to clarify whether there is a lag in understanding object-directedness as a property of actors consistent with the lag generally observed in infants' understanding of object-directedness of particular actions.

Intentionality in Action Combinations

Another approach to asking whether infants' intentional understanding is tied to particular action systems is to examine whether infants understand that different actions may reflect the same intention. For example, actions sequences can be organized in terms of an overarching goal such as in means–ends sequences (Woodward, 2005). In means–ends sequences, one action is used to set up the conditions for another action directed at a target object, for example, a cloth on which a toy is placed out of reach may be pulled to bring the toy within reach (Piaget, 1953). Woodward and Sommerville have studied infants' representations of such action combinations in a series of studies using a variant of the switch design (Sommerville & Woodward, 2005a,b; Woodward & Sommerville, 2000). They have found that by 12 months, infants understand the first action (e.g., pulling the cloth) to be directed at the ultimate goal object (the toy) rather than the intermediate object acted on (the cloth). Interestingly, although as a group 10-month-olds did not appear to understand the first action in a means–ends sequence as directed at the final goal, there was evidence that infants who themselves were performing such means–ends sequences did do so. Thus, as was the case for single actions, infants' action production was correlated with their action understanding.

In addition, just as transfer of intention across actors has been studied (as reviewed in the previous section), so transfer of intention across actions can also be studied. Phillips and colleagues (Phillips, Wellman, & Spelke, 2002, Experiments 3 and 4) familiarized 12- and 14-month-olds with an actor looking at and emoting positively towards one of two toys. They then present test trials in which the actor was shown holding either the same toy as attended during familiarization or the other toy. Only the older group discriminated these two test events, showing more recovery to the event in which the actor held the toy not previously attended. These results suggest that the older infants had an expectation about which toy the actor would hold based on her prior pattern of gaze and emotion expression. It is possible that recognizing the associations among different intentional actions may be simply a matter of detecting the statistical patterns of object-directed activity. That is, because people tend to grasp the same things that they look at, infants may acquire an expectation that grasping will follow looking at. However, it may also reveal that the understanding of intentional activity has become an abstraction across particular object-directed actions.

Conclusions From Looking-Time Work

The recent body of work using looking-time approaches has provided considerable insight into infants' intentional understanding. It is clear that from relatively early in the first year, infants do start to represent the intentional relations between actors and objects. Based on the evidence reviewed, a number of general conclusions may be drawn. First, there is no single insight into intentionality; different actions are represented as object-directed at different points during infancy. Second, infants appear to represent intentionality as a property of particular actions initially before representing it as a property of actors. Third, infants appear to tie intentionality to the proximal action that is directed at an object before becoming able to tie the intentionality to a sequence or combination of actions.

Together, these findings reveal that the understanding of intentionality during infancy comes gradually. It develops first for particular actions, or as Moore (2006b) termed them, "intentional islands," and only later becomes more abstract as it is applied to the actor rather than the action and to action combinations. This process appears wholly compatible with the kind of hierarchical constructivist concept formation processes that characterize many aspects of infant cognitive development (Cohen et al., 2002). Perhaps unlike many of these other areas of conceptual development, the understanding of intentional action also draws on action execution such that the first-person experience of intentional activity provides some of the relevant experience out of which intentional concepts are constructed. It is highly likely, therefore, that these concepts have multimodal mirror characteristics in that they may code own and others' action equivalently. However, this does not therefore entail that they are yet fully individualistic concepts of intentionality at the level of persons.

The way the mirror neuron system appears to operate is by associating the first-person intentional aspect of an action acquired from personal experience of action execution to the third-person observation of the action of others. This association occurs without necessarily linking a unified intentional understanding of the action to the other actor as an intentional agent who has a first-person form of experience of the action. The first-person aspect is, in a sense, projected onto the act, localized as co-occurring with the movement of the actor, but is not attributed to the actor. It gives meaning to the act, and anticipates what is about to happen, without linking it to the agent per se as an experiential process in an agent separate from self. As a result, the distinction between self and other is not required in order to understand simple object-directed actions such as reaching for or manipulating an object (e.g., Knoblich & Jordan, 2003).

This form of understanding may well occur with the infant in passive perception of the actions of other agents in similar contexts. The evidence reviewed above suggests that such understanding requires prior production of the actions by the agent, and perhaps perception of those actions through a common modality mediating self and other—such as vision or audition. Through such modalities, even complex acts and causal sequences of acts can be linked together and understood. Their association to a particular actor may also come to be appreciated. However, even with association to an individual actor, there still need not be an understanding of the actor as an intentional agent of the same kind as self. For that form of understanding, more is required. Furthermore, for intrinsically interactive intentional relations such as pointing and triadic eye gaze, which seem to require interaction contexts for learning, more is required than passive perception and projection from self to other of the intentional aspects associated with movements.

Active matching and perhaps imitation of the action is also required in order to generate shared understanding. In infancy, such active matching first occurs systematically during the triadic period characterizing the last few months of the first year.

INTERACTIVE APPROACHES TO ASSESSING INFANT UNDERSTANDING OF INTENTIONALITY

We have seen how infants begin to represent the intentional actions of others. There is now a great deal of evidence that during the second half of the first year, infants also become able both to respond to, and to influence, the object-directed activity of others. The onset of the phase of triadic interaction (Bakeman & Adamson, 1984) sees the infant able to follow and direct other people's attention, imitate their actions on objects, and use their emotional orientation towards objects to guide their own activity. Such social interactive abilities certainly look as if infants are now able to understand that others have intentional relations to objects. However, all of these forms of social interaction involve the infant coordinating her own action with that of the interactive partner. In the process, the infant's and the partner's intentional relations become aligned or shared. As such the infant can use the other's action to guide her own orientation to objects without necessarily representing the other's action as independently object-oriented. A simple example is provided by the case of gaze-following, in which infants will turn in the same direction as an interactive partner's gaze shift. To follow gaze successfully, all infants have to know is that the other's gaze is a good predictor of where interesting sights may be found. They do not need to represent the other as looking at something. And, indeed, research on gaze-following reveals clear differences in how infants process gaze in comparison to older children. For example, infants do not follow gaze based on eye direction until well into the second year (Moore & Corkum, 1998). Furthermore, whereas 12-month-olds appear most concerned with using gaze to find interesting sights, by 24 months, children appear more concerned with finding out what the interactive partner is looking at (Moore & Povinelli, 2007).

Tomasello et al. (2007) have recently suggested that infant-pointing provides evidence that even 12-month-olds have a concept of an intentional agent, though perhaps one that applies only in contexts of cooperative activity or shared intentionality. They are particularly impressed by infants' communicative use of pointing and eye gaze in joint attention, and feel that communication of this sort requires the concept of intentional agent. One of the experiments that they cite in support of an individualistic account of intentional agent is by Liszkowski, Carpenter, Henning, Striano, & Tomasello (2004). In this study, 12-month-old infants were presented with a novel interesting object in an interactive context with an adult who appears unaware of the object. The infant is sometimes induced to point at the object for the adult. The subsequent behavior of the adult in response to infant pointing was manipulated so that the adult: (a) looked at the infant and smiled; (b) looked at the object showing interest, but not at the infant; (c) did not respond to the point; or (d) engaged in joint attention by looking at the object and back at the infant with a positive emotional response. The experiment nicely showed that infants were dissatisfied with all the conditions other than the joint attention condition, and sometimes continued pointing in these other conditions. The question that arises here is how rich a motivation, as well as an understanding of mental states of others, should we ascribe to the infant.

Although we find much to agree with in Tomasello and colleagues' account of shared intentionality (Tomasello, Carpenter, Call, Behne, & Moll, 2005), we do not believe that their rich interpretation of infants' intentional understanding during the early triadic period of shared intentional activity is required (see also D'Entremont & Seamans, 2007). We do not doubt that infants of this age can understand variations in shared intentional relations in triadic interactions, but this understanding does not by itself require a notion of individuals as having their own particular intentional states. It is during this period that the infant is actively involved in sharing attention to objects with another person, so it is not surprising to find them able to appreciate the object-directed pointing behavior of others and to point themselves at objects for the other individual. Likewise, they can use and understand visual attention through eye gaze (or at least head orientation). Neither of these activities requires them to understand intentional relations as properties of individual agents. They need only determine whether there is a match or mismatch in object-directed activities of self and other, and take the appropriate actions necessary to produce the preferred state of matched or shared intentional relations.

The straightforward interpretation is that the infant wishes to share with the adult her experience of the novel object, and she points in order to draw the adult's attention to the object so that they can share the experience. As the various inadequate outcomes of her pointing show, she is dissatisfied when the point fails to achieve a shared intentional relation with the adult directed at the object. However, this dissatisfaction at each of the negative outcomes does not require that the infant have the conscious communicative intention of manipulating the individualistic "mental state" of the adult. She no doubt brings about such a change in individual mental attitude or intentional relation of the adult, but to view the infant's pointing as having this effect as its explicit, or even implicit, goal is to ascribe too rich an account to the infant, who may as yet have no understanding that individuals have distinct mental states. A more parsimonious reading of this same evidence is that she is manipulating the form of "shared activity" or intentional relations, from one in which there is less overlap to one in which there is more overlap between infant and adult. The target here is to draw the adult to share attention to a novel state of affairs in the world. It is not that the infant sees her target as one involving her own or the other's state of mind, either individually or collectively, but rather that she appreciates that previous shared intentionality between self and other toward the world about them has changed, and her aim is to redirect the other's behavior to reinstate the situation of shared intentionality toward changed aspects of that world.

In general, then, triadic forms of social interaction create a form of shared intentionality (Tomasello et al., 2005) whereby infant and interactive partner together participate in complementary actions directed at an object of shared interest. At this level of development, therefore, infants can be said to understand intentionality as a property of interactions, not of individual persons (Barresi & Moore, 1996). Perhaps the best illustration of the difference between interactive and individualistic forms of understanding comes from a recent study by Moll and Tomasello (2007). In this study, 14- and 18-month-old infants saw an adult play with two objects in one of two key conditions. In one condition (joint engagement), the infant and adult played together with the two objects and in the other condition (individual engagement), the adult played alone with the objects while the infant simply watched. For a third target object in both conditions, the adult left the room and the infant played with it. The adult then reentered the room and sat with the infant. With all three objects placed on a tray in front of the infant, the adult expressed excitement and animatedly asked the infant for an object while not indicating in any way which object was of such interest to her. The results showed that children of both ages in the interactive condition tended to give the adult the target object. However, only 18-month-olds in the individual engagement condition gave the target object to the adult. Thus, younger children were able to distinguish the objects based on whether they had previously used them in a shared interaction with the adult. However, they were not able to distinguish the objects based on the adult's individual experience or lack of experience with them. In contrast, by the middle of the second year, infants appear to be able to determine the object of another's interest even when their own experience with the objects would not allow a differentiation. Because his or her own experience with the objects differed from that of the adult in the individualistic case, it required a differentiation of self from other in order to determine the adult's interest. Only in the 18-month-olds could that individualistic kind of discrimination of intentional relations of self and other be made.

Understanding intentional relations as descriptive of independent individual agents requires the discrimination of the intentional relations of self and other. And to clearly discriminate the intentional relations of self and other, there has to be an awareness of the equivalence of self and other as separate agents with unique perspectives. As such, there are two sides to this discrimination (Moore, 2007). On the one hand, and perhaps most obvious, the child must understand that other people's intentional activity is accompanied by a first-person perspective that is different from their own. On the other hand, the child must be aware of the objective or third-person side of self—that the self has an objective nature as an agent just as others do. Overall, the evidence is quite consistent that children develop both sides of this understanding during the second year rather than during the first (see, e.g., Moore, 2006a).

Recently, Moore (2007) argued that the critical tests of children's understanding of intentionality as a property of individuals are those in which the intentional relation of the other person is explicitly in conflict with the child's own. This is because it is only in such cases that shared intentionality cannot provide the basis for the child's response. There are now various pertinent experimental demonstrations. A well-known study by Repacholi and Gopnik (1997) provides relevant evidence. Repacholi and Gopnik presented children of 14 and 18 months with two bowls, one containing crackers and the other raw broccoli. Children were first asked which food they liked, and they all chose the crackers. Another adult then sampled the two foods in turn. To the crackers she reacted with a clear expression of disgust, whereas to the broccoli she reacted with pleasure. The toddlers were then asked to give one of the foods to the same adult. In this situation a clear difference between the children's own preference and the adult's preference was set up, and the children could react based either on their own preference or on the adult's expressed preference. The results showed a developmental difference with only the 18-month-olds reliably giving the adult the broccoli for which she had previously expressed a liking. The 14-month-olds, in contrast, were more likely to give the adult the snack that they themselves liked—the crackers. So only at 18 months did children appear to represent the independent and individual preferences of self and other.

Research on visual perspective-taking also provides consistent evidence. Both classic and more recent research on level I perspective taking shows that children first become capable of understanding that another person may see something different from self between 18 and 24 months (e.g., Lempers, Flavell, & Flavell, 1977; Moll & Tomasello, 2006).* For example, Lempers and colleagues (1977) showed children a picture fixed to the inside face of a hollow wooden cube, so that the picture was only visible by looking into the cube. When asked to show the picture to the experimenter, 18-month-olds appeared to need to maintain the picture as part of their own immediate visual experience in order to show it to another person. So, many of the younger toddlers showed an interesting tendency to locate the cube open end up between the self and the adult and tilt it back and forth so that both could see in it. By 24 months, the majority of participants were quite able to show the picture even if to do so would mean they could no longer see it; they would orient the cube so that the other could see in it. This pattern shows that the younger toddlers needed to maintain their own visual relation to the picture in order to share it with another person; they did not appear to appreciate that the other person's visual relation could be independent of their own. In contrast, the older toddlers did seem to realize that the other person would see the picture even when they themselves did not. As such, they had a firmer grasp of the independence of their own and the other person's visual relation to the picture.

Research on self-awareness still lags behind that on understanding others. However, there are some well-established findings that point to 18 months as a time of transition (Moore, 2007). Most well known is the phenomenon of mirror self-recognition, whereby children show self-directed behavior when faced with a mirror image of themselves after having been surreptitiously marked with make-up or a sticker (e.g., Amsterdam, 1972; Bertenthal & Fischer, 1978; Lewis & Brooks-Gunn, 1979; Nielsen, Dissanayake, & Kashima, 2003). More recently,

---

* Recently, it has been suggested that younger infants show level I perspective taking in looking-time procedures (e.g., Sodian, Thoermer, & Metz, 2007). However, such results are ambiguous because in these procedures, there is no explicit conflict between the perspectives of the self and the actor. Thus, to succeed, infants could simply track the relations between the actor and the objects in much the same way they do in other looking-time procedures and gaze-following tasks (e.g., Moll & Tomasello, 2004; Woodward, 2003).

other aspects of awareness of the bodily self have been studied and found to show a similar developmental transition. For example, following an observation made by Piaget (1953), Moore, Mealiea, Garon, and Povinelli (2007) examined whether toddlers were aware of their bodies as objects that could be impediments to action. Toddlers were shown a toy shopping cart and asked to push the cart to their mothers on the other side of the room. The cart was rigged by attaching a small rug to the back, so that when the children approached the cart, they inevitably stood on the rug and thus were unable to push the cart forward. Successful performance on this task, as indicated by some sign of awareness that they had to move off the rug in order to push it, was observed after about 18 months and was correlated with mirror self-recognition. Similar findings have also been reported by Brownell and colleagues (Brownell, Zerwas, & Ramani, 2007). In general, then, there is strong evidence that children become aware of self from a third-person perspective after about the middle of the second year.

GENERAL CONCLUSIONS

In the present chapter we have attempted to provide an integrated account of how infants come to form the concept of an individual-embodied intentional agent, a concept that they can apply both to self and other. In order to have such a concept, the infants must not only appreciate the meaning of the actions of others. They must conceive of those actions as actions of an agent, who has both internal subjective experiences and external objective behavior. Furthermore, they must be able to ascribe this concept equally to self and other. In our original theory of intentional relations (Barresi & Moore, 1996) we assumed that matching of intentional relations of self and other during triadic interactions was required for eventually forming this concept of intentional agent, and that the concept was only acquired during the second year of life.

Since the discovery of mirror neurons, a great deal of progress has occurred both in research and theory. However, we believe that the situation has not changed fundamentally. The presence of mirror neurons (Gallese et al., 1996; Rizzolatti et al., 1996) suggests only that the meaning of actions of others can be understood, through a process of matching first-person information of action execution to third-person information of action observation. This understanding does not require a concept of an intentional agent as an individual person with both first and third-person characteristics. All that is required is the projection of the meaning of actions from self to others at a subpersonal level.

In order to understand self and other as intentional agents, with both first-person subjective and third-person objective properties, the similarity as well as difference between self and other must be explicitly cognized. During triadic interactions, the infant first becomes aware of similarity and dissimilarity in intentional relations of self and other. During this period, active matching occurs, so that the infant is often aware concurrently of the first-person or subjective aspect of an intentional relation from her own present or recent experience, while observing the third-person aspect of the same intentional relation in the actions of the other. From these experiences the infant begins to understand the notion of intentional activity as having both subjective and objective sides.

Gradually, infants come to appreciate that differences can occur between self and other in this intentional activity, and so cannot only project meaning into the action of others, but imagine their embodied intentional state from the inside. They can also begin to imagine their own embodied intentional state from the outside. In both cases, it is no longer a subpersonal form of understanding, but one at a personal level—at the level of an agent of intentional relations to objects, where such relations have an internal as well as external aspect.

We have reviewed here experiments that provide signposts along the route from understanding the meaning of actions at a subpersonal level in looking-time studies; through experiments in triadic interaction where shared intentionality is sustained by infant as well as adult activity, and the intentionality of others is only appreciated in this context of sharing; to experiments indicating that older infants in the second year can conceive of both themselves and others as individual intentional agents. The transformation in understanding of intentional relations by the infant during this period is subtle, and the present account highlights only what we view as the major trend. But it can be seen as a constructive process, with the mirror neuron system playing a contributory role. The role, we believe, is that of a necessary stepping stone between representations of self and other at a subpersonal level during the first year and the capacity to understand self and other as personal intentional agents during the second year of life.

ACKNOWLEDGMENTS

This chapter was written with the support of a grant from the Social Sciences and Humanities Research Council to the first author.

REFERENCES

Amsterdam, B. (1972). Mirror self-image reactions before age two. *Developmental Psychobiology, 5,* 297–305.

Bakeman, R., & Adamson, L. (1984). Coordinating attention to people and objects in mother–infant and peer–infant interactions. *Child Development, 5,* 1278–1289.

Barresi, J., & Moore, C. (1993). Sharing a perspective precedes the understanding of that perspective. *Behavioral and Brain Sciences, 16,* 513–514.

Barresi, J., & Moore, C. (1996). Intentional relations and social understanding. *Behavioral and Brain Sciences, 19,* 107–122.

Barresi, J., & Moore, C. (2008). The neuroscience of social understanding. In Zlatev, J., Racine, T. P., Sinha, C., & Itkonen, E. (Eds.), *The shared mind: Perspectives on intersubjectivity.* Amsterdam: John Benjamins.

Bertenthal, B., & Fischer, K. (1978). Development of self-recognition in the infant. *Developmental Psychology, 14,* 44–50.

Brownell, C. A., Zerwas, S., & Ramani, G. B. (2007). "So big": The development of body self-awareness in toddlers. *Child Development, 78,* 1426–1440.

Cohen, L. B., Chaput, H. H., & Cashon, C. H. (2002). A constructivist model of infant cognition. *Cognitive Development, 17,* 1323–1343.

D'Entremont, B., & Seamans, E. (2007). Do infants need social cognition to act socially? An alternative look at infant pointing. *Child Development, 78,* 723–728.

Gallese, V., Fadiga, L., Fogassi, L. and Rizzolatti, G. 1996. Action recognition in the premotor cortex. *Brain, 119,* 593–609.

Johnson, S. C., Ok, S.-J., & Luo, Y. (2007). The attribution of attention: 9-month-olds' interpretation of gaze as goal-directed action. *Developmental Science, 10,* 530–537.

Király, I., Jovanovic, B., Prinz, W., Aschersleben, G., & Gergely, G. (2003). The early origins of goal attribution in infancy. *Consciousness and Cognition, 12*(4), 752–769.

Knoblich, G., & Jordan, J. S. (2003). Action coordination in groups and individuals: Learning anticipatory control. *Journal of Experimental Psychology: Learning, Memory, and Cognition, 29,* 1006–1016.

Lempers, J. D., Flavell, E. R., & Flavell, J. H. (1977). The development in very young children of tacit knowledge concerning visual perception. *Genetic Psychology Monographs, 95,* 3–53.

LePage, J.-F., & Théoret, H. (2007). The mirror neuron system: Grasping others' actions from birth? *Developmental Science, 10,* 513–529.

Lewis, M., & Brooks-Gunn, J. (1979). *Social cognition and the acquisition of self.* New York: Plenum.

Liszkowski, U., Carpenter, M., Henning, A., Striano, T., & Tomasello, M. (2004). Twelve-month-olds point to share attention and interest. *Developmental Science, 7,* 297–307.

Masataka, N. (2003). From index-finger extension to index finger pointing: Ontogenesis of pointing in preverbal infants. In S. Kita (Ed.), *Pointing: Where language, culture, and cognition meet* (pp. 69–84). Mahwah, NJ: Lawrence Erlbaum.

Meltzoff, A. N. (2007). "Like me": A foundation for social cognition. *Developmental Science, 10*, 126–134.

Meltzoff, A. N., & Gopnik, A. (1993). The role of imitation in understanding persons and developing a theory of mind. In S. Baron-Cohen, H. Tager-Flusberg & D. J. Cohen (Eds.), *Understanding other minds: Perspectives from autism* (pp. 335–366). Oxford, U.K.: Oxford University Press.

Moll, H., & Tomasello, M. (2004). 12- and 18-month-old infants follow gaze to spaces behind barriers. *Developmental Science, 7*, F1–F9.

Moll, H., & Tomasello, M. (2006). Level 1 perspective-taking at 24 months of age. *British Journal of Developmental Psychology, 24*, 603–613.

Moll, H., & Tomasello, M. (2007). How 14- and 18-month-olds know what others have experienced. *Developmental Psychology, 43*, 309–317.

Moore, C. (1996). Theories of mind in infancy. *British Journal of Developmental Psychology, 14*, 19–40.

Moore, C. (1999). Intentional relations and triadic interactions. In P. D. Zelazo, J. Astington, W. & D. R. Olson (Eds.), *Developing theories of intention: Social understanding and selfcontrol* (pp. 43–61). Mahwah, NJ: Lawrence Erlbaum Associates.

Moore, C. (2006a). *The development of commonsense psychology in the first five years*. Mahwah, NJ: Lawrence Erlbaum Associates.

Moore, C. (2006b). Representing intentional relations and acting intentionally in infancy: Current insights and open questions. In G. Knoblich, I. Thornton, M. Grosjean, & M. Shiffrar, (Eds.), *Human body perception from the inside out* (pp. 427–442). New York: Oxford University Press.

Moore, C. (2007). Understanding self and other in the second year. In C. A. Brownell and C. B. Kopp (Eds.), *Transitions in early socioemotional development: The toddler years*. New York: Guilford Press.

Moore, C., & Corkum, V. (1994). Social understanding at the end of the first year of life. *Developmental Review, 14*, 349–372.

Moore, C. & Corkum, V. (1998). Infant gaze following based on eye direction. *British Journal of Developmental Psychology, 16*, 495–503.

Moore, C., Mealiea, J., Garon, N., & Povinelli, D. J. (2007). The development of the bodily self. *Infancy, 11*, 157–174.

Moore, C., & Povinelli, D. J. (2007). Differences in how 12- and 24-month-olds interpret the gaze of adults. *Infancy, 11*, 215–231.

Nielsen, M., Dissanayake, C., & Kashima, Y. (2003). A longitudinal investigation of self-other discrimination and the emergence of minor self-recognition. *Infant Behavior & Development, 26*, 213–226.

Piaget, J. (1976). *The child and reality. Problems of genetic psychology*. Harmondsworth, UK: Penguin Books.

Piaget, J. (1953). *The origins of intelligence in the child*. London: Routledge and Kegan Paul.

Repacholi, B. M., & Gopnik, A. (1997). Early reasoning about desires: Evidence from 14- and 18-month-olds. *Developmental Psychology, 33,* 12–21.

Rizzolatti, G., Fadiga, L., Gallese, V., & Fogassi, L. 1996. Premotor cortex and the recognition of motor actions. *Cognitive Brain Research,* 3, 131–141.

Sodian, B., & Thoermer, C. (2004). Infants' understanding of looking, pointing, and reaching as cues to goal-directed action. *Journal of Cognition & Development, 5,* 289–316.

Sodian, B., Thoermer, C., & Metz, U. (2007). Now I see it but you don't: 14-month-olds can represent another person's visual perspective. *Developmental Science, 10,* 199–204.

Tomasello, M., Carpenter, M., Call, J., Behne, T., & Moll, H. (2005). Understanding and sharing intentions: The origins of cultural cognition. *Behavioral and Brain Sciences, 28,* 675–691.

Tomasello, M., Carpenter, M., & Liszkowski, U. (2007). A new look at infant pointing. *Child Development, 78,* 705–722.

Woodward, A. L. (2003). Infants' developing understanding of the link between looker and object. *Developmental Science, 6,* 297–311.

Woodward, A. L., & Guajardo, J. J. (2002). Infants' understanding of the point gesture as an object-directed action. *Cognitive Development, 17,* 1061–1084.

Woodward, A. L., Sommerville, J., & Guajardo, J. J. (2001). How infants makes sense of intentional action. In B. Malle, L. Moses, & D. Baldwin, D. (Eds.), *Intentions and intentionality: Foundations of social cognition* (pp. 149–169). Cambridge, MA: MIT Press.

4

THEORY OF MIND AND EXECUTIVE FUNCTIONING
A Developmental Neuropsychological Approach

Jeannette E. Benson
Mark A. Sabbagh
Queen's University

Over the preschool years, young children develop an explicit "representational theory of mind" (RTM)—an understanding that mental states are representations, the contents of which are related to but also distinct from the real situations they represent. Research using the false-belief task (a widely accepted litmus test for explicit RTM understanding), has shown that children all over the world show a remarkably similar timetable for RTM development, with the achievement usually coming between children's fourth and fifth birthdays (Callaghan et al., 2005). The only exceptions to universal RTM competence have been seen in instances of brain injury (Apperly, Samson, Chiavarino, & Humphreys, 2004) or neurodevelopmental disability, such as autism (Baron-Cohen, 2005), in which RTM reasoning seems to be particularly impaired. Together, these findings raise the possibility that RTM development is paced, at least in part, by relatively specific neurological developments unfolding in the preschool years.

The goal of this paper is to offer some hypotheses about the nature of the neurodevelopmental changes associated with RTM development by considering research on the relation between RTM and children's developing response-conflict executive functioning (RC-EF). RC-EF enables individuals to negotiate situations in which habit or prior learning would typically impel them to respond in one manner, but the particular situation requires them to respond in an alternative way. In the following section we provide a brief overview of research describing specific characteristics of the relation between RC-EF and RTM development. We then describe and evaluate three possible explanations for this relation, and explore how each may provide insight into the neurodevelopmental factors associated with advances in RTM reasoning.

THE RELATION BETWEEN RTM DEVELOPMENT AND RC-EF

Specific Relationship

A now sizeable collection of studies has identified a robust relation between children's RC-EF and their RTM task performance (e.g., Carlson & Moses, 2001; Carlson, Moses, & Breton 2002; Hughes, 1998; Perner, Lang, & Kloo, 2002). In children, RC-EF is typically assessed using Stroop-like tasks that require the suppression of a more dominant response in favor of a less dominant one (e.g., day–night Stroop, grass–snow Stroop, Dimensional Change Card Sort; see Carlson, 2005, for a review of measures). These studies show that as preschoolers' performance on RC-EF tasks improves, so, too, does their RTM task performance. The relation typically survives stringent partial correlation analyses in which factors such as age, sex, and language abilities are statistically controlled; after controlling for these variables, the correlation between RTM and RC-EF is typically around $r = .40$ (e.g., Carlson & Moses, 2001). This suggests that the relation between RTM and RC-EF cannot be attributed to general, age-related improvements in both skills, or to markers of general intelligence.

Perhaps most noteworthy is that there are two senses in which the relation between RTM and RC-EF is a specific one (see Moses & Sabbagh, 2007 for a review). First, RTM is robustly related to RC-EF tasks, but not as robustly related to other aspects of executive functioning, such as delay of gratification or working memory (e.g., Carlson & Moses, 2001). Second, RC-EF is related to the development of RTM, but not to the development of other aspects of theory of mind, such as the understanding of intentions, desires, or knowledge (Moses, Carlson, Stieglitz, & Claxton, under review). Perhaps some of the most compelling evidence for this second sense comes from research testing the relation between RC-EF and reasoning about false photographs. This research shows that RC-EF tasks are associated with RTM, but not with performance on the closely matched false-photograph task. (Müller, Zelazo, & Imrisek, 2004; Sabbagh, Moses, & Shiverick, 2006). Taken together, these findings strongly suggest that RC-EF makes a specific contribution to RTM reasoning itself, and is not simply required for reasoning about mental states or representations in general.

Unidirectional Relationship

In addition to there being a concurrent relation between RC-EF and RTM development, several studies have now shown a robust longitudinal relation whereby early RC-EF skills predict later RTM skills, but not vice versa. Hughes (1998) found that 3- to 4-year-old children's early RC-EF skills predicted their performance on false-belief tasks a year later, even after controlling for age, verbal ability, and initial false-belief performance. In contrast, the relation did not hold in the opposite direction—that is, early RTM understanding did not predict later RC-EF abilities after relevant statistical controls.

Two additional studies replicated these results on different timescales. Flynn (2007) found 3- and 4-year-olds' early RC-EF to be predictive of false-belief performance 5 months later, after controlling for age, verbal ability, and initial RTM knowledge. Again, early RTM knowledge did not predict later RC-EF skills. Similarly, Carlson, Mandell, and Williams (2004) found that 24-month-olds' RC-EF was predictive of RTM after a 10-month period, when controlling for the combined effects of age, sex, verbal ability, maternal education, and initial theory-of-mind understanding (e.g., intentions and desires). This pattern of longitudinal correlations, seen now across varying timescales, is typically used as evidence for a unidirectional, potentially intrinsic relation between early RC-EF and later RTM skills.

There is, however, a caveat with respect to making this strong interpretation in that very few of these analyses controlled for Time 2 RC-EF in the longitudinal analyses. To the extent that early RC-EF is correlated with later RC-EF, any significant relations between early RC-EF and later RTM may exist because each are related to later RC-EF. Although neither Hughes (1998) nor Flynn (2007) address this issue, Carlson et al. (1998) report results of a limited partial correlation analysis in which early RC-EF performance remained predictive of later RTM reasoning after controlling for later RC-EF skills, though no other control variables were included. Taken together, these studies provide suggestive (though, not conclusive) evidence for a unidirectional and potentially intrinsic relation between RC-EF and RTM abilities in development.

RC-EF Is Not Sufficient for RTM Reasoning

Although it does appear that RC-EF plays some role in supporting RTM performance, some cross-cultural and clinical research suggests that the relation may not be entirely straightforward. For instance, Sabbagh, Xu, Carlson, Moses, and Lee (2006) tested Chinese (Beijing) and U.S. preschoolers on extensive batteries of both RC-EF and RTM tasks (see Carlson & Moses, 2001). Comparisons across the two groups showed that Chinese children were advanced on RC-EF relative to their U.S. counterparts; indeed, the Chinese 3.5-year-olds had executive skills on par with U.S. 4-year-olds. Yet, the two groups showed very similar performance on the RTM battery. This pattern of findings suggests that developmental advances in RC-EF alone are not sufficient to promote performance on RTM reasoning tasks; otherwise, the Chinese preschoolers would have outperformed the U.S. preschoolers on RTM tasks as well. Most interesting, however, within the group of Chinese preschoolers, RC-EF was predictive of false-belief task performance; the relation was roughly the same magnitude in the Chinese preschoolers as it was in the U.S. preschoolers. This latter finding suggests that even in cases where the timetable of RC-EF is advanced, individual differences in RC-EF still predict RTM task performance.

Similar findings were reported by Oh and Lewis (2008), who examined RTM and EF development in Korean preschoolers. In line with Sabbagh et al.'s findings, these researchers found notably advanced RC-EF performance in Korean children as compared to age-matched English preschoolers; on three of the four RC-EF measures administered, young 3-year-old Korean children obtained higher RC-EF scores than 4.5-year-old English participants. Again, this substantial executive performance advantage was not mirrored in false-belief task scores; Korean children's performance across four false-belief tasks was comparable to scores obtained by their English counterparts. These cross-cultural findings demonstrate that there are populations that show a relatively advanced developmental timetable of RC-EF skills, and that group advances in those skills do not themselves seem to give rise to advances in the timetable of RTM skills.

Findings from research on children with autism provide corroborating evidence that RC-EF skills may be necessary but not sufficient for RTM task performance. Although the autistic spectrum disorder is most commonly characterized by deficits in the realm of theory of mind (including the more specific RTM skills), a number of researchers have shown that those with autism also have EF difficulties (e.g., Ozonoff,

Pennington, & Rogers, 1991; Pellicano, 2007). That is, autistic participants obtain significantly lower scores on both types of tasks in comparison to controls. Moreover, individual-differences research has shown that EF and RTM performance are correlated in autistic populations (Joseph & Tager-Flusberg, 2004; Ozonoff et al., 1991; Pellicano, 2007; Zelazo, Jacques, Burack, & Frye, 2002). In one study of mildly impaired children (Zelazo et al., 2002), the zero-order correlation between RC-EF and RTM was particularly strong ($r = .82$); the only children with autism who did well on the RTM tasks were those who showed strong performance on the RC-EF tasks.

Despite this strong individual differences relationship, group analyses of children with autism show an intriguing pattern of impairment associations and dissociations. For instance, Pellicano (2007) reported that within an autistic sample, in all instances where RC-EF was impaired, so too was RTM. Although RTM impairments sometimes existed in the absence of RC-EF deficits, the reverse dissociation did not occur. Thus, while intact RC-EF skills appear to be necessary for RTM reasoning, it seems that atypical RTM performance in this special population is not, at least in its entirety, a result of underdeveloped RC-EF. Taken together, these findings show that in both atypical and typical cases, RC-EF may be necessary but not sufficient for RTM performance.

Summary

This brief review illustrates three basic characteristics of the relation between RC-EF and RTM. First, it is a relatively specific relation: RC-EF (but not usually other aspects of EF more generally) predicts RTM (but not other aspects of theory of mind or reasoning about representations in general). Second, the relation is a unidirectional, potentially intrinsic one whereby early RC-EF predicts later RTM, and not vice versa. Third, RC-EF is likely to be necessary but not sufficient for performance on RTM tasks. With these three characteristics in mind, we will now turn to some explanations of the relation between RC-EF and RTM, with particular attention given to the implications of these explanations for understanding the neuromaturational bases of RTM development.

NEURAL BASES OF RTM AND RC-EF

To begin this discussion, we should start with the caveat that little is known about the neural bases of either RTM or RC-EF in preschoolers. In contrast, much is known about the neural correlates of both RTM and RC-EF in adults. With respect to RC-EF, Ridderinkhof and colleagues (Ridderinkhof, Ullsperger, Crone, & Nieuwenhuis, 2004) recently reviewed the imaging work from tasks that have investigated RC-EF skills and highlighted the involvement of a relatively circumscribed region of the MPFC. This finding is intriguing for understanding the relation between RC-EF and RTM reasoning, because most studies also show a role for the MPFC in RTM reasoning (see Gallagher & Frith, 2003 for a review). However, the fact that both tasks rely, at least in part, on MPFC functioning may not be all that interesting. The MPFC is a relatively large cortical region, and a recent review by Kain and Perner (2005) has established that for adults, there are clear distinctions between areas of MPFC that are important for RC-EF and those responsible for RTM reasoning. Specifically, RC-EF task performance is typically associated with activation in areas of the MPFC that lie posterior to regions typically activated during RTM tasks. Of course, it is important to note that this separation does not necessarily characterize RC-EF and RTM activations in younger populations. Some studies have found more extensive, less specified regions of the prefrontal cortex to be recruited during executive tasks in children as compared to adults (Durston et al., 2002). It is possible, then, that RC-EF and RTM tasks share a common neural substrate during the preschool years. By developing a reasonable theoretical argument regarding the relation between RC-EF and RTM development, we hope that we will be able to develop hypotheses about the neural underpinnings of both skills.

EXPLANATIONS FOR THE RELATION BETWEEN RC-EF AND RTM

Cortical Neighbor Hypothesis

Perhaps the most straightforward developmental hypothesis based upon the adult imaging findings is that the neural systems underlying RC-EF and RTM development are proximal, but separable. That is, there may be one region of the MPFC that contributes to RC-EF, and a nearby but separate region that performs computations necessary for RTM reasoning. Developmentally speaking, then, it would be reasonable to argue that because of their proximity, the neural bases of RTM and RC-EF might be subject to the same endogenous neuromaturational factors that promote functional development within this broad region (e.g., Ozonoff et al., 1991). Although there are likely to be many such factors, one example of such an endogenous factor might be developmental changes in dopaminergic functioning. Dopamine is critical for the normal development of the frontal lobes (Diamond, 1996; 2001), and the MPFC is a primary site of dopamine projection. If critical neural substrates for both RTM and RC-EF are neighbors in the MPFC, then they might share relatively specific ontogenetic variance because of their common reliance on dopaminergic enervation. It follows that a reliance on common or proximal regions of the MPFC may lead two otherwise disparate processes to share some developmental variance because of dopamine's influence on this region as a whole.

Some support for a direct role for dopamine in the development of both RTM and RC-EF comes from studies of children with early and continuously treated phenylketonuria (PKU). PKU is a genetic metabolic disorder in which children lack the enzyme necessary for changing phenylalanine into tyrosine, a precursor of dopamine (Diamond, 2001). The result is that children with PKU have depleted dopamine, which in turn affects dopamine projections to the frontal lobes (Scriver, Kaufman, Eisensmith, & Woo, 1995). It is now well documented that children with early and continuously treated PKU have detectable difficulties in RC-EF tasks (Diamond, 1998), even in the face of relatively typical IQ scores. Although we know of no work investigating the performance of children with PKU on RTM tasks per se, there is some evidence that they may have a broad social–cognitive deficit. In particular, Dennis and colleagues (Dennis et al., 1999) showed that children with PKU have difficulty with the "Comprehension" subscale of the Weschler scales, which assesses children's ability to understand social roles, and explain others' behaviors. The fact that deficiencies in dopamine are leading to both of these difficulties is consistent with the possibility that they each rely on the MPFC, and share variance accordingly.

It is worth noting that unlike a number of competing theories, the cortical neighbor hypothesis posits no intrinsic connection between RC-EF and RTM. Several researchers have presented compelling theoretical analyses that have identified candidate structural relations between RTM and RC-EF tasks that could account for the correlation between the two constructs (e.g., Frye, Zelazo, & Palfai, 1995; Russell, 1997; Perner & Lang, 1999). The cortical neighbor hypothesis promotes a radically different view, namely that the correlation is epiphenomenal; it exists only because of a common reliance on endogenous factors that promote development in the cortical regions where both skills have their neural substrates.

The cortical neighbor hypothesis offers a clear explanation for the concurrent, specific relation between RTM and RC-EF. However, it faces a challenge from the longitudinal findings described above; data in support of a unidirectional, potentially intrinsic relation between

RC-EF and RTM run counter to the main prediction from the cortical neighbor hypothesis that the two skills are not intrinsically related.

It is worth noting, however, that the cortical neighbor hypothesis does provide a reasonable explanation of the cross-cultural and clinical findings. Specifically, common endogenous developmental factors could account for the shared variance in RTM and RC-EF development, even if the overall trajectory of each skill varies independently because of other domain-specific experiential and exogenous influences. Indeed, the role of experiential factors is well documented for both RTM and RC-EF development. With respect to RTM development, factors such as parent–child talk about mental states and having older siblings affects the developmental timetable of RTM development (see e.g., Carpendale & Lewis, 2004 for a review). With respect to RC-EF, a number of studies have suggested that being bilingual from an early age may be associated with an advanced trajectory of RC-EF development (e.g., Bialystok, 1999; Bialystok & Martin, 2004; Carlson & Meltzoff, 2008). Put simply, then, the individual differences relations between RC-EF and RTM reasoning might be attributable to endogenous factors (e.g., dopaminergic functioning in MPFC) while across-group variability in the developmental trajectories of the individual skills might be attributable to relatively independent exogenous experiential factors.

Common Computation Hypothesis

A number of theorists have suggested that the relation between RTM and RC-EF exists because performance on the two kinds of tasks relies, at least in part, on a common computation or cognitive process. There are two variations of this general hypothesis: the first suggests that RC-EF contributes to RTM task performance, while the second suggests that both RC-EF and RTM rely on a common subcomponent process.

RTM Task Performance Requires RC-EF   The relation between RTM and RC-EF task variability was first thought to reflect performance demands inherent to RTM tasks; standard RTM measures require children to resist their habitual tendency to reference what they know to be true in order to provide an alternate response (Carlson & Moses, 2001). As children become better able to negotiate the response demands of RTM tasks, they became better able to reveal their RTM knowledge. Initial empirical support for this hypothesis came from studies showing that children's performance on RTM tasks improves when researchers ease the RC-EF demands of the tasks (e.g., Wellman & Bartsch, 1988; Carlson et al., 1998). Likewise, RTM performance worsens when researchers increase the RC-EF demands of RTM tasks (Cassidy, 1998; Friedman & Leslie, 2005; Leslie, German, & Polizzi, 2005).

In addition to the possibility of response biases, some researchers have proposed that attributing false beliefs to others poses a special sort of RC-EF problem. As children gain some experience reasoning about beliefs through the preschool years, they may notice that people typically act in accordance with true beliefs. This may lead to a sort of "bias" to assume that others' beliefs are true. In those relatively rare contexts in which another's beliefs are not true (such as the false-belief task), RC-EF may be required to overcome a true-belief bias in order to correctly attribute a false belief (e.g., Leslie, Friedman, & German, 2004). Importantly, this theory suggests that children have acquired the knowledge that beliefs can theoretically be false; the difficulty lies in inhibiting the learned assumption that beliefs will match reality in order to ascribe false beliefs when appropriate, as is the case during false-belief tasks. Some evidence for RC-EF's role in overcoming the true-belief bias comes from Sabbagh et al. (2006), who showed that preschoolers' RTM skills are associated with their ability to reason about false signs—another kind of representation that may have a "true" bias (see also Parkin & Perner, 1996). In that same study, the ability to reason about false signs was also associated with RC-EF to the same magnitude as RTM skills.

A key feature of these performance accounts is that the computations associated with solving RTM tasks are separable from those associated with the ability to generate RTM concepts. This separation gives rise to the possibility of a distinction between RTM competence (i.e., the ability to generate representations of others' false beliefs) and performance on RTM tasks. Some researchers have proposed that RTM competence computations may be carried out by an early-emerging modular mechanism with direct correlates in neural architecture (e.g., Leslie's theory of mind module, Leslie, 1994; Leslie et al., 2004). However, performance accounts that include a strong commitment to early competence in RTM reasoning do not easily provide a compelling account of the finding that RC-EF is not alone sufficient for RTM task performance in young preschoolers. On the early competence view, once children acquire a certain level of RC-EF they should be able to demonstrate their skills in reasoning about false beliefs. Findings from the cross-cultural studies run counter to this prediction; they show that 3-year-old Chinese and Korean children who were at floor on RTM tasks had attained the same level of RC-EF functioning as had 4-year-old U.S. preschoolers, who showed strong RTM performance. Thus, performance theories that make strong claims about early competence seem unlikely.

It should be noted, though, that a strong commitment to early RTM competence is not a necessary feature of a performance account. The only necessary hypotheses are that the neural substrates of RC-EF are dissociable from the neural substrates of RTM reasoning, and that the distinct RC-EF networks are recruited alongside RTM networks when there are executive demands associated with RTM task performance. There is now some evidence that these hypotheses are supported, at least for adults. Saxe, Schulz, and Jiang (2006) compared the neural systems recruited during a location change false-belief task with those recruited during a task that involved the same sorts of executive demands as a typical false-belief task, but no mental state reasoning. Results indicated that brain regions associated with domain-general attention, response selection, and inhibitory control were activated during both the control and false-belief conditions. However, one additional brain region—the right temporo-parietal junction—was exclusively recruited during the false-belief condition. These findings lend support to the performance accounts' assertion that standard false-belief tasks recruit domain general RC-EF neural networks, in addition to more specific "mentalizing" brain regions (e.g., the right temporo-parietal junction).

From a developmental perspective, both versions of the performance account are generally in line with studies showing a specific, concurrent relation between RTM and RC-EF. It is worth noting, though, that the evidence is now tilting towards the notion that overcoming the true-belief bias may represent the more important contribution of RC-EF to RTM development, as opposed to negotiating superficial task demands. First, if RC-EF was critical for negotiating superficial task demands, there should be a relation between RC-EF and performance on the false-photograph task; however, as reviewed above, no research has detected such a relation. Second, RC-EF also predicts performance on false-belief "explanation" tasks in which children are asked to explain the behavior of a story character who acts in accordance with a false belief. This task is intriguing because children still presumably have to overcome the true-belief bias, but they do not need to suppress a prepotent motor response (e.g., Perner & Lang, 1999; Perner et al., 2002). Finally, a meta-analysis by Wellman, Cross, and Watson (2001) showed that even in false-belief tasks that were modified with the goal of lowering the executive response demands, young preschoolers did not demonstrate an explicit, systematic (i.e., responding different from chance) understanding of RTM. Taken together, these findings support the notion that RC-EF may be required to help children overcome a true-belief bias as they engage in RTM reasoning.

These performance accounts, and in particular the "true-belief bias" account, provide a compelling explanation for why there is a specific, concurrent relation between RC-EF and RTM reasoning. Moreover, they can account for the cross-cultural and clinical data through the proposal that RC-EF and RTM are essentially separate; although RC-EF is necessary for RTM reasoning in everyday settings, some other developments are necessary to render specific RTM computations. However, the performance account does not provide an easy explanation for the longitudinal data. That is, no performance account would predict the apparent unidirectional, intrinsic relation between RC-EF and RTM; if RC-EF is solely enabling children to reveal the extent of their RTM knowledge on RTM tasks, then there is no a priori reason to expect a predictive, intrinsic relationship between preschoolers' early RC-EF and their later RTM after taking relevant controls into account.

Shared Metacognitive Processes   The Performance Accounts posit that RC-EF and RTM are related because RTM tasks make RC-EF demands. Another possibility is that RC-EF and RTM tasks might each require some general metacognitive capabilities, which themselves rely on common neural substrates in the MPFC. Although this suggestion may seem counterintuitive, given the very different seeming functions of RTM and RC-EF, consider the problem posed by RC-EF tasks. In a typical RC-EF task, children are faced with two active representations of competing responses and have to decide which one is most appropriate. On some level, the decision-making process may invoke metacognitive processes for explicitly representing both competing response options and evaluating which one is most appropriate. Perner (1998) has suggested that these metacognitive processes might have a lot in common with those that are required to engage in RTM reasoning (see also Zelazo, 2004).

This view fits together well with a hypothesis offered by Gallagher and Frith (2003) regarding the role of MPFC in RTM reasoning. Based on their review of a wide range of adult studies on medial frontal contributions to mental state reasoning, they suggested that this region might be involved in "decoupling" mental states from the reality they represent (Leslie, 1987). One possibility is that decoupling is a metacognitive ability to explicitly separate mental states from reality and then evaluate those mental states with respect to a given task context. When put in this way, one might argue that decoupling is involved in both RTM tasks and RC-EF tasks. A similar idea was put forth by Sommer and colleagues (Sommer et al., 2007), who suggested that regions of the anterior cingulate cortex may be involved in cognitive processing that is detached from information in the immediate sensory environment. It seems reasonable that decoupling may be necessary for this process of disengaging oneself from the immediate sensory input to focus on internal mental representations.

From a neural perspective, this account makes a slightly different prediction from the performance accounts described above. Specifically, decoupling may be one computation that is shared by RC-EF and RTM tasks. If we make the assumption that this decoupling mechanism has a single neural substrate which is commonly recruited across task contexts, then we might assume that the neural bases of RTM and RC-EF would be partially overlapping. A decoupling mechanism would be only one overlapping part of the otherwise separable neurocognitive systems that underlie RC-EF and RTM.

In some ways, this explanation bears similarities to the cortical neighbor hypothesis in how they account for all three aspects of the RC-EF–RTM relation. The main difference is that, whereas the cortical neighbor hypothesis stipulates two independent neural circuits that rely on common endogenous factors, the common computation hypothesis stipulates a partially overlapping cortical region that can provide the basis for shared variance between the two tasks. Like the cortical neighbor hypothesis, then, this account provides a reasonable explanation for the specific, concurrent relation between RC-EF and RTM, and also for the cross-cultural and clinical data. Specifically, RC-EF and RTM may share variance as a result of a shared metacognitive process, but across populations the individual trajectories of RC-EF and RTM development could vary as a function of other domain-specific influencing factors.

Conversely, it is less clear that this account is compatible with the longitudinal, unidirectional relation between RC-EF and RTM. There does not seem to be any reason that this common computation account would predict advances in RC-EF skills to be related to the later (and not simply concurrent) development of RTM skills.

The Emergence Hypothesis

Although the common computation hypotheses are perhaps the most commonly proposed explanations for the relation between RC-EF and RTM reasoning, the collection of sometimes incongruent data on the relation between RC-EF and RTM has led some researchers to suggest an alternative "emergence" account. The emergence account asserts that RC-EF may actually facilitate the development of RTM concepts through the preschool years (Moses, 2001). Though there may be non-trivial RC-EF demands inherent to RTM reasoning, the emergence account suggests that these demands are not the primary cause of the ontogenetic relation between RC-EF and RTM task performance. Instead, the hypothesis is that RTM knowledge develops over time, and RC-EF skills, along with relevant experiential factors, make a significant contribution to that development.

RC-EF may facilitate children's RTM understanding in several ways. First, RC-EF skills may enable children to suppress their own perspectives in order to consider and reflect upon the mental states of others (Carlson & Moses, 2001). Second, having advanced RC-EF may enable children to competently engage in social interactions that, in turn, provide them with experiences relevant to furthering their RTM knowledge (Flynn, 2007; Hughes, 1998). Third, having the ability to ignore irrelevant stimuli may increase children's capacity to use information from both discourse with others and experience observing others' actions to develop the understanding that beliefs are sometimes false (Moses & Sabbagh, 2007). Finally, acquiring conceptual information that beliefs can be subjective and false may be particularly difficult because beliefs are meant to be true. Just as overcoming the "true-belief bias" may be important for reasoning about well-established belief concepts, it could be that overcoming the true-belief bias is particularly important for even acquiring the concept that beliefs can be false. Each of these provide a potential mechanism by which RC-EF might support the conceptual developments associated with RTM reasoning.

Perhaps more than any other account, the emergence account emphasizes the interaction between endogenous and exogenous (experiential) factors in shaping the developmental timetable of RTM reasoning. The straightforward prediction from the emergence hypothesis is that RC-EF may mediate the influence of exogenous factors on the development of RTM skills. Thus, early maturation of the neural systems associated with RC-EF should predict later RTM performance and related neural substrates, while the reverse should not hold true. Possibly the most striking prediction from this account is that relevant experiential factors (e.g., parental use of mental state terms and number of older siblings in the home) may affect the development of brain regions associated with RTM skills, with RC-EF maturation acting as a rate-limiting factor on that developmental pathway. Although theoretical models regarding how experience affects the neural substrates of high-level cognitive activity are only emerging (see Quartz & Sejnowski, 1997), a burgeoning body of research findings have clearly demonstrated the effects of experience on the neural substrates of both perceptual and conceptual developments (e.g., Neville, 2006; Maguire et al., 2003).

The emergence hypothesis offers a plausible explanation for many of the reported characteristics of the relation between RTM and RC-EF.

Most notably, the emergence account is the only account that straightforwardly explains the findings from longitudinal studies showing that RC-EF skills are unidirectionally predictive of later RTM skills. If RTM skills develop as a function of the interaction between RC-EF and experience over time, then we would expect early RC-EF abilities to predict variance in the subsequent development of RTM.

This account also provides some insight into the results suggesting that RC-EF is necessary but not sufficient for RTM understanding. According to this theory, RC-EF abilities enable children to capitalize on the types of experiences that provide them with information on others' mental states, thus fostering the development of a RTM. With respect to the cross-cultural findings, while the Chinese and Korean preschoolers were advanced in RC-EF, it may be that they had less exposure to the kinds of experiences that facilitate RTM development. In the case of the Chinese preschoolers, there is evidence to suggest that there are indeed substantial cultural differences in exposure to experiential factors that are related to RTM. For instance, when preschoolers have older siblings, they show adult-like performance on RTM tasks sooner than preschoolers who are the oldest child in the home (e.g., Ruffman, Perner, Naito, Parkin, & Clements, 1998). Because the Chinese government enforces a law that prohibits more than one child per household, Chinese children will likely not benefit from whatever advantages having older siblings confides. It is thus possible that the advantages afforded to Chinese children as a result of superior RC-EF abilities are offset by their relatively limited exposure to important experiential factors.

SUMMARY

We started with the notion that we might gain insight into the neurodevelopmental factors that affect RTM reasoning through a detailed analysis of the relation between RTM reasoning and children's developing RC-EF skills. To this end, we evaluated several hypotheses from the extant literature that have been marshaled to explain the relation between RC-EF and RTM reasoning, and explored their implications for understanding the neurodevelopmental bases of RTM. We would like to conclude by suggesting that although each of the hypotheses has been presented and evaluated separately, we do not have any strong reason to think that they would be mutually exclusive. For instance, it does seem clear that RTM tasks have nontrivial RC-EF demands, and although we do not feel that this captures the cause for the relation between RC-EF and RTM, it might be a mistake to suggest that negotiating these task demands plays no role. Further, it is possible that each of the explanatory hypotheses we offered may capture different aspects of the developmental process. These are important questions for future discussion.

Finally, we would like to note that a common thread that runs through the majority of these accounts of the relation between RC-EF and RTM reasoning is a role for experiential factors. In many cases, the neurodevelopmental implications of the relation between RC-EF and RTM points to the interplay of both endogenous neuromaturational factors and exogenous experiential factors. We feel that understanding the interaction between endogenous and exogenous factors in RTM development is perhaps the most promising and exciting avenue for future research in this area.

REFERENCES

Apperly, I. A., Samson, D., Chiavarino, C., & Humphreys, G. W. (2004). Frontal and temporo-parietal lobe contributions to theory of mind: Neuropsychological evidence from a false-belief task with reduced language and executive demands. *Journal of Cognitive Neuroscience, 16*, 1773–1784.

Baron-Cohen, S. (2005). *Autism and the origins of social neuroscience.* New York: Psychology Press.

Bialystok, E. (1999). Cognitive complexity and attentional control in the bilingual mind. *Child Development, 70*, 636–644.

Bialystok, E., & Martin, M. M. (2004). Attention and inhibition in bilingual children: Evidence from the Dimensional Change Card Sort task. *Developmental Science, 7*, 325–339.

Callaghan, T., Rochat, P., Lillard, A., Claux, M. L., Odden, H., Itakura, S., Tapanya, S., & Singh, S. (2005). Synchrony in the onset of mental state reasoning. *Psychological Science, 16*, 378–384.

Carlson, S. M. (2005). Developmentally sensitive measures of executive function in preschool children. *Developmental Neuropsychology, 28*, 595–616.

Carlson, S. M., Mandell, D. J., & Williams, L. (2004). Executive function and theory of mind: Stability and prediction from ages 2 to 3. *Developmental Psychology, 40*, 1105–1122.

Carlson, S. M., & Meltzoff, A. N. (2008). Bilingual experience and executive functioning in young children. *Developmental Science, 11*, 282–298.

Carlson, S. M., & Moses, L. J. (2001). Individual differences in inhibitory control and children's theory of mind. *Child Development, 72*, 1032–1053.

Carlson, S. M., Moses, L. J., & Breton, C. (2002). How specific is the relation between executive function and theory of mind? Contributions of inhibitory control and working memory. *Infant and Child Development, 11*, 73–92.

Carlson, S. M., Moses, L. J., & Hix, H. R. (1998). The role of inhibitory processes in young children's difficulties with deception and false belief. *Child Development, 69*, 672–691.

Carpendale, J. I. M., & Lewis, C. (2004). Constructing an understanding of mind: The development of children's social understanding within social interaction. *Behavioral and Brain Sciences, 27*, 79–151.

Cassidy, K. W. (1998). Three- and four-year-old children's ability to use desire- and belief-based reasoning. *Cognition, 66*, 1–11.

Dennis, M., Lockyer, L., Lazenby, A. L., Donnelly, R. E., Wilkinson, M., & Schoonheyt, W. (1999). Intelligence patterns among children with high-functioning autism, phenylketonuria, and childhood head injury. *Journal of Autism and Developmental Disorders, 29*, 5–17.

Diamond, A. (1996). Evidence for the importance of dopamine for prefrontal cortex functions early in life. *Philosophical Transactions of the Royal Society of London, 351*, 1483–1494.

Diamond, A. (1998). Evidence for the importance of dopamine for prefrontal cortex functions early in life. In A. C. Roberts, T. W. Robbins, & L. Weiskrantz (Eds.), *The prefrontal cortex: Executive and cognitive functions* (pp. 144–164). New York: Oxford University Press.

Diamond, A. (2001). A model system for studying the role of dopamine in the prefrontal cortex during early development in humans: Early and continuously treated phenylketonuria. In C. A. Nelson, & M. Luciana (Eds.), *Handbook of developmental cognitive neuroscience* (pp. 433–472). Cambridge, MA: MIT Press.

Durston, S., Thomas, K. M., Yang, Y., Ulug, A. M., Zimmerman, R. D., & Casey, B. J. (2002). A neural basis for the development of inhibitory control. *Developmental Science, 5*, 9–16.

Flynn, E. (2007). The role of inhibitory control in false belief understanding. *Infant and Child Development, 16*, 53–69.

Friedman, O., & Leslie, A. M. (2005). Processing demands in belief–desire reasoning: Inhibition or general difficulty? *Developmental Science, 8*, 218–225.

Frye, D., Zelazo, P. D., & Palfai, T. (1995). Theory of mind and rule-based reasoning. *Cognitive Development, 10*, 483–527.

Gallagher, H. L., & Frith, C. D. (2003). Functional imaging of 'theory of mind'. *Trends in Cognitive Sciences, 7*, 77–83.

Hughes, C. (1998). Finding your marbles: Does preschoolers' strategic behavior predict later understanding of mind? *Developmental Psychology, 34*, 1326–1339.

Joseph, R. M., & Tager-Flusberg, H. (2004). The relationship of theory of mind and executive functions to symptom type and severity in children with autism. *Development and Psychopathology, 16*, 137–155.

Kain, W., & Perner, J. (2005). What fMRI can tell us about the ToM–EF connection: False beliefs, working memory, and inhibition. In W. Schneider, R. Schumann-Hengsteler, & B. Sodian (Eds.), *Young children's cognitive development: Interrelationships among executive functioning, working memory, verbal ability and theory of mind* (pp.189–217). Mahwah, NJ: Lawrence Erlbaum Associates.

Leslie, A. M. (1987). Pretense and representation: The origins of "theory of mind." *Psychological Review, 94*, 412–426.

Leslie, A. M. (1994). ToMM, ToBy, and agency: Core architecture and domain specificity. In L. A. Hirschfeld, & S. A. Gelman (Eds.), *Mapping the mind: Domain specificity in cognition and culture* (pp. 119–148). New York: Cambridge University Press.

Leslie, A. M., Friedman, O., & German, T. P. (2004). Core mechanisms in "theory of mind." *Trends in Cognitive Sciences, 8*, 529–533.

Leslie, A. M., German, T. P., & Polizzi, P. (2005). Belief–desire reasoning as a process of selection. *Cognitive Psychology, 50*, 45–85.

Maguire, E. A., Spiers, H. J., Good, C. D., Hartley, T., Frackowiak, R. S. J., & Burgess, N. (2003). Navigation expertise and the human hippocampus: A structural brain imaging analysis. *Hippocampus, 13*, 250–259.

Moses, L. J. (2001). Executive accounts of theory-of-mind development. *Child Development, 72*, 688–690.

Moses, L. J., Carlson, S. M., Stieglitz, S., & Claxton, L. J. (Manuscript under review). Executive function, prepotency, and children's theories of mind.

Moses, L. J., & Sabbagh, M. A. (2007). Interactions between domain general and domain specific processes in the development of children's theories of mind. In M. J. Roberts (Ed.), *Integrating the mind: Domain general versus domain specific processes in higher cognition* (pp. 275–291). New York: Psychology Press.

Müller, U., Zelazo, P. D., & Imrisek, S. (2004). Executive function and children's understanding of false belief: How specific is the relation? *Cognitive Development, 20*, 173–189.

Neville, H. J. (2006). Different profiles of plasticity within human cognition. In Y. Munakata, & M. Johnson, (Eds.), *Processes of change in brain and cognitive development: Attention and Performance XXI* (pp. 287–314). London: Oxford University Press.

Oh, S., & Lewis, C. (2008). Korean preschoolers' advanced inhibitory control and its relation to other executive skills and mental state understanding. *Child Development, 79*, 80–99.

Ozonoff, S., Pennington, B. F., & Rogers, S. J. (1991). Executive function deficits in high-functioning autistic individuals: Relationship to theory of mind. *Journal of Child Psychology and Psychiatry, 32*, 1081–1105.

Parkin, L. J., & Perner, J. (1996). Wrong directions in children's theory of mind: What it means to understand belief as a representation. Unpublished manuscript, University of Sussex.

Pellicano, E. (2007). Links between theory of mind and executive function in young children with autism: Clues to developmental primacy. *Developmental Psychology, 43*, 974–990.

Perner, J. (1998). The meta-intentional nature of executive functions and theory of mind. In P. Carruthers, & J. Boucher (Eds.), *Language and thought: Interdisciplinary themes* (pp. 270–283). Cambridge, England: Cambridge University Press.

Perner, J., & Lang, B. (1999). Development of theory of mind and executive control. *Trends in Cognitive Sciences, 3*, 337–344.

Perner, J., Lang, B., & Kloo, D. (2002). Theory of mind and self control: More than a common problem of inhibition. *Child Development, 73*, 752–767.

Quartz, S. R., & Sejnowski, T. J. (1997). The neural basis of cognitive development: A constructivist manifesto. *Behavioural and Brain Sciences, 20*, 537–596.

Ridderinkhof, K. R., Ullsperger, M., Crone, E. A., & Nieuwenhuis, S. (2004). The role of the medial frontal cortex in cognitive control. *Science, 306*, 443–447.

Ruffman, T., Perner, J., Naito, M., Parkin, L., & Clements, W. (1998). Older (but not younger) siblings facilitate false belief understanding. *Developmental Psychology, 34*, 161–174.

Russell, J. (1997). How executive disorders can bring about an adequate theory of mind. In J. Russell (Ed.), *Autism as an executive disorder* (pp. 256–304). New York: Oxford University Press.

Sabbagh, M. A., Moses, L. J., & Shiverick, S. (2006). Executive functioning and preschoolers' understanding of false beliefs, false photographs, and false signs. *Child Development, 77*, 1034–1049.

Sabbagh, M. A., Xu, F., Carlson, S. M., Moses, L. J., & Lee, K. (2006). The development of executive functioning and theory of mind: A comparison of Chinese and U.S. preschoolers. *Psychological Science, 17*, 74–81.

Saxe, R., Schulz, L. E., & Jiang, Y. V. (2006). Reading minds versus following rules: Dissociating theory of mind and executive control in the brain. *Social Neuroscience, 1*, 284–298.

Scriver, C. R., Kaufman, S., Eisensmith, R. C., & Woo, L. C. (1995). The hyperphenylalaninemias. In C. R. Scriver, A. L. Beudet, W. S. Sly, & D. Valle (Eds.), *The metabolic and molecular bases of inherited disease* (7th ed., pp. 1015–1075). New York: McGraw-Hill.

Sommer, M., Döhnel, K., Sodian, B., Meinhardt, J., Thoermer, C., & Hajak, G. (2007). Neural correlates of true and false belief reasoning. *NeuroImage, 35*, 1378–1384.

Wellman, H. M., Cross, D., & Watson, J. (2001). Meta-analysis of theory-of-mind development: The truth about false belief. *Child Development, 72*, 655–684.

Wellman, H. M., & Bartsch, K. (1988). Young children's reasoning and beliefs. *Cognition, 30*, 239–277.

Zelazo, P. D. (2004). The development of conscious control in childhood. *Trends in Cognitive Sciences, 8*, 12–17.

Zelazo, P. D., Jacques, S., Burack, J. A., & Frye, D. (2002). The relation between theory of mind and rule use: Evidence from persons with autism-spectrum disorders. *Infant and Child Development, 11*, 171–195.

5

THE DEVELOPMENT OF ITERATIVE REPROCESSING

Implications for Affect and Its Regulation

William A. Cunningham
Ohio State University

Philip David Zelazo
University of Minnesota

INTRODUCTION

What is an emotion? How does emotion relate to cognition? How do we control our emotions? These questions have been debated vigorously at least since Socrates, and they have been central to psychology for over a century, but recent work in developmental social cognitive neuroscience provides a new framework for addressing them. In this chapter, we explore the development of emotion and its regulation in the context of our recent iterative reprocessing model (Cunningham & Zelazo, 2007; Zelazo & Cunningham, 2007). According to this model, emotion corresponds to an aspect of cognition—its evaluative, motivational aspect. This aspect of human information processing manifests itself in multiple dimensions—subjective experience, observable behavior, and physiological activity, among them. Although it is possible to have cognition that is more or less emotional, and more or less motivated, all cognition is motivated to some degree, and all emotional experience has a cognitive dimension. From this perspective, emotional experience is modulated in important ways by processes that have been studied under the rubric of "executive function," and research on the development of executive function has clear implications for the development of affective processing.

Here, we begin with an overview of the iterative reprocessing model, with a focus on the ways in which affective processing operates in adults. We then describe the development of these processes in childhood, linking age-related changes in children's emotional experience to the development of the brain.

THE ITERATIVE REPROCESSING MODEL

The iterative reprocessing (IR) model (Cunningham & Zelazo, 2007; Zelazo & Cunningham, 2007) examines emotion as a dynamic process that unfolds in time. Encountered or imagined stimuli (e.g., people, objects, or abstract concepts) elicit relatively rapid affective associations, but affective processing is normally modulated by an iterative cycle of reflective processes. With every iteration of this cycle, one's evaluation of a stimulus can be adjusted in light of an increasingly wide range of considerations. By a process of "reseeding," information resulting from higher-order reflective processes is fed back into lower-order processes, and the affective response is recalculated. This process allows for the attentional foregrounding of relevant (and backgrounding of irrelevant) attitude representations and contextual information in order to develop a more nuanced evaluation congruent with the current context, and/or current goals.

Because each iteration of the cycle allows for additional reflective processing, the IR Model implies a rough continuum from "automatic" affective responses, entailing few iterations and a limited set of cognitive operations, to more "reflective" affective responses, entailing more iterations and cognitive operations (see Cunningham & Johnson, 2007). The extent to which affect toward any particular stimulus is reprocessed (i.e., the number of iterations it receives) is likely to depend on a host of personal and situational factors, including differences in opportunity, cognitive ability, and developmental level. It is hypothesized, however, that an important determinant of the extent to which people engage in reprocessing is the dynamic tension between two competing motivational drives: (1) a drive to minimize the discrepancy between one's response and the hedonic environment (i.e., to minimize error), which tends to increase the likelihood of reflective processing; and (2) a drive to minimize processing demands, which tends to decrease the likelihood of reflective processing. The influence of these competing motivations varies as a function of situational demands, current goals, and individual differences in processing style—all of which shift the balance between reliance on initial iterations (yielding a "gut" reaction) and further reprocessing (permitting the construction of reactions that are more complex).

Initial iterations generally involve processing in subcortical brain regions such as the amygdala and the ventral striatum (especially the nucleus accumbens) and give rise to rapid evaluations based on innate biases (e.g., LeDoux, 1996; Öhman & Mineka, 2001) and learning (e.g., Armony & Dolan, 2002; Davis, 1997; Phelps et al., 2001; Whalen, 1998). When a stimulus is encountered, information about the stimulus triggers an unreflective motivational tendency to approach or avoid the stimulus, producing a series of physiological responses and reflexive reactions that are mediated by the hypothalamus (among other regions). These relatively undifferentiated physiological responses prepare the body for immediate action—fighting or fleeing—while additional neural processes continue to disambiguate the motivational implications of the stimulus. Importantly, although subcortical structures support automatic affect, such processes may maintain an ongoing and important role in generating current affect even as additional reflective processes are incorporated in affective processing.

Following this initial response to a stimulus, the physiological response is registered in the somatosensory cortex (and the insular cortex in particular), allowing for the representation of information about the current bodily state. Given connections among the insula, amygdala, and orbitofrontal cortex, the represented bodily state can be integrated into subsequent iterations of evaluative processing (Critchley, Wiens, Rotshtein, Öhman, & Dolan, 2004; Damasio, 1994, 1996; see also Rolls, 2005). During these subsequent iterations, more detailed information about the stimulus (from the sensory cortices) and the bodily state (from the insula) all provide input into the ongoing generation of the evaluation. These multiple inputs can serve to generate a more complex evaluation within a time period still typically labeled "automatic"—that is, several initial iterations may all occur within a few hundred milliseconds of stimulus perception (Oya, Kawasaki, Howard, & Adolphs, 2002).

After these initial iterations, however, amygdala projections to orbitofrontal cortex allow for a comparison of expected rewards and punishments with current experience. This allows context to play a top-down, regulatory role in shaping affect (e.g., Blair, 2004; Beer, Heerey, Keltner, Scabini, & Knight, 2003; Rolls, 2005; Rolls, Hornak, Wade, & McGrath,

1994). The orbitofrontal cortex receives input from multiple sensory modalities and may provide a common metric for representing and comparing disparate aspects of evaluative information (Montague & Berns, 2002; Rolls, 2005), including the evaluative connotations of self-generated mental representations (Cunningham, Johnsen, Mowrer, & Waggoner, 2008). Evidence suggests, however, that whereas posterior regions of medial orbitofrontal cortex may be relatively more involved in stimulus valuation, anterior regions of medial orbitofrontal cortex may be critical to the integration of these signals with goals, motivation, and the linking of affect to action (Cunningham, Kesek, & Mowrer, in press).

In many cases, affective responses to the situation mediated by the amygdala and orbitofrontal cortex will yield an evaluation sufficient to produce a behavioral response. In other cases, however, this joint processing will yield too much residual uncertainty or evidence of conflict (as when the stimulus is ambivalent or fails to provide expected rewards). According to the IR model, the presence of conflict (e.g., the simultaneous activation of strong approach and strong avoidance reactions) triggers anterior cingulate cortex activation (see Bush, Luu, & Posner, 2000; Carter et al., 1998), which may then initiate additional reprocessing of the stimulus in regions of lateral prefrontal cortex involved in cognitive control (see Bunge & Zelazo, 2006; MacDonald, Cohen, Stenger, & Carter, 2000; Ridderinkhof, Ullsperger, Crone, & Nieuwenhuis, 2004). Additional reflection proceeds, as needed, through increasingly higher-order regions within a hierarchy of prefrontal cortex regions: from ventrolateral prefrontal cortex to dorsolateral prefrontal cortex to rostrolateral prefrontal cortex (e.g., Badre & D'Esposito, 2007; Botvinick, 2008; Bunge & Zelazo, 2006; Koechlin, Ody, & Kounelher, 2003; see Figure 5.1).

More lateral prefrontal cortex mediated processing allows regulation of affect in a top-down fashion by deliberately amplifying or suppressing attention to certain aspects of the stimulus, changing the input to the system on subsequent iterations. This iterative reprocessing will not necessarily generate an altogether new affective state, but will likely modulate its current evaluation by modulating activity in lower-order regions (e.g., Cunningham, Johnson et al., 2004; Ochsner, Bunge, Gross, & Gabrieli, 2002; Ochsner et al., 2004).

More complex networks of processing permit more complex construals of a stimulus, in part simply because more information about a stimulus can be integrated into the construal during each iteration, and in part because these networks support the formulation and use of higher-order rules for deliberately selecting certain aspects of a stimulus or

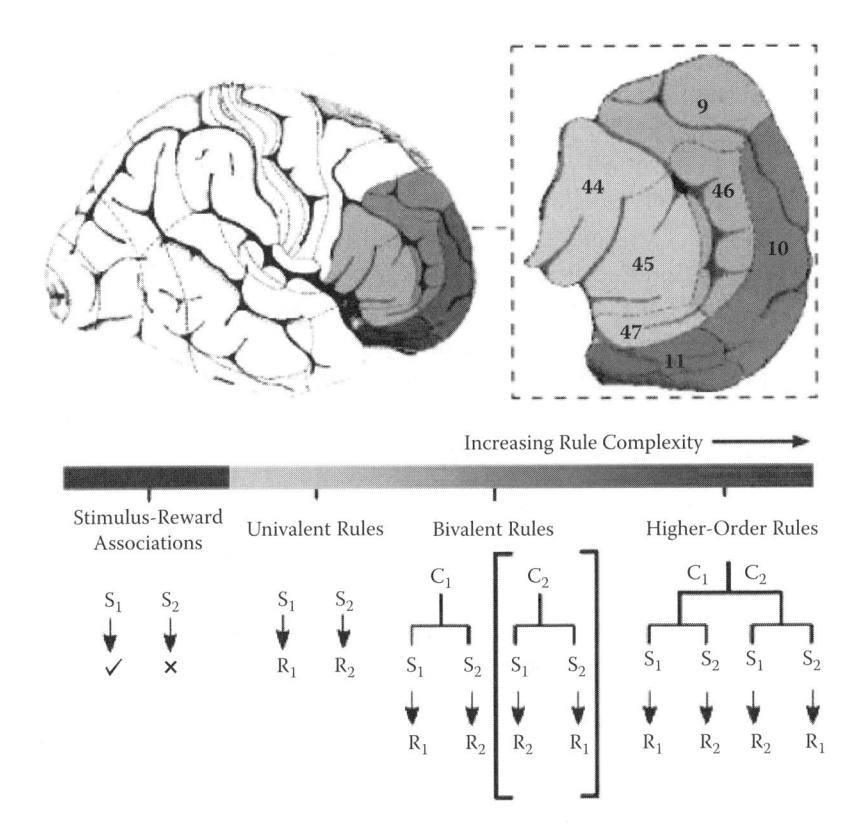

Figure 5.1 A hierarchical model of rule representation in prefrontal cortex. A lateral view of the human brain is depicted at the top of the figure, with regions of prefrontal cortex identified by the Brodmann areas (BA) that comprise them: orbitofrontal cortex (BA 11), ventrolateral prefrontal cortex (BA 44, 45, 47), dorsolateral prefrontal cortex (BA 9, 46), and rostrolateral prefrontal cortex (BA 10). The prefrontal regions are shown in various shades of gray, indicating which types of rules they represent. Rule structures are depicted below, with darker shades of gray indicating increasing levels of rule complexity. The formulation and maintenance in working memory of more complex rules depends on the reprocessing of information through a series of levels of consciousness, which in turn depends on the recruitment of additional regions of prefrontal cortex into an increasingly complex hierarchy of prefrontal activation. Note: S = stimulus; check = reward; cross = nonreward; R = response; C = context, or task set. Brackets indicate a bivalent rule that is currently being ignored. (From: Bunge, S., & Zelazo, P. D., 2006. *Current Directions in Psychological Science, 15*, 118–121. With permission.)

context to which to attend (Bunge & Zelazo, 2006). The selection function of prefrontal cortex is what foregrounds specific aspects of information (and backgrounds others), and it is these reweighted stimulus representations that are then used to reseed initial evaluative processing—for example, by influencing ongoing perception and processing of the stimulus. Prefrontal cortex may also play a role in keeping current goals and contextual demands/constraints in mind, which is important for fulfilling the competing goals of minimizing error while minimizing processing load (e.g., Cunningham, Van Bavel, & Johnsen, 2008). This characterization of the prefrontal cortex is consistent with its hypothesized role in allowing for higher levels of reflective consciousness via reprocessing (Zelazo, 2004) and in the monitoring and control of cognition and behavior (e.g., Carver & Scheier, 2001; Shallice, 1982; Stuss & Benson, 1986). Taken together, the dynamic interactions among different brain regions identified in the IR model are argued to support a flexible and complex process of evaluation that unfolds in time and exists on a continuum from relatively automatic (and simple) to relatively reflective (and complex).

EMOTIONAL EXPERIENCE

The IR model also provides a framework for understanding how emotional experience is generated by computations of valence in a hierarchical cognitive system (Cunningham & Van Bavel, in press). Previous hedonic states represent how the organism was doing in the distant or immediate past. The current hedonic state represents an appraisal of how the organism is doing right now in the current situation—an idea that is conceptually similar to the idea of core affect (Russell, 2003). Predicted hedonic states represent how the organism is likely to do in the future, including how perceived changes in the environment are likely to change one's hedonic state; this notion is similar to that of anticipated affect (Russell, 2003). The delta function (the result of a comparator process) computes and represents whether things are getting (or have gotten) better or worse by comparing the current with previous and/or predicted hedonic states (cf. Frank & Claus, 2006).

These relatively simple states, together with more complex comparisons among states, offer an account of the basic emotions. For example, fear corresponds to a predicted negative state, whereas sadness follows when a comparison between the current hedonic state and a previous hedonic state results in a decrease in valence or perpetuation of a negative state. The experience of more complex emotions requires iterative reprocessing in order to construct appropriately complex representations of a situation. For example, the guilt that follows from a current hedonic state (e.g., of sadness) being less negative than predicted requires a higher order chain of comparisons.

THE DEVELOPMENT OF ITERATIVE REPROCESSING

The IR model was developed as an integration of a social cognitive neuroscience framework for thinking about attitudes and evaluation (Cunningham & Johnson, 2007) and a developmental cognitive neuroscience model of the development of prefrontally mediated reflective processing (Bunge & Zelazo, 2006; Zelazo, 2004). As such, the model is explicitly informed by a growing body of research on the development of prefrontal cortex in childhood (see Zelazo, Carlson, & Kesek, 2008, for review). The development of prefrontal cortex plays a well established role in the development of executive function, but we argue here that the consequences of prefrontal cortex development extend beyond the range of behaviors normally studied under the rubric of executive function and include consequences for emotion regulation, and even the nature of affective experience (Zelazo & Cunningham, 2007).

THE DEVELOPMENT OF NEURAL SYSTEMS INVOLVING PREFRONTAL CORTEX

Although it continues to develop throughout childhood and adolescence (e.g., Giedd et al., 1999; Gogtay et al., 2004), prefrontal cortex function first emerges early—probably toward the end of the first year of life (e.g., Chugani & Phelps, 1986; Diamond & Goldman-Rakic, 1989). The growth of prefrontal cortex beyond infancy has been documented using a variety of measures, and a number of consistent patterns have been noted. For example, whereas myelination starts postnatally in prefrontal cortex, and then increases monotonically over the course of childhood (e.g., Klingberg, Vaidya, Gabrieli, Moseley, & Hedehus, 1999; Yakovlev & Lecours, 1967), gray matter volume in prefrontal cortex shows a pattern of early increases followed by gradual decreases that start in late childhood and continue into adulthood (e.g., Gogtay et al., 2004; Huttenlocher, 1990; O'Donnell, Noseworthy, Levine, & Dennis, 2005). These changes occur at different rates in different regions of prefrontal cortex (Giedd et al., 1999; O'Donnell et al., 2005).

In addition to these structural changes, there are general changes in the patterns of neural activation that occur during performance on measures of executive function, including an increasing reliance on more anterior regions of prefrontal cortex associated with age and executive function development (i.e., frontalization; Lamm, Zelazo, & Lewis, 2006; Rubia et al., 2000). For example, Lamm et al. (2006) used high-density (128-channel) electroencephalography (EEG) to measure event-related potentials (ERPs) on the scalp as children and adolescents performed a go/no-go task, and collected a number of behavioral measures of executive function. The N2 component of the ERP, an index of cognitive control, was source-localized to the cingulate cortex and to orbitofrontal cortex. However, the source of the N2 in older children and in children who performed better on the executive function tasks (regardless of age) was more anterior than that of younger children and children who performed poorly.

MECHANISMS UNDERLYING THE DEVELOPMENT OF EXECUTIVE FUNCTION: REFLECTION AND RULE USE

There is still considerable debate about how best to understand the development of prefrontally mediated EF in psychological terms, but there is growing support for the suggestion that the development of EF in childhood is due in part to age-related increases in the complexity of the representations that children are able to formulate, maintain in working memory, and use to control their behavior in a top-down, goal-directed fashion (Zelazo, Müller, Frye, & Marcovitch, 2003). These increases in complexity, in turn, have been explained by corresponding increases in the extent to which children can reflect on their representations (Zelazo, 2004).

On this account, each degree of reflection allows a stimulus to be considered in relation to a wider set of contextual considerations. For example, children may go beyond merely seeing a ball and appreciating its affordances to noting that it is called "a ball," to thinking about the relation between this ball and the ball they lost yesterday, etc. Such reflections increase the range of aspects of the situation to which they may respond, and allow them to formulate more complex systems of verbally mediated if/then rules linking stimuli and responses. That is, reflection on rules formulated at one level of complexity is required to formulate higher-order rules that subsequently control the selection and application of these rules. Rather than taking rules for granted, and simply assessing whether their antecedent conditions are satisfied, reflection involves making those rules themselves an object of consideration and considering them in relation to other rules at the same level of complexity. The top-down selection of certain rules, all within a complex system of rules, then results in the goal-directed amplification and diminution of attention to potential influences on thought (inferences) and action when multiple possible influences are present. This, in turn, allows for greater cognitive flexibility in situations where behavior might otherwise be determined by the bottom-up activation of rules that have been primed through previous experience.

Bunge (2004) and Bunge and Zelazo (2006) summarized evidence that different regions of prefrontal cortex are involved in representing rules at different levels of complexity—from simple stimulus-reward associations (orbitofrontal cortex), to sets of conditional rules (ventrolateral prefrontal cortex and dorsolateral prefrontal cortex), to explicit consideration of task sets (rostrolateral prefrontal cortex). Figure 5.1 depicts the functional correspondence between regions of prefrontal cortex and rule systems at different levels of complexity (Bunge & Zelazo, 2006). On this account, as individuals engage in reflective reprocessing and formulate more complex rule systems, additional regions of prefrontal cortex are integrated into an increasingly elaborate hierarchical network of prefrontal activations. As Bunge and Zelazo (2006) noted, developmental research suggests that the order of acquisition of rule types shown in Figure 5.1 corresponds to the order in which corresponding regions of prefrontal cortex mature. With development, children construct neural systems involving the hierarchical coordination of more regions of prefrontal cortex—a hierarchical coordination that develops in a bottom-up fashion, with higher levels in the hierarchy operating on the products of lower levels through thalamocortical circuits.

In the service of EF, this hierarchical network is constructed in the same way that it is constructed in the service of reflective evaluation, described earlier. Information is first processed via circuits connecting the thalamus and orbitofrontal cortex. Orbitofrontal cortex generates learned approach-avoidance (stimulus–reward) rules. If these relatively unreflective processes do not provide an unimpeded response to the situation (i.e., if there is some degree of conflict), then anterior cingulate cortex—serving as a performance monitor (e.g., Ridderinkhof et al., 2004)—signals the need for further reflection, and the information is then reprocessed via circuits connecting the thalamus and ventrolateral prefrontal cortex. Further processing—as required, for example, when prepotent response tendencies elicited by bivalent rules need to be ignored—occurs via circuits connecting the thalamus to dorsolateral prefrontal cortex. Thalamocortical circuits involving rostrolateral prefrontal cortex play a role in the explicit consideration of task sets or perspectives at each level in the hierarchy. The efficiency with which these circuits function develops markedly during childhood, but continues to change into adulthood, as shown by both behavioral measures and measures of neural function (see Zelazo, Carlson, & Kesek, 2008, for review).

IMPLICATIONS FOR THE DEVELOPMENT OF EMOTION AND EMOTION REGULATION

Having reviewed our current conceptualization of affect and the development of executive function, we now outline the implications for the development of emotion and emotion regulation. Specifically, we examine how development can lead to increases in the types of emotions available, and how these emotions unfold in time.

Investigations of newborn emotional states (e.g., Wolff, 1987) suggest that babies are born with a relatively simple affective system allowing for a discrimination between positive current states and negative current states. These states can be conceptually linked to basic motivational systems where infants approach stimuli eliciting positive current states and avoid stimuli eliciting negative current states. Such stimuli include appetitive reinforcers and punishers, such as tastes (Lipsitt, 1986), but also information. P. R. Zelazo and colleagues (e.g., Weiss, Zelazo, & Swain, 1988), for example, used a head-turning procedure in which newborn infants were placed between two speakers and sounds were presented laterally, sometimes on the left and sometimes on the right. Following repeated presentation of a word ("titi") at a particular fundamental frequency, babies heard a version of this word that differed in fundamental frequency by one of five degrees of discrepancy (ranging from a change of 0–28%). Results indicated maximal orienting to the stimuli that were moderately discrepant in fundamental frequency (changes of 14% and 21%), as well as active avoidance of (turning away from) stimuli that did not change.

As infants explore their environments (aided by increases in motor control), they rapidly begin to learn to pair particular cues (potentially mediated through amygdala-striatal circuits) with expected changes in current affective state. The onset, at around 6 weeks of age, of social smiling to familiar caregivers is a good example. With encoding of stimulus-outcome relation, and with the increased ability to represent predicted hedonic states, children begin to anticipate pleasure in the company of particular individuals. At the same time, however, these comparisons also give rise to the experience of fear and frustration. Lewis and colleagues (e.g., Alessandri, Lewis, & Sullivan, 1990; Lewis, Alessandri, & Sullivan, 1990) provided infants as young as 6 weeks with an opportunity to learn a simple operant response (pulling a string) associated with a consequence (a toy moving). During acquisition, infants displayed interest and enjoyment, but were less likely to do so in response to the consequence alone. Moreover, infants at this age showed evidence of frustration during extinction. In this context, frustration likely reflects a degree of sensitivity to a failure to obtain a predicted hedonic state.

As prefrontal executive development continues, infants begin to be able to represent absent stimuli in working memory, and to imagine these stimuli changing over time. As a result, they can generate more explicit representations of predicted hedonic states, allowing for richer experiences of joy and sadness, as well as the emergence of anxiety. A good example of this is the onset of stranger anxiety at around the end of the first year of life (Kagan, 1981)—a time of numerous behavioral changes that have been linked to rapid improvements in prefrontally mediated working memory (e.g., Baird et al., 2002; Liston & Kagan, 2002). The increases in reflection believed to underlie these changes (e.g., see Marcovitch & Zelazo, 2009) allow for the anticipation of a possible increase in the negativity of one's hedonic state, corresponding to anxiety. Orbitofrontal cortex is believed to be critical for the integration of information about previous, predicted, and current hedonic states, and more complex emotions requiring this integration appear as orbitofrontal cortex continues to develop (especially in more anterior regions).

Lastly, as prefrontal cortex continues to develop, allowing for further self-reflection and the understanding of other minds, children's hedonic circuits will be more informed by social contexts and others' emotions, allowing an increase in the range of situations that elicit emotions such as empathy, guilt, shame, and pride. Lewis, Sullivan, Stanger, and Weiss (1989), for example, have documented the development during the third year of life of self-conscious emotions such as guilt and shame. Specifically, only after children demonstrate self-recognition in mirrors (arguably implying self-awareness of their bodies) do they display these more social emotions. Unlike more basic emotions, these emotions require a representation of self, other, and self relative to other—representations that require a theory of mind. Consistent with this idea, many preschool age children have difficulty discriminating between the subjective experience of emotion and the overt expression of an emotion (Harris, Donnelly, Guz, & Pitt-Watson, 1986).

Development of prefrontal cortex during the preschool years allows children to take into consideration contextual affective rules that facilitate effective, socially appropriate emotion regulation. A good example comes from Saarni's (1984) classic disappointment paradigm, in which children are given an undesirable gift. Carlson and Wang (2007) found improvements between 4 and 6 years of age in children's ability to hide their disappointment, and individual differences in emotion regulation were related to individual differences in other indices of executive function. With further development, however, context (including the imagined context of others' minds) can play an increasingly large role in affective processing. Critically, from our point of view, the further development of prefrontal cortex does not simply provide new affective states per se, but rather allows for richer contexts that can be used to inform current emotional states and the affective subtleties of abstract experiences, aesthetic experiences, and complex mixed emotions such as *schadenfreude*.

One prediction that follows from this conceptualization is that as children develop, they should be more likely to engage in reflection during affective processing and emotion regulation. This should result in increasing reliance on more anterior regions of prefrontal cortex (i.e., frontalization; cf. Rubia et al., 2000), measurable by methods such as functional magnetic resonance imaging [fMRI] and electroencephalography). One EEG index of this prefrontally mediated processing is an event-related potential (ERP) called the lateralized late positive potential (LPP), which has been shown to be larger in amplitude following negative than positive stimuli in adults (Cunningham et al., 2005), but the IR model predicts that there will be a cascading sequence of neural markers as evaluation unfolds in time. Todd, Lewis, Meusel, and Zelazo (2008) investigated the time course of 4- to 6-year-old children's ERP responses when these children were presented with pictures of their mothers' and strangers' happy and angry faces. ERPs were scored following face presentation and following a subsequent cue signaling a "go/no-go" response. Responses to face presentation showed early perceptual components that were larger following strangers' faces, suggesting increased automatic processing of novel faces; a mid-latency frontocentral negativity that was greatest following angry mothers' faces, indicating increased attentional monitoring or affect regulation evoked by an angry parent; and a right-lateralized late positive potential that was largest following angry faces, suggesting extended processing of negatively valenced social stimuli. Following the go/no-go response cue, a mid-latency negativity, commonly thought to measure effortful attention, was larger in "no-go" than "go" trials, and showed a right lateralized response that was greater to angry faces, possibly reflecting increased effortful control. All of this suggests that facial affect may elicit an interactive, hierarchically organized set of processes associated with social-emotional processing. Such processes include relatively implicit or automatic levels of stimulus evaluation and response as measured by the early components, as well as more elaborated, temporally enduring processes, as indexed by the LPP. Mid-latency components (such as the N2 component) may tap either cortical processes signaling recruitment of more distributed networks for more extended social-emotional processing, or for the regulation of affective and behavioral. Overall, however, the findings from Todd et al. (2008) provide evidence that, by 4 to 6 years, networks for elaborative processing and regulation of important socio-affective information are subserved by overlapping but differentiated networks, some of which are right-lateralized, consistent with data on adult brains.

It is important to note that just because adults (and older children) have the ability to generate more complex evaluations, this does not mean that they always will. Specifically, as noted earlier, the degree to which more reflective processing and additional iterative processes will emerge is constrained by competing drives to minimize complexity and to reduce error. However, even with a desire to reduce error, situational constraints often limit one's ability to perform the necessary processing. Reflective processing (unlike automatic processing) requires time, attention, and resources. Thus, manipulations that reductions in the likelihood of reflective processing (e.g., imposing response deadlines that require quick responses) will result in affective responses that are more "childlike" and that are based on the most salient, superficial aspects of the situation. On this account, response deadlines will interrupt the cycles of reprocessing involved in reflection, resulting in evaluations based on less complex representations of a situation, and associated, as well, with decreases in activation in anterior regions of lateral prefrontal cortex. Older children and adults should look like younger children when required to respond quickly, resulting in relatively immature reactions and relatively immature patterns of neural activation. Given that prefrontally mediated reprocessing is effortful, manipulations such as divided attention would also be predicted to result in decreases in reflective evaluation and decreases in activation in anterior regions of lateral prefrontal cortex. Interestingly, however, these manipulations should be less likely to influence more "basic emotions" (ones that are the result of more automatic hedonic processing; Cunningham & Van Bavel, in press) than more complex social emotions such as guilt and pride.

CONCLUSION

Although only in its infancy, the developmental social cognitive neuroscience approach to the study of affect and emotion seems well positioned to generate new understandings of fundamental processes involved in human subjective experience. Specifically, our IR model of evaluation and affect proposes a neurally plausible account of the computational processes of evaluation and affect. This model takes seriously the nested hierarchies of representations that are used to understand the world, to make predictions about it, and to shape our behaviors. This integration of developmental cognitive neuroscience and social cognitive neuroscience provides critical predictions for adult and child emotional processing and regulation. Because these processes unfold in developmental time, the model makes explicit predictions about the developmental trajectory and nature of emotion during brain development. In addition, however, observation of the development of emotion should also yield a richer and more integrated model of affective experience in adulthood.

ACKNOWLEDGMENTS

We thank Michael Chandler and Andy Jahn for helpful comments on an earlier draft. Both authors contributed equally to this chapter.

REFERENCES

Alessandri, S. M., Sullivan, M. W., & Lewis, M. (1990). Violation of expectancy and frustration in early infancy. *Developmental Psychology, 26*, 738–744.

Armony, J. L., & Dolan, R .J. (2002). Modulation of spatial attention by fear conditioned stimuli: An event–related fMRI study. *Neuropsychologia, 40*, 807–826.

Badre, D., & D'Esposito, M. (2007). Functional magnetic resonance imaging evidence for a hierarchical organization of the prefrontal cortex. *Journal of Cognitive Neuroscience, 19*, 2082–2099.

Baird, A. A., Kagan, J., Gaudette, T., Walz, K., Hershlag, N., & Boas, D. (2002). Frontal lobe activation during object permanence: Evidence from near infrared spectroscopy. *NeuroImage, 16*, 1120–1126.

Beer, J. S., Heerey, E. H., Keltner, D., Scabini, D., & Knight, R. T. (2003). The regulatory function of self-conscious emotion: Insights from patients with orbitofrontal damage. *Journal of Personality and Social Psychology, 85*, 594–604.

Blair, R. J. R. (2004). The roles of orbitalfrontal cortex in the modulation of antisocial behavior. *Brain and Cognition, 55*, 198–208.

Bunge, S. A. (2004). How we use rules to select actions: A review of evidence from cognitive neuroscience. *Cognitive, Affective & Behavioral Neuroscience, 4*, 564–579.

Bunge, S., & Zelazo, P. D. (2006). A brain-based account of the development of rule use in childhood. *Current Directions in Psychological Science, 15*, 118–121.

Bush, G., Luu, P., & Posner, M. I. (2000). Cognitive and emotional influences in anterior cingulate cortex. *Trends in Cognitive Sciences, 4*, 215–222.

Carlson, S. M., & Wang, T. S. (2007). Inhibitory control and emotion regulation in preschool children. *Cognitive Development, 22*, 489–510.

Carter, C. S., Braver, T. S., Barch, D. M., Botvinick, M. M., Noll, D., & Cohen, J. D. (1998). Anterior cingulate cortex, error detection, and the online monitoring of performance. *Science, 280*, 747–749.

Carver, C. S., & Scheier, M. F. (2001). *On the self-regulation of behaviour.* Cambridge, UK: Cambridge University Press.

Chugani, H., & Phelps, M. (1986, February 21). Maturational changes in cerebral function in infants determined by 18^{FDG} positron emission tomography. *Science, 231*, 840–843.

Critchley, H. D., Wiens, S., Rotshtein, P., Öhman, A., & Dolan, R. J. (2004). Neural systems supporting interoceptive awareness. *Nature Neuroscience, 7*, 189–195.

Cunningham, W. A., Espinet, S. D., DeYoung, C. G., & Zelazo, P. D. (2005). Attitudes to the right—and left: Frontal ERP asymmetries associated with stimulus valence and processing goal. *NeuroImage, 28*, 827–834.

Cunningham, W. A., & Johnson, M. K. (2007). Attitudes and evaluation: Toward a component process framework. In E. Harmon-Jones & P. Winkielman (Eds.), *Social neuroscience: Integrating biological and psychological explanations of social behavior* (pp. 227–245). New York: Guilford Press.

Cunningham, W. A., Johnsen, I. R., Mowrer, S. M., & Waggoner, A. (2008). Orbitofrontal cortex provides cross-modal valuation of self-generated stimuli. Manuscript under review.

Cunningham, W. A., Johnson, M. K., Raye, C. L., Gatenby, J. C., Gore, J. C., & Banaji, M. R. (2004). Separable neural components in the processing of Black and White faces. *Psychological Science, 15*, 806–813.

Cunningham, W. A., Kesek, A., Mowrer, S.M. (in press). Distinct orbitofrontal regions encode stimulus and choice valuation. *Journal of Cognitive Neuroscience.*

Cunningham, W. A., & Van Bavel, J. J. (2009). Varieties of emotional experience: Differences in object or computation? *Emotion Review, 1*, 56–57.

Cunningham, W. A., Van Bavel, J. J., & Johnsen, I. R. (2008). Affective flexibility: Evaluative processing goals shape amygdala activity. *Psychological Science, 19*, 152–160.

Cunningham, W. A., & Zelazo, P. D. (2007). Attitudes and evaluations: A social cognitive neuroscience perspective. *Trends in Cognitive Sciences, 11*, 97–104.

Damasio, A. R. (1994). *Descartes' error.* New York: Putnam.

Davis, M. (1997). Neurobiology of fear responses: The role of the amygdala. *The Journal of Neuropsychiatry and Clinical Neurosciences, 9*, 382–402.

Diamond, A., & Goldman-Rakic P.S. (1989). Comparison of human infants and rhesus monkeys on Piaget's AB task: Evidence for dependence on dorsolateral prefrontal cortex. *Experimental Brain Research, 74*, 24–40.

Frank, M. J., & Claus, E. D. (2006). Anatomy of a decision: Striato-orbitofrontal interactions in reinforcement learning, decision making and reversal. *Psychological Review, 113*, 300–326.

Giedd, J. N., Blumenthal, J., & Jeffries, N. O. et al. (1999). Brain development during childhood and adolescence: A longitudinal MRI study. *Nature Neuroscience, 2*, 861–863.

Gogtay, N., Giedd, J. N., Lusk, L., Hayashi, K. M., Greenstein, D., Vaituzis, A. C. et al. (2004). Dynamic mapping of human cortical development during childhood through early adulthood. *Proceedings of the National Academy of Sciences U S A, 101*(21), 8174–8179.

Harris, P. L. Donnelly, K., Guz, G. R., & Pitt-Watson, R. (1986) Children's understanding of the distinction between real and apparent emotion. *Child Development, 57*, 895–909.

Huttenlocher, P. R. (1990). Morphometric study of human cerebral cortex development. *Neuropsychologia, 28*, 517–527.

Kagan, J. (1981). *The second year: The emergence of self-awareness*. Cambridge: Harvard University Press.

Klingberg, T., Vaidya, C. J., Gabrieli, J. D. E., Moseley, M. E., & Hedehus, M. (1999). Myelination and organization of the frontal white matter in children: A diffusion tensor MRI study. *Neuroreport, 10*, 2817–2821.

Koechlin, E., Ody, C., & Kounelher, F. (2003). The architecture of cognitive control in the human prefrontal cortex. *Science, 302*, 1181–1185.

Lamm, C., Zelazo, P. D., & Lewis, M. D. (2006). Neural correlates of cognitive control in childhood and adolescence: Disentangling the contributions of age and executive function. *Neuropsychologia, 44*, 2139–2148.

LeDoux, J. E. (1996). *The emotional brain*. New York: Simon & Schuster.

Lewis, M., Alessandri, S., & Sullivan, M. W. (1990). Violation of expectancy, loss of control, and anger expressions in young infants. *Developmental Psychology, 26*, 745–751.

Lewis, M., Sullivan, M. W., Stanger, C., & Weiss, M. (1989). Self development and self-conscious emotions. *Child Development, 60*, 146–156.

Lipsitt, L. P. (1986). Toward understanding the hedonic nature of infancy. In L. P. Lipsitt & J. H. Cantor (Eds.), *Experimental child psychologist: Essays and experiments in honor of Charles C. Spiker* (pp. 97–109). Hillsdale, NJ: Lawrence Erlbaum Associates.

Liston, C., & Kagan, J. (2002). Memory enhancement in early childhood. *Nature, 419*, 896.

MacDonald, A. W., Cohen, J. D., Stenger, V. A., & Carter, C. S. (2000). Dissociating the role of the dorsolateral prefrontal and anterior cingulate cortex in cognitive control. *Science, 288*, 1835–1838.

Marcovitch, S., & Zelazo, P. D. (2009). A hierarchical competing systems model of the emergence and early development of executive function. *Developmental Science, 12*, 1–18.

Montague, P. R., & Berns, G. S. (2002). Neural economics and the biological substrates of valuation. *Neuron, 36*, 265–284.

Ochsner, K. N., Bunge, S. A., Gross, J. J., & Gabrieli, J. D. E. (2002). Rethinking feelings: An fMRI study of the cognitive regulation of emotion. *Journal of Cognitive Neuroscience, 14*, 1215–1299.

Ochsner, K. N., Ray, R. D., Robertson, E. R., Cooper, J. C., Chopra, S., Gabrieli, J. D. E., & Gross, J. J. (2004). For better or for worse: Neural systems supporting the cognitive down- and up-regulation of negative emotion. *NeuroImage, 23*, 483–499.

O'Donnell, S., Noseworthy, M. D., Levine, B., & Dennis, M. (2005). Cortical thickness of the frontopolar area in typically developing children and adolescents. *Neuroimage, 24*(4), 948–954.

Öhman, A., & Mineka, S. (2001). Fears, phobias, and preparedness: Toward an evolved module of fear and fear learning. *Psychological Review, 108*, 483–522.

Oya, H., Kawasaki, H., Howard, M. A., & Adolphs, R. (2002). Electrophysiological responses in the human amygdala discriminate emotion categories of complex visual stimuli. *Journal of Neuroscience, 22*, 9502–9512.

Phelps, E. A., O'Connor, K. J., Gatenby, J. C., Grillon, C., Gore, J. C., & Davis, M. (2001). Activation of the human amygdala to a cognitive representation of fear. *Nature Neuroscience, 4*, 437–441.

Ridderinkhof, K. R., Ullsperger, M., Crone, E. A., & Nieuwenhuis, S. (2004). The role of the medial frontal cortex in cognitive control. *Science, 306*, 443–447.

Rolls, E. T. (2005). *Emotion explained*. Oxford University Press: New York.

Rolls, E. T., Hornak, J., Wade, D., & McGrath, J. (1994). Emotion-related learning in patients with social and emotional changes associated with frontal lobe damage. *Journal of Neurology, Neurosurgery & Psychiatry, 57*, 1518–1524.

Rubia, K., Overmeyer, S., Taylor, E., Brammer, M., Williams, S. C. R., & Simmons, A. et al. (2000). Functional frontalisation with age: Mapping neurodevelopmental trajectories with fMRI. *Neuroscience and Biobehavioral Reviews, 24*, 13–19.

Russell, J. A. (2003). Core affect and the psychological construction of emotion. *Psychological Review, 110*, 145–172.

Saarni, C. (1984). An observational study of children's attempts to monitor their expressive behavior. *Child Development, 55*, 1504–1513.

Shallice, T. (1982). Specific impairments of planning. *Philosophical Transactions of the Royal Society of London, B, 298*, 199–209.

Stuss, D. T., & Benson, D. F. (1986). *The frontal lobes*. New York: Raven Press.

Todd, R., Lewis, M., Muesel, L.-A., & Zelazo, P. D. (2008). The time course of emotion processing in early childhood: ERP responses to affective stimuli in a Go-Nogo task. *Neuropsychologia, 46*, 595–613.

Yakovlev, P. I., & Lecours, A. R. (1967). The myelogenetic cycles of regional maturation of the brain. In A. Minkowski (Ed.), *Regional development of the brain in early life* (pp. 3–70). Oxford, England: Blackwell.

Weiss, M. J., Zelazo, P. R., Swain, I. U. (1988). Newborn response to auditory stimulus discrepancy. *Child Development, 59*, 1530–1541.

Whalen, P. J. (1998). Fear, vigilance, and ambiguity: Initial neuroimaging studies of the human amygdala. *Current Directions in Psychological Science, 7*, 177–188.

Wolff, P. H. (1987). *The development of behavioral states and the expression of emotions in early infancy: New proposals for investigation.* Chicago, IL: University of Chicago Press.

Zelazo, P. D. (2004). The development of conscious control in childhood. *Trends in Cognitive Sciences, 8*, 12–17.

Zelazo, P. D., Carlson, S. M., & Kesek, A. (2008). Development of executive function in childhood. In C. A. Nelson, & M. Luciana (Eds.), *Handbook of developmental cognitive neuroscience*, 2nd ed. (pp. 553–574). Cambridge, MA: MIT Press.

Zelazo, P. D., Müller, U., Frye, D., & Marcovitch, S. (2003). The development of executive function in early childhood. *Monographs of the Society for Research in Child Development, 68*(3), vii-137.

Zelazo, P. D., & Cunningham, W. (2007). Executive function: Mechanisms underlying emotion regulation. In J. Gross (Ed.), *Handbook of emotion regulation* (pp. 135–158). New York: Guilford Press.

6

BRAIN MECHANISMS IN THE TYPICAL AND ATYPICAL DEVELOPMENT OF SOCIAL COGNITION

Susan B. Perlman
Brent C. Vander Wyk
Kevin A. Pelphrey
Yale University

DEFINING SOCIAL COGNITION

Definitions of social cognition are diverse and wide ranging. Therefore, we look to various scientific disciplines to delineate this multidisciplinary construct. Social psychologists generally describe social cognition as a range of phenomena including moral reasoning, attitude formation, stereotyping, and related topics (Kunda, 1999). This may include the processes underlying the perception, memory, and judgment of social stimuli, the effects of social, cultural, or affective factors on the processing of information, or even the behavioral and interpersonal consequences of these cognitive processes. In developmental psychology, the depiction of social cognition and its development has focused most frequently upon the study of "theory of mind," the awareness that other individuals maintain thoughts, beliefs, and desires that are different from our own and that their actions can be explained with reference to their individual mental states (Premack & Woodruff, 1978; Frith & Frith, 1999). The discipline of neuroscience, however, defines social cognition more narrowly. Neuroscientists tend to view this versatile construct as the ability to perceive the intentions, actions, and dispositions of others (Brothers, 1990). In this chapter, we use the term *social cognition* to refer to the fundamental abilities to perceive, categorize, remember, analyze, reason with, and behave toward other conspecifics (Adolphs, 2001; Pelphrey, Adolphs, & Morris, 2004).

Little is known about the neural correlates of social cognition in children or about the changes in brain function that underlie normative development in this domain. Understanding brain maturation in relation to changes in social cognition will allow the field to construct normative developmental curves for the functioning of circuits supporting different aspects of social cognition. Further, knowledge of the development of the "social brain" may aid researchers in their search for genetic and possible environmental factors (and gene × environment interactions) related to suboptimal social cognitive development.

Functional brain correlates or neurobiological markers of social cognition may prove useful in the early identification of children at risk for difficulties in social information processing (e.g., children with autism). To the extent that this research can elucidate developmental trajectories of the neural circuitry supporting pivotal early social cognitive abilities, it can inform the design of more effective programs for the identification and remediation of children at risk for difficulties in these areas. Further, techniques for understanding the neural correlates of social cognition may provide a means for assessing the efficacy of treatment and reveal whether behavioral improvements correspond to compensatory changes in brain function or the normalization of developmental pathways.

Our chapter focuses on two important and interrelated aspects of social cognition: *social perception* (the initial stages of evaluating intentions and dispositions of others by analysis of gaze direction, body movement, and other types of biological motion) and *action understanding* (the ability to appreciate other people's actions in terms of the mental states that produce such behavior). In the text following, we outline a network of brain regions implicated in social cognition in typically developing adults. We then present work from our own laboratory focusing on the superior temporal sulcus (STS) region and its role in social cognition in both autism and typically developing populations. We discuss mechanisms of abnormal social perception processes in autism, and suggest how early insults to the brain mechanisms involved in these processes might influence subsequent development of higher-order social cognition abilities. Finally, we put forward a new hypothesis regarding the role of connectivity among key regions as a mechanism for social cognitive development in typically and atypically developing populations.

Neuroanatomical Substrates of Social Cognition

Our research begins with the assumption that human beings (and other primates), in response to the unique computational demands of highly social environments, have evolved a "social brain" that is comprised of brain mechanisms supporting several highly important, socially relevant abilities including (1) individuating others, (2) perceiving agents and their actions, (3) perceiving the emotional states of others, (4) analyzing the intentions and motivations of others, (5) sharing attention and intentions, and (6) representing another person's perceptions and beliefs. Groundbreaking work in nonhuman primates identified several important regions involved in processing socially relevant stimuli. These included the superior temporal sulcus (STS), the amygdala, and the orbitofrontal cortices (for a review see Brothers, 1990). More recent work, employing a variety of neuroimaging techniques in humans, has elaborated upon this core network of social brain areas. Figure 6.1 provides a sketch of several of the critical brain regions, whereas their functions are elaborated in the sections that follow.

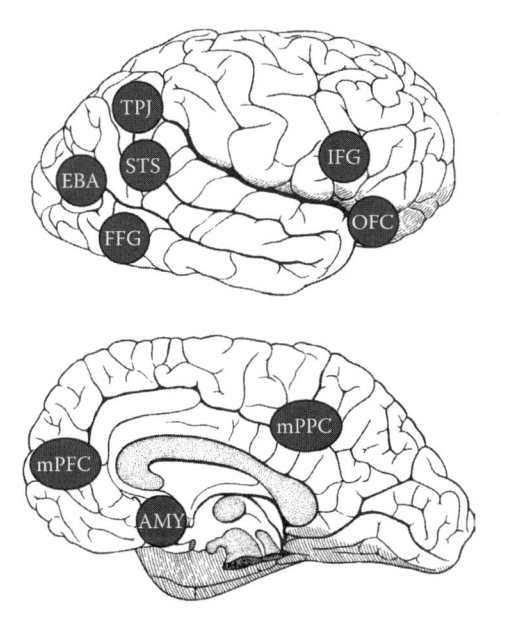

Figure 6.1 Brain regions that have been implicated in various aspects of social cognition. AMY, amygdala; EBA, extrastriate body area; FFG, fusiform gyrus; IFG, inferior frontal gyrus; mPFC, medial prefrontal cortex; mPPC, medial parietal cortex; OFC, orbital frontal cortex; STS, superior temporal sulcus region; TPJ, temporal parietal junction. (From Pelphrey and Perlman (2009), Neuroimaging in Developmental Clinical Neuroscience, eds. Judith M. Rumsey and Monique Ernst. Cambridge, UK: Cambridge University Press.

Ventral Occipital Temporal Regions: Representing Other Beings   Several areas located in the ventral occipitotemporal cortex (VOTC) support the basic perception of socially relevant information. These include the lateral fusiform gyrus (FFG), which contains the "fusiform face area" (FFA), and the extrastriate body area (EBA). The former has a clear role in face perception and recognition (e.g., Kanwisher, McDermott, & Chun, 1997; Puce, Allison, Asgari, Gore, & McCarthy, 1996) while the latter has been implicated in the visual perception of human bodies (e.g., Downing, Jiang, Shuman, & Kanwisher, 2001). These brain regions appear to provide the basic representation of their specific visual category which supports both bottom-up visual processing of social information and top-down imagery of those categories.

Limbic Regions: Perceiving Emotion   Emotion perception is a critical aspect of social interactions. For instance, to identify the best manner in which to interact with an individual, we must accurately perceive their emotional state. A number of areas within the extended limbic system (Morgane & Mokler, 2006) subserve various aspects of emotion perception; chief among them is the amygdala. The amygdala, which consists of a set of bilateral subcortical nuclei, functions in multiple aspects of emotion (e.g., Kluver & Bucy, 1997; Davis & Whalen, 2001). One such role is to rapidly generate appropriate reactions within the self to emotionally salient stimuli given a course sensory input (LeDoux, 2000). From a social perspective, another human face could serve as a critical stimulus, and in particular, one conveying emotional expression. Indeed, the amygdala plays a role in the analysis of facial expressions (e.g., Morris et al., 1996). Dysfunction in amygdala activation to faces, due to atypical development or damage, correlates with dysfunction in emotional face processing (Adolphs, Baron-Cohen, & Tranel, 2002; Calder et al., 1996). The amygdala has reciprocal connections with many of the cortical areas involved in other aspects of face processing (Amaral, Price, Pitkanen, & Carmichael, 1992, Amaral, Behniea, & Kelly, 2003) and may modulate online activation to emotional stimuli through priming of refined cortical representations or by biasing the allocation of attentional resources (Fox, Russo, Bowles, & Dutton, 2001). Additionally, amygdala activation appears to influence learning (Cahill, 2000) in posterior areas dedicated to processing visual stimuli, including ventral temporal–occipital pathways.

Medial Prefrontal Cortex: Reasoning About the Self and Others   The medial prefrontal cortex (mPFC) is implicated in a diverse set of social cognition processes. It has been shown to be active during reasoning about others' beliefs, making inferences or attributions about others' mental states (e.g., Frith & Frith, 1999; Castelli, Frith, Happe, & Frith, 2002) or representing semantic knowledge about the psychological aspects of others (e.g., Mitchell, Banaji, & MaCrae, 2005). However, the mPFC has also been shown to be active in a number of self-oriented tasks such as autobiographical memory (Shannon & Buckner, 2004), self-reflection (e.g., Kelley et al., 2002; Northoff et al., 2006; Saxe & Powell, 2006), and the attribution of emotion to the self and others (e.g., Ochsner et al., 2004). The specific function of the mPFC is difficult to specify because of the range of stimuli and tasks to which it becomes active. However, the mPFC is often characterized as being sensitive to the perceived similarity of the self and other people. This argument is employed to support the hypothesis that the mPFC is involved in reasoning about or simulates other people's thoughts in higher order theory-of-mind related tasks. Results from other domains have shown that degree of activity in frontal-executive areas is related to the processing load (e.g., Braver et al., 1997). It is possible that under typical circumstances the medial prefrontal system is presented with a richer, more complex set of information when thinking about oneself or those similar to oneself. Thus, the self-versus-other findings may be related to social-reasoning load rather than selfness per se.

Posterior Parietal Regions: Representing the Thoughts of Others. Several studies have reported greater activity during tasks that require thinking about others' mental states than tasks requiring thinking about other aspects of a person or object in and around the temporal parietal junction (TPJ) (e.g., Saxe & Kanwisher, 2003). And like the mPFC, the TPJ is also active when reflecting on one's own mental states (Happe, 2003) such as when you read stories where the self is a protagonist in the story. Activation is usually seen in both mPFC and the TPJ during classic social cognition tasks, making differentiating their functional role challenging. Saxe and Wexler (2005) argued that the specific job of the TPJ is to integrate a set of attributed mental states into a coherent whole. However, recent evidence has implicated this region in attention allocation more broadly (Mitchell, 2008). The two interpretations of the role of the TPJ (it is engaged in mental state representation or it is engaged in attentional modulation) seem completely at odds with one another. However, the views may be potentially reconcilable if one considers the role of attention as binding information together. In order to attribute a coherent set of mental states to others, the social cognition system may have to keep active a number of conceptual states that provide little to no mutual support for one another. That is, keeping track of the fact that an individual believes in ghosts provides little support for maintaining or activating the knowledge that they believe that dogs are great. In contrast, conceptual knowledge—for example, birds have wings and can fly—may be much more interdependent. If this is the case, then it would be no surprise that binding together information with weak inherent coherence would require greater activation of attentional resources than simply representing other facts about a person or object. Thus, it is possible that the TPJ aids in building an integrated mind by binding together conceptual knowledge, along with other knowledge represented in areas in social cognition network. The mPFC could then use the bound information to simulate and reason. Whether self-reflection, such as this, is built on theory of mind or vice versa is unclear.

The precuneus is often identified in tasks involving high-level social cognition. Recent work has argued for a role of the precuneus in encoding and processing beliefs in a moral judgment tasks (Young & Saxe, 2008). However, the precuneus shows greater activity for a variety of tasks that involve internally directed focus, and depressed activity for many tasks that require externally directed focus (for a review see Cavanna & Trimble, 2006). Thus, the activation found in this region may not be about social cognition per se, but rather it may reflect an artefact of the common design of high level social cognition experiments (i.e., passively ruminating about others minds in an MRI scanner). However, the precuneus connects with many of the regions involved in social cognition, so its ability to modulate an internal versus external focus may be critical in the normal development of social cognition and especially social reasoning.

Temporal Regions: Perceiving Biological Motion and Analyzing Intentions The STS region, particularly the posterior STS, has been implicated in the perception of dynamic biological motion including eye, hand, and other body movements (e.g., Bonda, Petrides, Ostry, & Evans 1996; Pelphrey, Mitchell et al., 2003). Additionally, the posterior STS is differentially sensitive to the "intent" of the motion (Pelphrey, Morris, & McCarthy, 2004; Pelphrey et al., 2003). Specifically, it exhibits differences in activation to actions that are congruent with contextual factors relative to actions that are incongruent with the context. The STS probably plays this role because intents are used, chiefly, for predicting what people will do in the future. Since future actions are intrinsically dynamic biological motion, it seems reasonable that the STS, whose job it is to represent such motion, could be used for this role. However, it is unknown whether regions that are doing intention representation for anticipation are distinct from areas that simply represent the current state of perceived biological motion.

STS REGION AND SOCIAL PERCEPTION IN TYPICALLY DEVELOPING ADULTS

Much of our own research has focused on brain regions involved in social perception and action understanding. As mentioned previously, social perception refers to the initial stages of evaluating the actions and intentions of others by analysis of biological motion cues, including gaze direction and body movement. In this context, biological motion refers to the visual perception of an entity engaged in a recognizable activity. This definition includes the observation of humans walking and making eye and mouth movements, but the term can also refer to the visual system's ability to recover information about another person's motion from sparse input. The latter is well illustrated by the discovery that point-light displays (moving images created by placing lights on the major joints of a walking person and filming them in the dark), while being relatively impoverished visual stimuli, contain the information necessary to identify the agent of motion and the kind of motion produced by the agent (Johansson, 1973). Indeed, demonstrations that neurologically normal people not only recognize the activity of walking from dynamic point-light displays, but can also perceive relatively complex social attributes of the person including emotional and mental states (Heberlein, Adolphs, Tranel, & Damasio, 2004) serve as an excellent illustration of social perception. Social perception is part of a domain of processes variably referred to as theory of mind, social attention, social cognition, mentalizing, and mind reading. We view social perception as an ontogenetic and phylogenetic prequel to more sophisticated aspects of social cognition such as theory of mind.

Using functional magnetic resonance imaging (fMRI), we have conducted a series of studies with the goal of further specifying the role of the posterior STS region in the network of brain regions comprising the social brain. As detailed below, this work initially focused on evaluating STS function when typically developing, adult subjects were engaged in the visual analysis of other people's actions and intentions. These studies tested the hypothesis that the STS region plays an important role in social perception via its involvement in interpreting the actions and social intentions of other people from an analysis of biological-motion cues (Allison, Puce, & McCarthy, 2000). This hypothesis was based on the small amount of available human neuroimaging evidence as well as elegant, long-standing research in nonhuman primates demonstrating the sensitivity of neurons in the STS to various socially relevant visual cues, including head and gaze direction (e.g., Perrett et al., 1985). From the results of these studies, it soon became clear that knowledge of the role of the STS region in social perception could be highly relevant for understanding the nature of social perception dysfunction in autism. We therefore conducted fMRI studies to examine the STS region in adults with autism. More recently, we have begun to expand this line of research to take a developmental perspective in charting the typical and atypical development of the STS region in children with and without autism.

Detecting Life: Perceiving Biological Motion

As we navigate the social world, a critical component of social perception is the detection and recognition of other dynamic agents in the environment. In particular, we must distinguish between living, dynamic, animate beings with whom there is a potential for social interaction, and static, lifeless objects that require a different mode of reasoning and interaction (Johnson, 2006). As Baron-Cohen (1995) has proposed, the detection of other animate beings, particularly other humans, may be an essential building block along the pathway to the development of more complex social cognition abilities. Dynamic agents are often revealed via the perception of biological motion. Troje and Westhoff (2006) hypothesized that humans (and other vertebrates) have an early developing visual-perceptual mechanism that that is tuned to detect the motion of the limbs of an animal in locomotion. They further speculated that this mechanism serves as a general detection system for articulated terrestrial animals: a perceptual "life detector."

Given such a perceptual mechanism, what might be its neural basis? Neuroimaging work has provided increasing evidence that humans develop specialized pathways for the perception of biological motion (e.g., Bonda et al., 1996; Puce, Allison, Bentin, Gore, & McCarthy, 1998; Pelphrey, Mitchell et al., 2003; Peelen, Wiggett, & Downing, 2006; Carter & Pelphrey, 2006). In particular, many studies have implicated the posterior STS region in the visual perception of biological motion (for reviews, see Allison et al., 2000; Pelphrey & Morris, 2006). For example, Bonda and colleagues (1996) reported that the perception of point-light displays representing goal-directed hand actions and body movements selectively activates the STS region relative to random motion. Later, Puce and colleagues (Puce, Allison, Bentin, Gore, & McCarthy, 1998) demonstrated that the STS region responds more strongly to observed mouth and eye movements than it does to various nonbiological-motion controls.

Many of the early neuroimaging studies of biological motion perception used point-light displays as stimuli, leaving open the possibility the response from the posterior STS region was being driven by the fact that biological motion was more familiar, recognizable, and nameable than the random motion used as a control condition. It was possible that coordinated and meaningful nonbiological motion might also activate the STS region. This called into question the specificity of this region for processing biological motion. To address this issue, we conducted an event-related fMRI study to compare the response from the STS region to four different types of motion conveyed via animated virtual-reality characters (Pelphrey, Mitchell et al., 2003). Participants viewed walking, a biological motion conveyed by a robot ("Robot") or a human ("Human"). They also viewed a nonmeaningful but complex nonbiological motion in the form of a disjointed mechanical figure ("Mechanical") and a complex, meaningful, and nameable nonbiological motion involving the movements of a grandfather clock ("Clock"). Our design addressed the critical question of whether the posterior STS region is specialized for the perception of biological motion. As shown in the top panel of Figure 6.2, we observed strong and equivalent activity in the right hemisphere posterior STS region to the Human and Robot conditions. This result ruled out the possibility that the STS region was merely responding to the presence of a human form. Overall, the response to biological motion was far greater than that to the moving clock and the mechanical figure. Critically, not every brain region showed this pattern of effects. For instance, MT or V5 (MT/V5), which is known to respond to various kinds of motion (Watson et al., 1993; Zeki et al., 1991), responded strongly to all four types of motion (see bottom panel of Figure 6.2). From these results, we concluded that biological motion selectively activates the STS region. Thus, we began to view the STS region as a node of the neural system supporting social perception via its role in the "detection of life" and more broadly in the representation of observed human actions.

We employed this same paradigm to chart the development of biological motion perception in the STS region. We identified a similar network of brain regions in a sample of 7- to 10-year-old typically developing children that had greater responses evoked by biological than by nonbiological motion, including the STS region and portions of the purported human mirror neuron system, such as the inferior frontal gyri, the precentral gyri, and middle and superior frontal gyri (Carter & Pelphrey, 2006). Additionally, we found a developmental change that

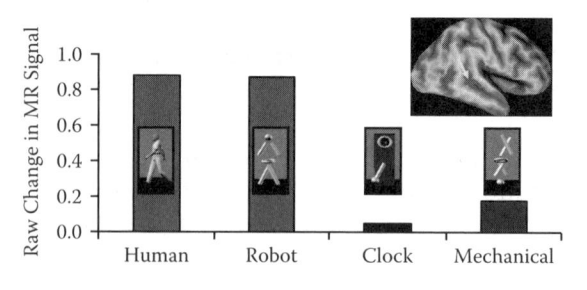

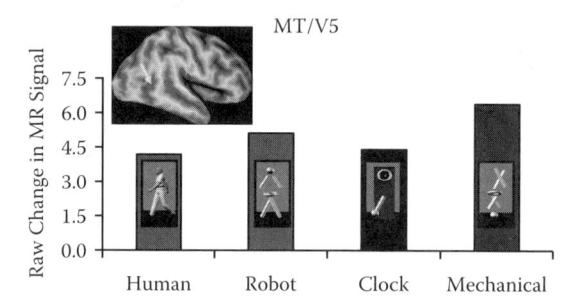

Figure 6.2 (Top panel) Experiment to determine brain responses to biological versus nonbiological motion. There were four experimental conditions: a human, a robot, a mechanical assembly, and a grandfather clock. The four figures were always present, and on each trial one of the four figures moved for 2 sec. Trials were separated by a 16 sec ITI, during which all four figures were present on the screen and none were moving. (Bottom panel) Peak amplitudes from all voxels in the right hemisphere STS by condition. (Modified from Pelphrey, Mitchell et al. (2003a), courtesy of the Society for Neuroscience.)

suggested increasing specificity for biological motion with age in the STS region. The magnitude of the biological greater than nonbiological difference score was positively correlated with age in the right STS region ($r = .64$, $p < .03$).

Analyzing the Intentions of Others

Our prior studies demonstrated that the posterior STS region is involved selectively in the perception of biological motion. We next sought to determine whether the STS region is involved in representing the intentions of other individuals with respect to objects in the visual field. That is, we wanted to know whether the STS region derives higher-level, mentalistic descriptions from motion for use in action interpretation and other inferences. In this case, we examined brain responses to the movements of a character's eyes that were either consistent or inconsistent with the participant's expectation about what the virtual character "ought" to do in a particular context (Pelphrey, Singerman, Allison, & McCarthy, 2003).

Inside the MRI scanner, our subjects watched an animated character as a small checkerboard appeared and flickered within her visual field (see Figure 6.3). On congruent trials, the character looked toward the checkerboard (Figure 6.3, top), acting in accordance with the subject's presumed expectation. On incongruent trials, the character looked away from the checkerboard at a different part of her visual field (Figure 6.3, bottom), violating expectations. We had suspected that the STS region would be sensitive to these differences in intentionality and that this region would therefore differentiate between the congruent and incongruent conditions. This would suggest that this area is involved in monitoring expectations about the goals of others. Activity in the STS region was greater for incongruent than for congruent gaze shifts, demonstrating a need for different levels of processing for observed goal-directed and nongoal-directed observed actions. We note that the pattern of effects (incongruent > congruent) is not specific to eye movements because it is also observed when subjects view congruent and incongruent reaching-to-grasp movements of the hand

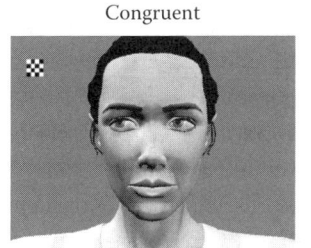

Figure 6.3 Experiment to determine brain activation in response to expected and unexpected gaze shifts on the part of another person. Incongruent trials evoked greater right hemisphere STS activity than did congruent trials, demonstrating the sensitivity of the STS region to the intentions conveyed by eye-gaze shifts. (Modified from Pelphrey & Morris, 2006. *Current Directions in Psychological Science, 15*(3), 136–140. Courtesy of Blackwell Publishing.)

and arm (Pelphrey et al., 2004). Thus, we concluded that the STS region participates in social perception beyond its role in the simple detection of biological motion: it is also involved in the visual analysis of other people's actions and intentions.

STS REGION AND SOCIAL PERCEPTION DYSFUNCTION IN AUTISM

Autism is a behaviorally defined, relatively common, pervasive neurodevelopmental disorder characterized by a triad of deficits: (1) impairments in social interactions, (2) delays in and absences of communicative skills, and (3) restricted interests and stereotyped, repetitive behaviors, as well as a distinctive developmental course with an onset no later than the first 2–3 years of life (DSM-IV; American Psychiatric Association, 1994). Leo Kanner originally described the disorder in 1943 upon seeing it in a set of 11 young patients in his clinic. His characterization of the disorder highlighted two common threads: "... (a) the children's inability from the beginning of life to relate themselves to people and situations in the ordinary way, and (b) an anxiously obsessive desire for the preservation of sameness" (Kanner, 1971, p. 140).

The social deficits and a characteristic developmental course appear to be unique, pathognomic features of autism spectrum disorders (Kanner, 1943; Wing & Gould, 1979). While repetitive behaviors and restricted interests as well as language deficits are seen in other disorders (e.g., obsessive compulsive disorder and specific language impairment, respectively), only autism presents with a set of deficits that includes social dysfunction.

Our findings from fMRI studies of typically developing individuals provided information concerning brain mechanisms for social perception, focusing particularly on the involvement of the STS region as component in the network of brain regions involved in analyzing other people's actions, goals, and intentions. These studies set the conceptual and methodological stage for studies of social perception dysfunction in autism and the typical and atypical development of the social brain. The findings from our basic neuroimaging work examining the role of the STS in social perception, coupled with our observations of the nature of social perception deficits in autism, led us to the hypothesis that STS dysfunction might play a role in the social perception deficits that are central to autism.

Neural Basis of Social Perception Deficits in Autism

Individuals with autism exhibit characteristic and early appearing deficits in using gaze information to understand the intentions and mental states of others as well as to coordinate joint attention (Baron-Cohen, 1995; Dawson, Meltzoff, Osterling, Rinaldi, & Brown, 1998; Leekam, Hunnisett, & Moore, 1998; Leekam, Lopez, & Moore, 2000; Loveland & Landry, 1986; Mundy, Sigman, Ungerer, & Sherman, 1986; Baron-Cohen et al., 1999; Frith & Frith, 1999). Joint attention can be taught in some cases, but using gaze information to infer mental states and intentions is consistently impaired even in high-functioning adults with autism (Baron-Cohen, Wheelwright, Hill, Raste, & Plumb, 2001). Note that it is not that these individuals cannot detect gaze direction, but rather that they cannot use such information to infer others' mental states and behaviors. For example, Baron-Cohen, Campbell, Karmiloff-Smith, Grant, & Walker (1995) found that children with autism do not use the direction of an individual's gaze to determine what that individual wants or knows, even though they succeed in following gaze and can accurately determine where an individual is looking.

Based on our prior findings and knowledge of these deficits, we decided to explore the role of the STS in gaze-processing dysfunction in autism (Pelphrey, Morris, & McCarthy, 2005). We utilized our checkerboard paradigm. We knew that typically developing individuals should show an increased response for incongruent versus congruent gaze shifts in the STS region, demonstrating effects of intentionality (Pelphrey, Singerman et al., 2003; Pelphrey et al., 2004; Mosconi, Mack, McCarthy, & Pelphrey, 2005). Although we found activity in the same brain regions in individuals with autism (Figure 6.4, top two panels), we saw no such differentiation in brain activity based on condition (Figure 6.4, bottom two panels). This suggests an absence of contextual influence on the STS region and a lack of sensitivity to the intentional and goal-directed structure of actions, which is a possible brain mechanism underlying the gaze-processing deficits reported behaviorally in autism. Consistent with our findings, hypoactivation of the STS and reduced functional connectivity between the STS and portions of the inferior occipital gyrus (visual area V3) have also been reported in individuals with autism during tasks involving attribution of intentions to moving geometric figures (Castelli et al., 2002).

One of the most interesting aspects of this study was the observation that dysfunction in the STS region was strongly and specifically correlated with the level of social impairment exhibited by individual subjects. As a group, activity in the STS region in the subjects with autism did not

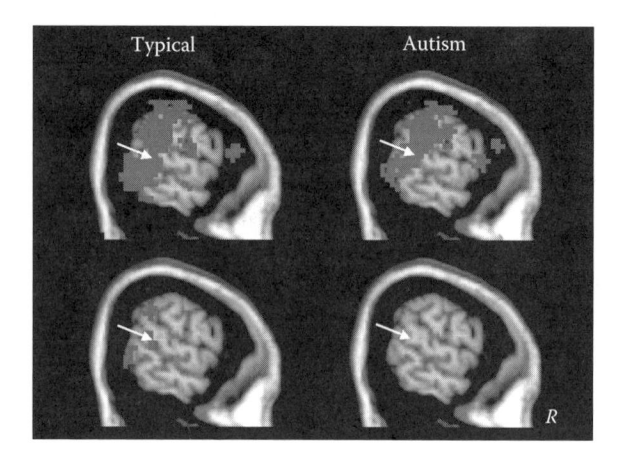

112 • Developmental Social Cognitive Neuroscience

Figure 6.4 Results from random effects analyses to compare people with and without autism. (A and B) Activation maps indicating regions with significant eye gaze-evoked activity (collapsing across congruent and incongruent gaze shifts) in the group of neurologically normal subjects (A) and subjects with autism (B). (C and D) Activation maps indicating regions where the average response at expected peak amplitude to incongruent gaze shifts was greater than the average response to congruent gaze shifts in neurologically normal subjects (C) and the lack of these activations in subjects with autism (D). The right hemisphere is shown. (Modified from Pelphrey et al., 2005. *Brain, 128*(5), 1038–1048. Courtesy of Oxford University Press, Oxford, UK.)

differ significantly for incongruent and congruent gaze shifts. However, just as autism is heterogeneous in severity, there were clear individual differences in the degree of STS dysfunction. To explore whether these individual differences were related to the severity of autism, we computed correlations between the scores on several algorithmic domains of the Autism Diagnostic Interview–Revised (ADI-R; Lord, Rutter, & Le Couteur, 1994), which is used to support the diagnosis of autism, and the magnitude of the incongruent versus congruent differentiation in the right STS region. We assumed that lower levels of incongruent versus congruent differentiation (i.e., incongruent minus congruent difference scores) in the STS region would indicate greater cortical dysfunction. Higher scores on aspects of the ADI-R can indicate greater severity of autism. Strikingly, the magnitude of the incongruent versus congruent difference score was strongly negatively correlated with scores in the reciprocal social interaction domain ($r = -0.78$, $p = 0.004$), but was not significantly correlated with impairments in the communication domain or the restricted, repetitive and stereotyped patterns of behavior domain. Nor was the measure of STS dysfunction correlated with levels of general intellectual function (as measured by an IQ test). These findings suggest that the degree of neurofunctional impairment in the right STS region is related to the severity of specific core features of the autism phenotype. Our findings suggest a brain mechanism that can help to explain why individuals with autism fail to link the perceptual representation of eyes moving and the concurrent representation regarding a character's goals, motives, and desires (i.e., the contents of the actor's mind) to determine the intentions of another person.

How Might Early Alterations in STS Development Affect Later Developments in Social Cognition?

Recent, exciting behavioral work combined with initial fMRI studies of children with autism have raised the intriguing possibility that early alterations in STS functional development in children with autism might dramatically alter the normative course of social cognitive development in these children. Blake, Turner, Smoski, Pozdol, and Stone (2003) demonstrated that school-age children with autism were less proficient in visually perceiving biological motion in a point-light display relative to their typically developing counterparts. Further evidence from a case study of a 15-month-old child who would later receive a diagnosis of autism (Klin et al., 2004) suggested that the visual perception of biological motion was more attuned to the physical contingencies of the stimulus rather than its social context. Klin and Jones (2008) presented the young subject with point-light displays of an actor portraying social situations common in infancy (playing peek-a-boo, saying hello, clapping to nursery rhymes). The animations appeared on one half of the screen in an upright manner with an inverted version of the animation presented on the other half of the screen (played in reverse to eliminate the effects of mirroring). The sound track from the social interaction was played while the experimenters recorded the child's point of regard. The researchers found that the infant with autism looked equally at both the inverted and upright figure, in contrast to typically developing infants who infer the social context from the sound track and look to the upright animation. The only exception to this effect was in an animation of a person clapping along to the "pat-a-cake" nursery rhyme. During this trial, the infant preferred looking to the upright figure whose clapping motion was synchronous with the clapping of the sound track. The authors suggest that infants and children with autism lack the ability to perceive social context in biological motion and are instead more drawn to the physical synchrony of the actor.

To investigate this hypothesis further, Klin and Jones (2008) examined the visual scanpath of the child with autism when viewing videos of a female actor during face-to-face interaction with the infant. Infants typically fixate on the eyes of their caregiver by the age of 3 months (Haith, Bergman, & Moore, 1977) and it is well documented that individuals with autism lack spontaneous fixation upon the eye region when viewing faces (Klin, Jones, Schultz, Volkmar, & Cohen, 2002; Pelphrey, Sasson, Reznick, Paul, Goldman, & Piven, 2002). The child with autism fixated mostly upon the mouth of the actor, which is the area of the face that moves in synchrony with the sound, rather than upon the eye region. The authors suggest that children with autism may glean the informational content of their social interactions by matching the voice of the person to their moving mouth and, in effect, "reading lips" rather than fixating upon the eyes, which convey the most social and emotional content (Emery, 2000). In fact, Klin et al. (2002) found that focus on the mouth region of a speaking face was the best predictor of adaptive social functioning in individuals with autism.

While it has been hypothesized that STS deficits are present in individuals with autism at an early age (Zilbovicius, Meresse, Chabane, Brunelle, Samson, & Boddaert, 2006), still under investigation are the mechanisms by which this occurs. As stated above, it may be that children with autism possess both abnormal STS structure and functioning beginning at birth or an early age due to genetic factors underlying their disability. Atypical STS functioning may inhibit young children from directing their focus towards the eyes of their caregivers and peers during social interaction, preventing them from developing adequate joint attention abilities. A lack of spontaneous engagement in joint attention could then become a mechanism for abnormal theory of mind development. It must be noted, however, that the precise relationship between STS functioning, joint attention, and theory of mind may be multidirectional. It may be the case that STS functioning in individuals with autism is normal from birth, but develops along an atypical trajectory due to a lack of input from social cues (i.e., experience with the social and emotional information conveyed by the eye region of a face). The observed lack of fixation upon the eyes of faces may be due to outside factors unrelated to STS development, but in fact lead to abnormal theory of mind understanding through its affects on input to the STS. Future studies will be needed to address both bidirectionality of this complicated relationship and early STS development in autism.

We recently initiated a study to examine the degree to which the STS region is selective for biological as compared to nonbiological motion in 7- to 10-year-old children with autism (Pelphrey & Carter, 2008). In a behavioral study, Blake and colleagues (2003) demonstrated that children 8 to 10 years old with autism are significantly impaired at recognizing biological motion from point-light displays relative to IQ-matched neurologically normal children. We sought to identify the brain correlates of these biological motion perception deficits in children with autism. We employed our prior design with four different motion conditions: a walking man, a walking robot, a disjointed mechanical figure with the same components as the robot, and a grandfather clock (Pelphrey, Mitchell et al., 2003). In this way, we were able to control for whether the figure was biological, whether the motion was biological, and whether the motion was organized. As reviewed above, we had previously shown that the STS region in neurologically normal adults and children was activated more by the biological motion conditions (human and robot) than by the nonbiological motion conditions (clock and mechanical figure). When we used this same paradigm with 7- to 10-year-old high-functioning children with autism, we found that they did not have different STS activity for biological and non-biological motion, nor did the specificity of the response to biological versus non-biological motion vary as a function of age. Critically, not every brain region showed these patterns of effects: the motion-sensitive visual area MT/V5 was equally activated by both motion types in children with and without autism. This functional dissociation suggests that the STS and other social brain regions are specific for biological motion in typically developing children but not in children with autism. We did not observe reduced activity in all brain areas investigated, indicating that the effect we observed was not being driven by a general reduction in Blood-oxygen-level-dependent (BOLD) signal.

These findings are particularly intriguing in light of theoretical perspectives that emphasize the importance of biological motion in the development of theory of mind abilities. For example, Frith and Frith (1999) suggested that the ability to distinguish between biological and nonbiological figures and their actions is one of the likely evolutionary and developmental precursors to theory of mind. Thus, early biological motion detection abilities could allow children to develop the ability to use knowledge about the actions and intentions of others to infer mental states and thus to develop full-fledged theory of mind abilities. The findings from our fMRI study of biological motion perception in children with and without autism suggest that children with autism do not possess one of the basic building blocks believed to underlie theory of mind. It is noteworthy that these individuals also show pronounced deficits on theory of mind tasks (Baron-Cohen, Leslie, & Frith, 1985; Perner, Frith, Leslie, & Leekam, 1989). This raises the important point that key well-documented deficits in autism, such as difficulties in theory of mind (Baron-Cohen et al., 1985), might actually result from disruption in or of more basic abilities such as altered responses to biological and nonbiological motion. Then, with this basic difference, early in development, the later failure to develop theory of mind could reflect an altered course of experiment, from which intact abilities become directed toward other, nonsocial, nontheory-of-mind relevant ends. Through the lens of a developmental perspective, with an understanding of the development progression, those aspects of autism often noted as potential cognitive mechanisms may be culminations of altered developmental pathways as opposed to key mechanisms or aspects of the pathogenesis.

Incorporating a Developmental Perspective on the Social Brain

To date, we have begun to dissect some of the neurobiological mechanisms underlying specific aspects of social perception dysfunction in autism. Advances in our basic understanding of the social brain have informed and have been informed by our studies of the brain mechanisms underlying social perception dysfunction in autism. Given some initial successes in this area we have recently turned or attention to efforts aimed at incorporating a developmental perspective on the study of the social brain in children with and without autism. Methodological advances in techniques for scanning young children have provided the opportunity to address a number of fascinating theoretical questions including: (1) Do later-developing social cognition abilities colonize the systems engaged by earlier-developing abilities? Or, as development proceeds, are new regions recruited into a more basic underlying network? (2) How do the components of the social brain come to interact with each other and with other brain circuits involved in such activities as executive function and language? (3) What are the constraints for specific regions to take on aspects of social perception? (e.g., proximity and connectivity to motion and speech decoding?). (4) How are these developmental processes disrupted in autism?

Clearly there are a diverse set of brain regions that subserve the complex set of tasks that fall under the umbrella of social cognition. Given principles of division of labor in the brain, we should expect that each of these regions will show functional specialization. However, social cognition will emerge out of the joint activity and connectivity of each brain region as it contributes to processing. In the following text we describe a tentative model of the components of social cognition and discuss some of the developmental implications.

Early in development, the main source of social input comes from looking at faces and actions of others. Panel (a) of Figure 6.5 depicts this early social network. External input is processed for its basic social components—emotionality in the limbic areas, static face and body information in the VOTC, and dynamic biological motion in the STS.

Brain Mechanisms in the Typical and Atypical Development of Social Cognition • 117

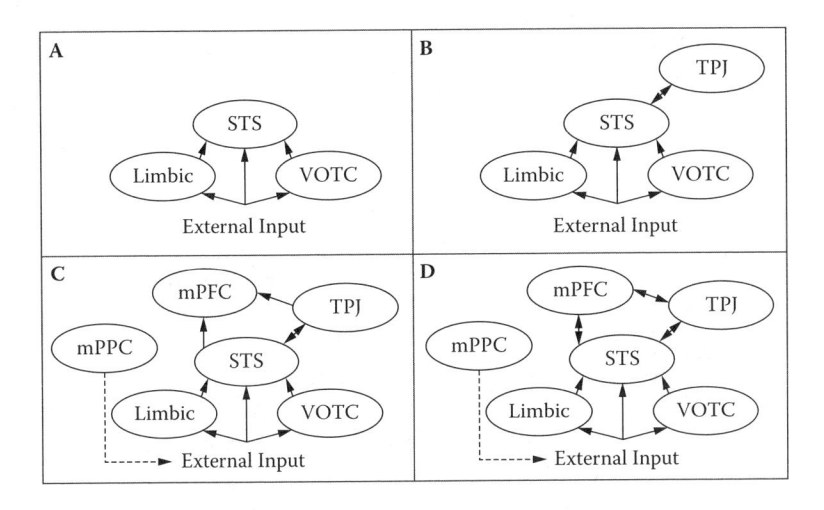

Figure 6.5 Hypothesized model of social cognitive development emphasizing connectivity among key regions as a mechanism for the increasing development of social cognition.

This information can be used to anticipate other actions and guide social exploration, but only if it is integrated across these individual domains and sensory modalities. We know that in school-age children and adults, the integration of information for the understanding of action takes place in the STS, but we do not know when this begins to happen (Mosconi et al., 2005).

This early circuit relies on sensory input that is available to even young infants and probably does not require much sophistication on the part of the child in terms of their ability to represent complex environmental knowledge or knowledge about others' mental states (Senju, Johnson, & Csibra, 2006). However, it is the very simplicity that provides the typically developing child with their first abilities to perceive and anticipate the actions of others and reap the positive rewards of being able to do so. With this early success, the child (and their social cognitive system) would be motivated to dedicate more resources to the problem of predicting and understanding others.

Indeed, this simple circuit, while probably remaining critically important throughout life, can only do very temporally limited predictions. For example, smiling and reaching may provide information enough to anticipate actions over the next few seconds, it probably cannot be used to predict what a person will be doing an hour from now. To predict this longer scale of actions, more sophisticated constructs about others need to be built. At this point, the TPJ is recruited (Panel [b]) to allow the attribution of more abstract mental states. To do this, the TPJ probably also communicates with other areas representing language and semantic knowledge (not shown). It also must communicate with the STS. First, projections from the STS to the TPJ keep the latter informed about the current state of action as it is perceived. Reciprocal connections from the TPJ back to the STS would allow the system to take advantage of the knowledge about how intentions (and more sophisticated representations) can unfold into action.

Further development of the system (Panel [c]) would bring online prefrontal areas. It is well known that prefrontal regions and connection to the prefrontal regions from the posterior portions of the brain continue to develop well into adulthood (Diamond, 2000). Thus, in early stages, the prefrontal contribution to social cognitive tasks is probably limited and impoverished. However, as these regions begin to develop, this would allow the child to begin reasoning and simulation of social situations, albeit with limited capacity. Additionally, mPPC regions such as the precuneus may begin to allow the child to switch between internal and external foci, instead of being driven completely by external stimuli. Behaviorally, this likely manifests in and is strengthened by pretense, which could be thought of as a precursor to adult-like abilities to deliberately ruminate and ponder. Indeed, pretense may be critical for adult level performance for two reasons. First, pretense allows the child to generate situations, including social situations, which can be reprocessed for additional learning. Second, by reducing the dominance of external factors, the frontal-executive control of posterior areas, which may be relatively weak, can be developed.

Finally, in the final stage (Panel [d]), the child develops adult-like control over posterior areas representing social information. These regions can still provide bottom-up processing of social information, but they now can also be activated by top-down mechanisms. One consequence of this is that social events that are separate in time can be brought together for the purposes of deliberate social reasoning. For example, we might be able to recollect an expression we perceived our friend make earlier in the day (via top-down reactivation of limbic and FFA regions) and couple it with body language we are perceiving now (via the STS) and reach a conclusion about their mental states.

Our model of social cognitive development can also account for the abnormal development of social cognition in atypical cases. In this model, we view the ontogeny of the development of social cognition in a "building block-like" manner. That is, each component of the system builds upon earlier developing components, assimilating later developing abilities into preexisting cognitive capacities. If an early component of the system, such as the ability to perceive biological motion cues, develops in an atypical fashion due to genetic factors or lack of adequate input, added levels may also not receive ample input due to decreased connectivity, or the input they receive may be impoverished or distorted. These later stages may be assimilated into an already compromised foundation, derailing the appropriate developmental trajectory for later the development of higher-order social cognition, such as the development of theory of mind. Future research will be needed to determine appropriate interventions to aid the atypically developing child in retaining not only appropriate social and cognitive behaviors, but adaptive brain functioning as well.

ACKNOWLEDGMENTS

The studies reviewed herein were supported in part by grants from the National Institute of Mental Health, the John Merck Scholars Fund, the National Institute of Child Health and Human Development, the Veterans Affairs Administration, Autism Speaks, and the National Institute for Neurological Disorders and Stroke. Kevin Pelphrey is supported by a Career Development Award from the National Institutes of Health, NIMH Grant MH071284. We gratefully acknowledge our collaborators, especially Gregory McCarthy, James Morris, and Truett Allison.

REFERENCES

Adolphs, R. (2001). The neurobiology of social cognition. *Current Opinion in Neurobiology, 2*(1), 231–239.

Adolphs, R., Baron-Cohen, S., & Tranel, D. (2002). Impaired recognition of social emotions following amygdale damage. *Journal of Cognitive Neuroscience, 14,* 1264–1274.

Allison, T., Puce, A., & McCarthy, G. (2000). Social perception from visual cues: Role of the STS region. *Trends in Cognitive Sciences, 4,* 267–278.

Amaral, D. G., Behniea, H., & Kelly, J. L. (2003). Topographic organization of projections from the amygdala to the visual cortex in the macaque monkey. *Neuroscience, 118,* 1009–1120.

Amaral, D. G., Price, J. L., Pitkanen, A., & Carmichael, S. T. (1992). Anatomical organization of the primate amygdaloid complex. In J. Aggleton (Ed.), *The amygdala* (pp. 1–67). Wiley-Liss: New York.

Baron-Cohen, S. (1995). *Mindblindness: An essay on autism and theory of mind.* Cambridge, MA., MIT Press.

Baron-Cohen, S., Campbell, R., Karmiloff-Smith, A., Grant, J., & Walker, J. (1995). Are children with autism blind to the mentalistic significance of the eyes? *British Journal of Developmental Psychology, 168,* 158–163.

Baron-Cohen, S., Leslie, A. M., & Frith, U. (1985). Does the autistic child have a "theory of mind"? *Cognition, 21*, 37–46.

Baron-Cohen, S., Ring, H. A., Wheelwright, S., Bullmore, E. T., Brammer, M. J., Simmons, A. et al. (1999). *European Journal of Neuroscience, 11*, 1891–1898.

Baron-Cohen, S., Wheelwright, S., Hill, J., Raste, Y., & Plumb, I. (2001). The "reading the mind in the eyes" test revised version: A study with normal adults, and adults with asperger syndrome or high-functioning autism. *Journal of Child Psychology and Psychiatry and Allied Disciplines, 42*, 241–251.

Blake, R., Turner, L. M., Smoski, M. J., Pozdol, S. L., & Stone, W. (2003). Visual recognition of biological motion is impaired in children with autism. *Psychological Science, 14*(2), 151–157.

Bonda, E., Petrides, M., Ostry, D., & Evans, A. (1996). Specific involvement of human parietal systems and the amygdala in the perception of biological motion. *Journal of Neuroscience, 16*(11), 3737–3744.

Braver, T. S., Cohen, J. D., Nystrom, L. E., Jonides, J., Smith, E. E., & Noll, D. C. (1997). A parametric study of prefrontal cortex involvement in human working memory. *NeuroImage 5*(1), 49–62.

Brothers, L. (1990). The neural basis of primate social communication. *Motivation and Emotion, 14*(2), 81–91.

Cahill, L. (2000). Modulation of long-term memory in humans by emotional arousal: Adrenergic activation and the amygdala. In J. Aggleton (Ed.), *The amygdala: A functional analysis* (pp. 83–116). Oxford, UK: Oxford University Press.

Calder, A. J., Young, A. W., Rowland, D., Perrett, D. I., Hodges, J. R., & Etcoff, N. L. (1996). Facial emotion recognition after bilateral amygdale damage: Differentially severe impairment of fear. *Cognitive Neuropsychology, 13*(5), 699–745.

Carter, E. J., & Pelphrey, K. A. (2006). School-aged children exhibit domain-specific responses to biological motion. *Social Neuroscience, 1*(3-4), 396–411.

Castelli, F., Frith, C., Happe, F., & Frith, U. (2002). Autism, Asperger syndrome and brain mechanisms for the attribution of mental states to animated shapes. *Brain, 125*(8), 1839–1849.

Cavanna, A. E., & Trimble, R. (2006). The precuneus: A review of its functional anatomy and behavioural correlates. *Brain, 129*(3), 564–583.

Dawson, G., Meltzoff, A. N., Osterling, J., Rinaldi, J., & Brown, E. (1998). Children with autism fail to orient to naturally occurring social stimuli. *Journal of Autism and Developmental Disorders, 28*(6), 479–485.

Davis, M., & Whalen, P. J. (2001). The amygdala: Vigilance and emotion. *Molecular Psychiatry, 6*(1), 13–34.

Diamond, A. (2000). Close interrelation of motor development and cognitive development and of the cerebellum and prefrontal cortex. *Child Development, 71*(1), 44–56.

Downing, P. E., Jiang, Y., Shuman, M., & Kanwisher, N. (2001). A cortical area selective for visual processing of the human body. *Science, 293*(5539), 2470–2473.

Emery, N. J. (2000). The eyes have it: The neuroethology, function and evolution of social gaze. *Neuroscience and Biobehavioral Review, 24,* 581–604.

Fox, E., Russo, R., Bowles, R., & Dutton, K. (2001). Do threatening stimuli draw or hold attention in visual attention in subclinical anxiety? *Journal of Experimental Psychology: General, 130,* 681–700.

Frith, C. D., & Frith, U. (1999). Interacting minds—a biological basis. *Science, 286,* 1692–1695.

Haith, M. M., Bergman, T., & Moore, M. (1977). Eye contact and face scanning in early infancy. *Science, 218,* 179–181.

Happe, F. (2003). Theory of mind and the self. *Annals of the New York Academy of Sciences, 1001,* 134–144.

Heberlein, A. S., Adolphs, R., Tranel, D., & Damasio, H. (2004). Cortical regions for judgment of emotions and personality traits from point-light walkers. *Journal of Cognitive Neuroscience, 16*(7), 1143–1158.

Johansson, G. (1973). Visual perception of biological motion and a model for its analysis. *Perception & Psychophysics, 14,* 201–211.

Johnson, M. H. (2006). Biological motion: A perceptual "life detector"? *Current Biology, 16,* 376–377.

Kanner, L. (1943). Autistic disturbances of affective contact. *Nervous Child, 2,* 217–250.

Kanner, L. (1971). Follow-up study of eleven autistic children originally reported in 1943. *Journal of Autism and Developmental Disorders, 1*(2), 119–145.

Kanwisher, N., McDermott, J., & Chun, M. M. (1997). The fusiform face area: A module in human extrastriate cortex specialized for face perception. *Journal of Neuroscience, 17*(11), 4302–4311.

Klin, A., Chawarska, K., Paul, R., Rubin, E., Morgan, T., Wiesner, L. et al. (2004). Autism in a 15-month-old child. *American Journal of Psychiatry, 161*(11), 1981–1988.

Klin, A., & Jones, J. (2008). Altered face scanning in impaired recognition of biological motion in a 15-month-old infant with autism. *Developmental Science, 11*(1), 40–46.

Klin, A., Jones, W., Schultz, R. T., Volkmar, F. R., & Cohen, D. J. (2002). Visual fixation patterns during viewing of naturalistic social situations as predictors of social competence in individuals with autism. *Archives of General Psychiatry, 59*(9), 809–816.

Kluver, H., & Bucy, P. C. (1997). Preliminary analysis of functions of the temporal lobes in monkeys. *The Journal Neuropsychiatry and Clinical Neuroscience, 9*(4), 606–620.

Kunda, Z. (1999). *Social cognition: Making sense of people.* Cambridge, MA, MIT Press.

Kelley, W. M., MaCrae, C. N., Wyland, C. L., Caglar, S., Inati, S., & Heatherton, T. F. (2002). Finding the Self? An Event-Related fMRI Study. *Journal of Cognitive Neuroscience, 5*(14), 785–794.

LeDoux, J. E. (2000). Emotion circuits in the brain. *Annual Review of Neuroscience, 23*, 155–184.

Leekam, S.R., Hunnisett, E., & Moore, C. (1998). Targets and cues: Gaze following in children with autism. *Journal of Child Psychology and Psychiatry and Allied Disciplines, 39*, 951–962.

Leekam, S. R., Lopez, B., & Moore, C. (2000). Attention and joint attention in preschool children with autism. *Developmental Psychology, 36*(2), 261–273.

Lord, C., Rutter, M., & Le Couteur, A. (1994). Autism diagnostic interview-revised: A revised version of a diagnostic interview for caregivers of individuals with possible pervasive developmental disorders. *Journal of Autism and Developmental Disorders, 24*(5), 659–685.

Loveland, K. A., & Landry, S. H. (1986). Joint attention in language in autism and developmental language delay. *Journal of Autism and Developmental Disorders, 16*(3), 1335–349.

Mitchell, J. P. (2008). Activity in right temporo-parietal junction is not selective for theory-of-mind. *Cerebral Cortex, 18*(2), 262–271.

Mitchell, J. P., Banaji, M. R., & MaCrae, C. N. (2005). General and specific contributions of the medial prefrontal cortex to knowledge about mental states. *NeuroImage, 28*(4), 757–762.

Morgane, P. J., & Mokler, D. J. (2006). The limbic brain: Continuing resolution. *Neuroscience & Biobehavioral Reviews, 30*(2), 119–125.

Morris, J. S., Frith, C. D., Perrett, D. I., Rowland, D., Young, A.W., Calder, A.J., & Dolan, R.J. (1996). A differential neural response in the human amygdala to fearful and happy expressions. *Nature, 383*(6603), 812–815.

Mosconi, M. W., Mack, P. B., McCarthy, G., & Pelphrey, K. A. (2005). Taking an "intentional stance" on eye-gaze shifts: A functional neuroimaging study of social perception in children. *NeuroImage, 27*(1), 247–252.

Mundy, P., Sigman, M., Ungerer, J., & Sherman, T. (1986). Defining the social deficits of autism: The contribution of non-verbal communication measures. *Journal of Child Psychology and Psychiatry, 27*(5), 657–669.

Northoff, G., Heinzel, A., de Greck, M., Bermpohl, F., Dobrowolny, H., & Panksepp, J. (2006). Self-referential processing in our brain: A meta-analysis of imaging studies on the self. *NeuroImage, 31*(1), 440–457.

Ochsner, K. N., Knierim, K., Ludlow, D. H., Hanelin, J., Ramachandran, T., Glover, G. et al. (2004). Reflecting upon feelings: An fMRI study of neural systems supporting the attribution of emotion to self and other. *Journal of Cognitive Neuroscience, 16*(10), 1746–1772.

Peelen, M., Wiggett, A., & Downing, P. (2006). Patterns of fMRI activity dissociate overlapping functional brain areas that respond to biological motion. *Neuron, 49*(6), 815–822.

Pelphrey, K. A., Adolphs, R., & Morris, J. P. (2004). Neuroanatomical substrates of social cognition dysfunction in autism. *Mental Retardation and Developmental Disabilities Research Reviews, 10*(4), 259–271.

Pelphrey, K. A., & Carter, E. J. (2008). Charting the typical and atypical development of the social brain. *Proceedings of the New York Academy of Sciences, 1145*, 283–299.

Pelphrey, K. A., Mitchell, T. V., McKeown, M. J., Goldstein, J., Allison, T., & McCarthy, G. (2003). Brain activity evoked by the perception of human walking: Controlling for meaningful coherent motion. *Journal of Neuroscience, 23*(17), 6819–6825.

Pelphrey, K. A., & Morris, J. P. (2006). Brain mechanisms for interpreting the actions of others from biological-motion cues. *Current Directions in Psychological Science, 15*(3), 136–140.

Pelphrey, K. A., Morris, J. P., & McCarthy, G. (2004). Grasping the intentions of others: The perceived intentionality of an action influences activity in the superior temporal sulcus during social perception. *Journal of Cognitive Neuroscience, 16*, 1706–1716.

Pelphrey, K. A., Morris, J. P., & McCarthy, G. (2005). Neural basis of eye gaze processing deficits in autism. *Brain, 128*(5), 1038–1048.

Pelphrey, K. A., Sasson, N. J., Reznick, J.S., Paul, G., Goldman, B. D., & Piven, J. (2002). Visual scanning of faces in autism. *Journal of Autism and Developmental Disorders, 32*(4), 249–261.

Pelphrey, K. A., Singerman, J. D., Allison, T., & McCarthy, G. (2003). Brain activation evoked by perception of gaze shifts: The influence of context. *Neuropsychologia, 41*(2), 156–170.

Perner, J., Frith, U., Leslie, A. M., & Leekam, S. R. (1989). Exploration of the autistic child's theory of mind: Knowledge, belief, and communication. *Child Development, 60*(3), 689–700.

Perrett, D. I., Smith, P. A. J., Potter, D. D., Mistlin, A. J., Head, A. S., Milner, A. D. et al. (1985). Visual cells in the temporal cortex sensitive to face view and gaze direction. *Proceedings of the Royal Society of London, B 223*, 239–317.

Premack, D., & Woodruff, G. (1978). Does the chimpanzee have a theory of mind? *Behavioral and Brain Sciences, 3*, 615–36.

Puce, A., Allison, T., Asgari, M., Gore, J. C., & McCarthy, G. (1996). Differential sensitivity of human visual cortex to faces, letterstrings, and textures: A functional magnetic resonance imaging study. *Journal of Neuroscience, 16*(16), 5205–5215.

Puce, A., Allison, T., Bentin, S., Gore, J. C., & McCarthy, G. (1998). Temporal cortex activation in humans viewing eye and mouth movements. *Journal of Neuroscience, 18*, 2188–2199.

Saxe, R., & Kanwisher, N. (2003). People thinking about people: The role of the temporo-parietal junction in "theory of mind". *NeuroImage, 19*(4), 1835–1842.

Saxe, R., & Powell, L. J. (2006). It's the thought that counts: Specific brain regions for one component of theory of mind. *Psychological Science, 17*(8), 692–699.

Saxe, R., & Wexler, A. (2005). Making sense of another mind: The role of the right temporal-parietal junction. *Neuropsychologia, 43*(10), 1311–1399.

Senju, A., Johnson, M. H., & Csibra, G. (2006). The development and neural basis of referential gaze perception. *Social Neuroscience, 1*, 220–234.

Shannon, B. J., & Buckner, R. L. (2004). Functional-anatomic correlates of memory retrieval that suggest nontraditional processing roles for multiple distinct regions within posterior parietal cortex. *The Journal of Neuroscience, 24*(45), 10084–10092.

Troje, N., & Westhoff, C. (2006). The inversion effect in biological motion perception: Evidence for a "life detector"? *Current Biology, 16*(8), 821–824.

Watson, J. D., Myers, R., Frackowiak, R. S., Hanjal, J. V., Woods, R. P., Mazziotta, J. V. et al. (1993). Area V5 of the human brain: Evidence from a combined study using positron emission tomography and magnetic resonance imaging. *Cerebral Cortex, 3*, 79–94.

Wing, L., & Gould J. (1979). Severe impairments of social interaction and associated abnormalities in children: Epidemiology and classification. *Journal of Autism and Developmental Disorders, 9*(1), 11–29.

Young, L., & Saxe, R. (2008). The neural basis of belief encoding and integration in moral judgment. *NeuroImage, 40*(4), 1912–1920.

Zeki, S., Watson, J. D., Lueck, C. J., Friston, K. J., Kennard, C., & Frackowiak, R. S. (1991). A direct demonstration of functional specialization in human visual cortex. *Journal of Neuroscience, 11*, 641–649.

Zilbovicius, M., Meresse, I., Chabane, N., Brunelle, F., Samson, Y., & Boddaert, N. (2006). Autism, the superior temporal sulcus and social perception. *Trends in Neurosciences, 29*(7), 359–366.

7

AUTISM AND THE EMPATHIZING–SYSTEMIZING (E-S) THEORY

Simon Baron-Cohen
Cambridge University

Classic autism and Asperger syndrome both share three core diagnostic features: (a) difficulties in social development, (b) the development of communication, and (c) unusually strong, narrow interests and repetitive behavior (APA, 1994). Since communication is always social, it might be more fruitful to think of autism and Asperger syndrome as sharing features in two broad areas: social communication and narrow interests/repetitive actions. As for distinguishing features, a diagnosis of Asperger syndrome requires that the child spoke on time and has an average IQ or above.

Today, the notion of an autistic spectrum is no longer defined by any sharp separation from "normality" (Wing, 1997). The clearest way of seeing this "normal" distribution of autistic traits is by looking at the results from the Autism Spectrum Quotient (or AQ) (Baron-Cohen, Hoekstra, Knickmeyer, & Wheelwright, 2006; Baron-Cohen, Wheelwright, Skinner, Martin, & Clubley, 2001). This is a screening instrument in the form of a questionnaire, either completed by a parent about his or her child, or by self-report (if the adult is "high-functioning"). There are 50 items in total, and when administered to a large population the results resemble a "normal distribution." Most people without a diagnosis fall in the range 0–25; most with a diagnosis of an autism spectrum condition fall between 26 and 50. It has been shown that 80% score above 32, and 99% score above 26. Given that 93% of the general population fall in the average range of the AQ, and 99% of the autistic population fall in the extreme (high end) of the scale, the AQ neatly separates these groups.

In the general population, males score slightly (but statistically significantly) higher than females. Since autism spectrum conditions are far more common in males than in females (classic autism occurs in four males for every one female, and AS occurs in nine males for every one female; Rutter, 1978), this may suggest that the number of autistic traits a person has is linked to a sex-linked biological factor—genetic or hormonal, or both (Baron-Cohen, Knickmeyer, & Belmonte, 2005; Baron-Cohen, Lutchmaya, & Knickmeyer, 2004). These two aspects—the autistic spectrum and the possibility of sex-linked explanations—have been at the core of my research and theorizing over recent years.

THE MINDBLINDNESS THEORY

In my early work I explored the theory that children with autism spectrum conditions are delayed in developing a theory of mind (ToM): the ability to put oneself into someone else's shoes, to imagine their thoughts and feelings (Baron-Cohen, 1995; Baron-Cohen, Leslie, & Frith, 1985). When we "mind read" or mentalize, we not only make sense of another person's behavior (Why did their head swivel on their neck? Why did their eyes move left?), but we also imagine a whole set of mental states (they have seen something of interest, they know something or want something) and we can predict what they might do next.

The mindblindness theory proposes that children with autism and Asperger syndrome are delayed in the development of their ToM, leaving them with degrees of mindblindness. As a consequence, they find other people's behavior confusing and unpredictable, even frightening. Evidence for this comes from difficulties they show at each point in the development of the capacity to mindread:

- A typical 14-month-old shows joint attention (such as pointing or following another person's gaze), during which they not only look at another person's face and eyes, but pay attention to what the other person is interested in (Scaife & Bruner, 1975). Children with autism and Asperger syndrome show reduced frequency of joint attention, in toddlerhood (Swettenham et al., 1998).
- The typical 24-month-old engages in pretend play, using their mindreading skills to be able to understand that in the other

person's mind, they are just pretending (Leslie, 1987). Children with autism and Asperger syndrome show less pretend play, or their pretence is limited to more rule-based formats (Baron-Cohen, 1987).

- The typical 3-year-old child can pass the Seeing Leads to Knowing Test: understanding that merely touching a box is not enough to know what is inside (Pratt & Bryant, 1990). Children with autism and Asperger syndrome are delayed in this (Baron-Cohen & Goodhart, 1994).
- The typical 4-year-old child passes the False Belief Test, recognizing when someone else has a mistaken belief about the world (Wimmer & Perner, 1983). Most children with autism and Asperger syndrome are delayed in passing this test (Baron-Cohen et al., 1985).
- Deception is easily understood by the typical 4-year-old child (Sodian & Frith, 1992). Children with autism and Asperger syndrome tend to assume everyone is telling the truth, and may be shocked by the idea that other people may not say what they mean (Baron-Cohen, 1992; Baron-Cohen, 2007a). The typical 9-year-old can figure out what might hurt another's feelings and what might therefore be better left unspoken. Children with Asperger syndrome are delayed by around 3 years in this skill, despite their normal IQ (Baron-Cohen, O'Riordan, Jones, Stone, & Plaisted, 1999).
- The typical 9-year-old can interpret another person's expressions from their eyes, to figure out what they might be thinking or feeling (see Figure 7.1). Children with Asperger syndrome tend to find such tests far more difficult (Baron-Cohen, Wheelwright, Scahill, Lawson, & Spong, 2001), and the same is true when the adult test of Reading the Mind in the Eyes is used (Figure 7.2). Adults with autism and Asperger syndrome score

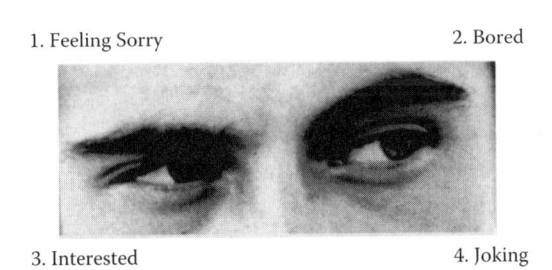

Figure 7.1 The child version of the Reading the Mind in the Eyes Test.

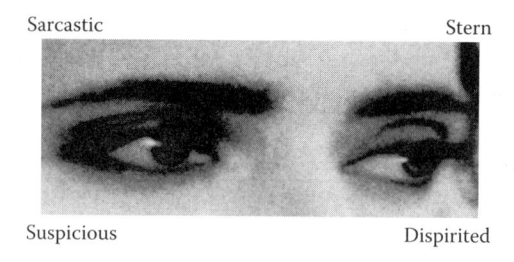

Figure 7.2 The adult version of the Reading the Mind in the Eyes Test.

below average on this test of advanced mindreading (Baron-Cohen, Wheelwright, Hill, Raste, & Plumb, 2001).

A strength of the mindblindness theory is that it can make sense of the social and communication difficulties in autism and Asperger syndrome, and that it is universal in applying to all individuals on the autistic spectrum. Its shortcoming is that it cannot account for the nonsocial features. A second shortcoming of this theory is that whilst mind reading is one component of empathy, true empathy also requires an emotional response to another person's state of mind (Davis, 1994). Many people on the autistic spectrum also report that they are puzzled by how to respond to another person's emotions (Grandin, 1996). A final limitation of the mindblindness theory is that a range of clinical conditions show forms of mindblindness, such as patients with schizophrenia (Corcoran & Frith, 1997), or narcissistic and borderline personality disorders (Fonagy, 1989), and in some studies children with conduct disorder (Dodge, 1993), so this may not be specific to autism and Asperger syndrome.

Two key ways to revise this theory have been to explain the nonsocial areas by reference to a second factor, and to broaden the concept of ToM to include an emotional reactivity dimension. Both of these revisions were behind the development of the next theory.

THE EMPATHIZING–SYSTEMIZING (E-S) THEORY

This newer theory explains the social and communication difficulties in autism and Asperger syndrome by reference to delays and deficits in empathy, while explaining the areas of strength by reference to intact or even superior skill in systemizing (Baron-Cohen, 2002).

ToM is just the cognitive component of empathy. The second component of empathy is the response element: having an appropriate emotional reaction to another person's thoughts and feelings. This is referred to as affective empathy (Davis, 1994). On the empathy quotient (EQ), a questionnaire either filled out by an adult about themselves or by a parent about their child, both cognitive and affective empathy are assessed. On this scale, people with autism spectrum conditions score lower than comparison groups.

According to the empathizing–systemizing (E-S) theory, autism and Asperger syndrome are best explained not just with reference to empathy (below average) but also with reference to a second psychological factor (systemizing), which is either average or even above average. So it is the discrepancy between E and S that determines if you are likely to develop an autism spectrum condition.

To understand this theory we need to turn to this second factor, the concept of systemizing. (See Table 7.1) Systemizing is the drive to analyze or construct systems. These might be any kind of system. What defines a system is that it follows rules, and when we systemize we are trying to identify the rules that govern the system, in order to predict how that system will behave (Baron-Cohen, 2006). These are some of the major kinds of system:

- Collectible systems (e.g., distinguishing between types of stones or wood),
- Mechanical systems (e.g., a video-recorder or a window lock),
- Numerical systems (e.g., a train timetable or a calendar),
- Abstract systems (e.g., the syntax of a language, or musical notation),
- Natural systems (e.g., the weather patterns or tidal-wave patterns),
- Social systems (e.g., a management hierarchy, or a dance routine with a dance partner)
- Motoric systems (e.g., throwing a Frisbee or bouncing on a trampoline).

In all these cases, you systemize by noting regularities (or structure) and rules. The rules tend to be derived by noting if A and B are associated in a systematic way. The evidence for intact or even unusually strong systemizing in autism and Asperger syndrome is that, in one study, such children performed above the level that one would expect on a physics test (Baron-Cohen, Wheelwright et al., 2001). Children with Asperger syndrome as young as 8–11 years old scored higher than a comparison group who were older (typical teenagers).

A second piece of evidence comes from studies using the systemizing quotient (SQ). The higher your score, the stronger your drive to systemize. People with high functioning autism or Asperger syndrome score higher on the SQ compared to people in the general population

Table 7.1 Systemizing in Classic Autism and/or Asperger Syndrome (in italics)

- Sensory systemizing
  - Tapping surfaces, or letting sand run through one's fingers
  - *Insisting on the same foods each day*
- Motoric systemizing
  - Spinning round and round, or rocking back and forth
  - *Learning knitting patterns or a tennis technique*
- Collectible systemizing
  - Collecting leaves or football stickers
  - *Making lists and catalogues*
- Numerical systemizing
  - Obsessions with calendars or train timetables
  - *Solving math problems*
- Motion systemizing
  - Watching washing machines spin round and round
  - *Analyzing exactly when a specific event occurs in a repeating cycle*
- Spatial systemizing
  - Obsessions with routes
  - *Developing drawing techniques*
- Environmental systemizing
  - Insisting on toy bricks being lined up in an invariant order
  - *Insisting that nothing is moved from its usual position in the room*
- Social systemizing
  - Saying the first half of a phrase or sentence and waiting for the other person to complete it
  - *Insisting on playing the same game whenever a child comes to play*
- Natural systemizing
  - Asking over and over again what the weather will be today
  - *Learning the Latin names of every plant and their optimal growing conditions*
- Mechanical systemizing
  - Learning to operate the VCR
  - *Fixing bicycles or taking apart gadgets and reassembling them*
- Vocal/auditory/verbal systemizing
  - Echoing sounds
  - *Collecting words and word meanings*
- Systemizing action sequences
  - Watching the same video over and over again
  - *Analysing dance techniques*
- Musical systemizing
  - Playing a tune on an instrument over and over again
  - *Analyzing the musical structure of a song*

(Baron-Cohen, Richler, Bisarya, Gurunathan, & Wheelwright, 2003). The above tests of systemizing are designed for children or adults with Asperger syndrome, not classic autism. However, children with classic autism perform better than controls on the Picture Sequencing Test where the stories can be sequenced using physical-causal concepts (Baron-Cohen, Leslie, & Frith, 1986). They also score above average on a test of how to figure out how a Polaroid camera works, even though they have difficulties figuring out people's thoughts and feelings (Baron-Cohen et al., 1985; Perner, Frith, Leslie, & Leekam, 1989). Both of these are signs of their intact or even strong systemizing.

The strength of the E-S theory is that it is a two-factor theory that can explain the cluster of both the social and nonsocial features in autism spectrum conditions. Below average empathy is a simple way to explain the social-communication difficulties, while average or even above-average systemizing is a way of explaining the narrow interests, repetitive behavior, and resistance to change/need for sameness. This is because when you systemize, it is easiest to keep everything constant, and only vary one thing at a time. That way, you can see what might be causing what, rendering the world predictable.

When this theory first came out, one criticism of it was that it might only apply to high-functioning individuals with autism or Asperger syndrome. While their obsessions (with computers or math, for example) could be seen in terms of strong systemizing (Baron-Cohen, Wheelwright, Stone, & Rutherford, 1999), surely this didn't apply to low-functioning individuals? However, when we think of a child with autism, many of the classic behaviors can be seen as a reflection of their strong systemizing. Some examples are listed here:

Like the weak central coherence (WCC) theory (Frith, 1989), the E-S theory is about a different cognitive style (Happe, 1996). Like that theory, it also posits excellent attention to detail (in perception and memory), since when you systemize you have to pay attention to the tiny details. This is because each tiny detail in a system might have a functional role. Excellent attention to detail in autism has been repeatedly demonstrated (Jolliffe & Baron-Cohen, 2001; Mottron, Burack, Iarocci, Belleville, & Enns, 2003; O'Riordan, Plaisted, Driver, & Baron-Cohen, 2001; Shah & Frith, 1983, 1993). The difference between these two theories is that while the WCC theory sees people with autism spectrum conditions as drawn to detailed information (sometimes called local processing) for negative reasons (an alleged inability to integrate), the E-S theory sees this same quality (excellent attention to detail) as being highly purposeful; it exists in order to understand a system. Attention to detail is occurring for positive reasons in the service of achieving an ultimate understanding of a system (however small and specific that system might be).

Whereas the WCC theory predicts that people with autism or Asperger syndrome will be forever lost in the detail and never achieve an understanding of the system as a whole (since this would require a global overview), the E-S theory predicts that over time, the person may achieve an excellent understanding of a whole system, given the opportunity to observe and control all the variables in that system. The existence of talented mathematicians with AS like Richard Borcherds is proof that such individuals can integrate the details into a true understanding of the system (Baron-Cohen, 2003). It is worth noting that the executive dysfunction (ED) theory (Ozonoff, Pennington, & Rogers, 1991; Rumsey & Hamberger, 1988; Russell, 1997) has even more difficulty in explaining instances of good understanding of a whole system, such as calendrical calculation, or indeed why the so-called obsessions in autism and AS should center on systems at all.

So, the low-functioning person with classic autism shaking a piece of string thousands of times close to his eyes is seen by the ED theory as showing a perseveration arising from some neural dysfunction that would normally enable the individual to shift attention; the E-S theory sees the same behavior as a sign that the individual understands, or is attempting to understand, the physics of that string movement. He may be able to make it move in exactly the same way every time. When he makes a long, rapid sequence of sounds, he may know exactly that acoustic pattern, and get some pleasure from the confirmation that the sequence is the same every time. Much as a mathematician might feel an ultimate sense of pleasure that the "golden ratio" (a + b /a = a/b) always comes out as 1.61803399, so the child, even with low functioning autism who produces the same outcome every time with their repetitive behavior appears to derive some emotional pleasure at the predictability of the world. This may be what is clinically described as "stimming" (Wing, 1997). Autism was originally described as involving "resistance to change" and "need for sameness" (Kanner, 1943), and here we see that important clinical observation may be the hallmark of strong systemizing.

One final advantage of the E-S theory is that it can explain what is sometimes seen as an inability to generalize in autism spectrum conditions (Plaisted, O'Riordan, & Baron-Cohen, 1998; Rimland, 1964; Wing, 1997). According to the E-S theory, this is exactly what you would expect if the person were trying to understand each system as a unique system. A good systemizer is a splitter, not a lumper, since lumping things together can lead to missing key differences that enable you to predict how these two things behave differently.

EXTREME MALE BRAIN THEORY

The E-S theory has been extended into the extreme male brain (EMB) theory of autism (Baron-Cohen, 2002). This is because there are clear sex differences in empathizing (females performing better on many such tests) and in systemizing (males performing better on tests of this), such that autism and Asperger syndrome can be seen as an extreme of the typical male profile, a view first put forward by the pediatrician Hans Asperger (Asperger, 1944). To see how this theory is effectively just an extension of the E-S theory, one needs to understand that that theory posits two independent dimensions (E for empathy and S for systemizing) in which individual differences are observed in the population. When you plot these, five different "brain types" are seen:

- Type E (E>S): individuals whose empathy is stronger than their systemizing
- Type S (S>E): individuals whose systemizing is stronger than their empathy
- Type B (S=E): individuals whose empathy is as good (or as bad) as their systemizing. (B stands for "balanced")
- Extreme Type E (E>>S): individuals whose empathy is above average, but who are challenged when it comes to systemizing
- Extreme Type S (S>>E): individuals whose systemizing is above average, but who are challenged when it comes to empathy

The E-S model predicts that more females have a brain of Type E, and more males have a brain of Type S. People with autism spectrum conditions, if they are an extreme of the male brain, are predicted to be more likely to have a brain of Extreme Type S. If one gives people in the general population measures of empathy and systemizing (the EQ and SQ), the results fit this model reasonably well. The largest subgoup of males (54%) do have a brain of Type S, the largest subgroup of females (44%) have a brain of Type E, and the majority of people with autism and Asperger syndrome (65%) have an extreme of the male brain (Goldenfeld, Baron-Cohen, & Wheelwright, 2005).

Apart from the evidence from the SQ and EQ, there is other evidence that supports the EMB theory. Regarding tests of empathy, on the Faux Pas Test, where a child has to recognize when someone has said something that could be hurtful, typically girls develop faster than boys, and children with autism spectrum conditions develop even slower than typical boys (Baron-Cohen, O'Riordan et al., 1999). On the Reading the Mind in the Eyes Test, on average women score higher than men, and people with autism spectrum conditions score even lower than typical males (Baron-Cohen, Jolliffe, Mortimore, & Robertson, 1997). Regarding tests of attention to detail, on the Embedded Figures Test, where one has to find a target shape as quickly as possible, on average males are faster than females, and people with autism are even faster than typical males (Jolliffe & Baron-Cohen, 1997).

Recently, the extreme male brain theory has been extended to the level of neurology, with some interesting findings emerging (Baron-Cohen et al., 2005). Thus, in regions of the brain that on average are smaller in males than in females (e.g., the anterior cingulate, superior temporal gyrus, prefrontal cortex, and thalamus), people with autism have even smaller brain regions than typical males. In contrast, in regions of the brain that on average are bigger in males than in females (e.g., the amygdala, cerebellum, overall brain size/weight, and head circumference), people with autism have even bigger brain regions than typical males. Also, the male brain on average is larger than in females, and people with autism have been found to have even larger brains than typical males. Not all studies support this pattern but some do, and it will be important to study such patterns further.

In summary, the EMB theory is relatively new and may be important for understanding why more males develop autism and Asperger syndrome than do females. It remains in need of further examination. It extends the E-S theory which has the power to explain not just the social-communication deficits in autism spectrum conditions, but also the uneven cognitive profile, repetitive behaviour, islets of ability, savant skills, and unusual narrow interests that are part of the atypical neurology of this subgroup in the population. The E-S theory has implications for intervention, as is being tried by "systemizing empathy," presenting emotions in an autism-friendly format (Baron-Cohen, 2007b; Golan, Baron-Cohen, Wheelwright, & Hill, 2006). Finally, the E-S theory destigmatizes autism and Asperger syndrome, relating these to individual differences we see in the population (between the sexes, and within the sexes), rather than as categorically distinct or mysterious.

ACKNOWLEDGMENTS

Parts of this chapter appeared in an article by the author published in The Psychologist (2008). The author was supported by the MRC (UK) and the NLM Family Foundation during the period of this work.

REFERENCES

APA (1994). *DSM-IV Diagnostic and Statistical Manual of Mental Disorders (4th ed.)*. Washington, DC: American Psychiatric Association.

Asperger, H. (1944). Die "Autistischen Psychopathen" im Kindesalter. *Archiv fur Psychiatrie und Nervenkrankheiten, 117*, 76–136.

Baron-Cohen, Jolliffe, T., Mortimore, C., & Robertson, M. (1997). Another advanced test of theory of mind: Evidence from very high-functioning adults with autism or Asperger Syndrome. *Journal of Child Psychology and Psychiatry, 38*, 813–822.

Baron-Cohen, S. (1987). Autism and symbolic play. *British Journal of Developmental Psychology, 5*, 139–148.

Baron-Cohen, S. (1992). Out of sight or out of mind: Another look at deception in autism. *Journal of Child Psychology and Psychiatry, 33*, 1141–1155.

Baron-Cohen, S. (1995). *Mindblindness: An essay on autism and theory of mind*. MIT Press/Bradford Books: Boston.

Baron-Cohen, S. (2002). The extreme male brain theory of autism. *Trends in Cognitive Science, 6*, 248–254.

Baron-Cohen, S. (2003). *The Essential Difference: Men, Women and the Extreme Male Brain*. Penguin: London.

Baron-Cohen, S. (2006). The hyper-systemizing, assortative mating theory of autism. *Progress in Neuropsychopharmacology and Biological Psychiatry, 30*, 865–872.

Baron-Cohen, S. (2007a). I cannot tell a lie. *In Character, 3*, 52–59.

Baron-Cohen, S. (2007b). Transported into a world of emotion. *The Psychologist, 20*, 76–77.

Baron-Cohen, S., & Goodhart, F. (1994). The "seeing leads to knowing" deficit in autism: The Pratt and Bryant probe. *British Journal of Developmental Psychology, 12*, 397–402.

Baron-Cohen, S., Hoekstra, R. A., Knickmeyer, R., & Wheelwright, S. (2006). The Autism-Spectrum Quotient (AQ)—Adolescent version. *Journal of Autism & Developmental Disorders, 36*, 343–350.

Baron-Cohen, S., Knickmeyer, R., & Belmonte, M. K. (2005). Sex differences in the brain: Implications for explaining autism. *Science, 310*, 819–823.

Baron-Cohen, S., Leslie, A. M., & Frith, U. (1985). Does the autistic child have a "theory of mind'? *Cognition, 21*, 37–46.

Baron-Cohen, S., Leslie, A. M., & Frith, U. (1986). Mechanical, behavioural and Intentional understanding of picture stories in autistic children. *British Journal of Developmental Psychology, 4*, 113–125.

Baron-Cohen, S., Lutchmaya, S., & Knickmeyer, R. (2004). *Prenatal testosterone in mind: Amniotic fluid studies*. Cambridge, MA: MIT/Bradford Books.

Baron-Cohen, S., O'Riordan, M., Jones, R., Stone, V., & Plaisted, K. (1999). A new test of social sensitivity: Detection of faux pas in normal children and children with Asperger syndrome. *Journal of Autism and Developmental Disorders, 29*, 407–418.

Baron-Cohen, S., Richler, J., Bisarya, D., Gurunathan, N., & Wheelwright, S. (2003). The Systemising Quotient (SQ): An investigation of adults with Asperger Syndrome or High-Functioning Autism and normal sex differences. *Philosophical Transactions of the Royal Society, 358,* 361–374.

Baron-Cohen, S., Wheelwright, S., Hill, J., Raste, Y., & Plumb, I. (2001). The "Reading the Mind in the Eyes" test-revised version: A study with normal adults, and adults with Asperger syndrome or high-functioning autism. *Journal of Child Psychology and Psychiatry, 42,* 241–252.

Baron-Cohen, S., Wheelwright, S., Scahill, V., Lawson, J., & Spong, A. (2001). Are intuitive physics and intuitive psychology independent? *Journal of Developmental and Learning Disorders, 5,* 47–78.

Baron-Cohen, S., Wheelwright, S., Skinner, R., Martin, J., & Clubley, E. (2001). The Autism Spectrum Quotient (AQ): Evidence from Asperger Syndrome/High Functioning Autism, Males and Females, Scientists and Mathematicians. *Journal of Autism and Developmental Disorders, 31,* 5–17.

Baron-Cohen, S., Wheelwright, S., Stone, V., & Rutherford, M. (1999). A mathematician, a physicist, and a computer scientist with Asperger Syndrome: Performance on folk psychology and folk physics test. *Neurocase, 5,* 475–483.

Corcoran, R., & Frith, C. (1997). Conversational conduct and the symptoms of schizophrenia. *Cognitive Neuropsychiatry, 1,* 305–318.

Davis, M. H. (1994). *Empathy: A social psychological approach.* Colorado: Westview Press.

Dodge, K. A. (1993). Social-cognitive mechanisms in the development of conduct disorder and depression. *Annual Review of Psychology, 44,* 559–584.

Fonagy, P. (1989). On tolerating mental states: Theory of mind in borderline personality. *Bulletin of the Anna Freud Centre, 12,* 91–115.

Frith, U. (1989). *Autism: Explaining the enigma.* Oxford: Basil Blackwell.

Golan, O., Baron-Cohen, S., Wheelwright, S., & Hill, J. J. (2006). Systemising empathy: Teaching adults with Asperger Syndrome to recognise complex emotions using interactive multimedia. *Development and Psychopathology, 18,* 589–615.

Goldenfeld, N., Baron-Cohen, S., & Wheelwright, S. (2005). Empathizing and systemizing in males, females and autism. *Clinical Neuropsychiatry, 2,* 338–345.

Grandin, T. (1996). *Thinking in pictures.* Vancouver, WA: Vintage Books.

Happe, F. (1996). *Autism.* UCL Press.

Jolliffe, T., & Baron-Cohen, S. (1997). Are people with autism or Asperger's Syndrome faster than normal on the Embedded Figures Task? *J. Child Psychol. Psychiatry, 38,* 527–534.

Jolliffe, T., & Baron-Cohen, S. (2001). A test of central coherence theory: Can adults with high-functioning autism or Asperger Syndrome integrate fragments of an object. *Cognitive Neuropsychiatry, 6,* 193–216.

Kanner, L. (1943). Autistic disturbance of affective contact. *Nervous Child, 2,* 217–250.

Leslie, A. M. (1987). Pretence and representation: The origins of "theory of mind." *Psychological Review, 94*, 412–426.

Mottron, L., Burack, J. A., Iarocci, G., Belleville, S., & Enns, J. T. (2003). Locally orientated perception with intact global processing among adolescents with high-functioning autism: Evidence from multiple paradigms. *Journal of Child Psychology and Psychiatry, 44*, 904–913.

O'Riordan, M., Plaisted, K., Driver, J., & Baron-Cohen, S. (2001). Superior visual search in autism. *Journal of Experimental Psychology: Human Perception and Performance, 27*, 719–730.

Ozonoff, S., Pennington, B., & Rogers, S. (1991). Executive function deficits in high-functioning autistic children: Relationship to theory of mind. *Journal of Child Psychology and Psychiatry, 32*, 1081–1106.

Perner, J., Frith, U., Leslie, A. M., & Leekam, S. (1989). Exploration of the autistic child's theory of mind: Knowledge, belief, and communication. *Child Development, 60*, 689–700.

Plaisted, K., O'Riordan, M., & Baron-Cohen, S. (1998). Enhanced visual search for a conjunctive target in autism: A research note. *Journal of Child Psychology and Psychiatry, 39*(777–783).

Pratt, C., & Bryant, P. (1990). Young children understand that looking leads to knowing (so long as they are looking into a single barrel). *Child Development, 61*, 973–983.

Rimland, B. (1964). *Infantile Autism: The syndrome and its implications for a neural theory of behaviour.* New York: Appleton-Century-Crofts.

Rumsey, J., & Hamberger, S. (1988). Neuropsychological findings in high-functioning men with infantile autism, residual state. *Journal of Clinical and Experimental Neuropsychology, 10*, 201–221.

Russell, J. (1997). How executive disorders can bring about an inadequate theory of mind. In J. Russell (Ed.), *Autism as an executive disorder.* Oxford: Oxford University Press.

Rutter, M. (1978). Diagnosis and definition. In M. Rutter & E. Schopler (Eds.), *Autism: A reappraisal of concepts and treatment* (pp. 1–26). New York: Plenum Press.

Scaife, M., & Bruner, J. (1975). The capacity for joint visual attention in the infant. *Nature, 253*, 265–266.

Shah, A., & Frith, U. (1983). An islet of ability in autism: A research note. *Journal of Child Psychology and Psychiatry, 24*, 613–620.

Shah, A., & Frith, U. (1993). Why do autistic individuals show superior performance on the block design test? *Journal of Child Psychology and Psychiatry, 34*, 1351–1364.

Sodian, B., & Frith, U. (1992). Deception and sabotage in autistic, retarded, and normal children. *Journal of Child Psychology and Psychiatry, 33*, 591–606.

Swettenham, J., Baron-Cohen, S., Charman, T., Cox, A., Baird, G., Drew, A., Rees, L., & Wheelwright, S. (1998). The frequency and distribution of spontaneous attention shifts between social and nonsocial stimuli in autistic, typically developing, and non-autistic developmentally delayed infants. *Journal of Child Psychology and Psychiatry, 9*, 747–753.

Wimmer, H., & Perner, J. (1983). Beliefs about beliefs: Representation and constraining function of wrong beliefs in young children's understanding of deception. *Cognition, 13*, 103–128.

Wing, L. (1997). *The Autistic Spectrum*. London: Pergamon.

III
Social Cognition in Adolescence

8

THE NEURAL FOUNDATIONS OF EVALUATIVE SELF-KNOWLEDGE IN MIDDLE CHILDHOOD, EARLY ADOLESCENCE, AND ADULTHOOD

Jennifer H. Pfeifer
University of Oregon

Mirella Dapretto
University of California, Los Angeles

Matthew D. Lieberman
University of California, Los Angeles

The self has been an object of scientific curiosity for decades in psychology, and many centuries longer in philosophy. During that time research has arguably emphasized the study of mature self-concepts, but the developing self has become an important topic as well—in part due to the association between self-concepts and critical developmental outcomes such as psychological well-being, academic achievement, or engagement in risky behavior (Harter, 1999). Naïve developmental theories about the self abound, such as the expectation that babies are essentially lacking in self-knowledge while teenagers are preoccupied with the self—despite research suggesting adolescents are neither completely self-absorbed nor without reason for their enhanced self-focus (Vartanian, 2000), and infants are self-aware to some extent (Meltzoff & Moore, 1977). But how, indeed, do we come to possess a self that organizes, guides, and motivates our expectations and behaviors? This question can be asked on many levels, such as how perceptions of oneself develop in context (at home, at school, across cultures) or what kind of mental representations of the self can be held by children of different ages. A relatively new approach afforded by technology, under the banner of "developmental social cognitive neuroscience," examines the neural systems supporting the uniquely human capacity for personal identity throughout development.

In this chapter we focus specifically on change in the brain regions associated with processing representations, descriptions, or perceptions of oneself ("Who am I? What are my likes and dislikes?"). We subsequently will refer to these phenomena as evaluative self-knowledge or self-evaluations (see Harter, 1999). This focus excludes many other aspects of self-related processing, such as self-recognition or visual self-awareness; self-control or self-regulation; and agency or self-generated actions and intentions. Recent neuroimaging work is beginning to address the neural systems supporting these phenomena as well (for a review, see Lieberman, 2007), but here we emphasize the significant and independent body of research that has examined the process of reflecting on the self's attributes, abilities, and preferences. In the following section, we briefly describe the functional importance and development of evaluative self-knowledge. We next review relevant developmental and social cognitive neuroscience research addressing three aspects of self-evaluations: general evaluative self-knowledge, domain-specific self-concepts, and taking the perspective of others on the self. The chapter concludes by proposing a developmental model of the neural systems supporting self-evaluations and discussing promising directions for future research. We propose that taking a developmental social cognitive neuroscience approach to the self may help to provide new insights about the social or cognitive sources and mechanisms of self-development in typically developing children, biologically rooted justifications for the powerful effects of self-concepts during development, and foundations for understanding social developmental disorders associated with atypical self-perception, such as autism spectrum disorders (ASD).

FUNCTION AND DEVELOPMENT OF EVALUATIVE SELF-KNOWLEDGE

One way to begin our account of evaluative self-knowledge development is with the global sense of worth we ascribe to ourselves, also known as self-esteem. Very early in childhood, children do begin to exhibit individual differences in self-esteem that manifest themselves in aspects of behavior, such as confidence and independence, but it is not until middle childhood that behavioral representations of self-esteem become grounded in competence and skills across contexts to form a hierarchically organized self (Haltiwanger, 1989; Harter, 1990a; Harter, 1999). For example, part of how I feel about myself as a person may be determined by my self-evaluations in the academic domain, which is a function of additional subordinate self-concepts in multiple disciplines (math, science, reading, etc.). The importance of various domains to the self was proposed over a century ago to weight the relative contribution of such evaluations towards global self-esteem (James, 1890); for example, if I do not value academic abilities at all, my lack thereof will trouble me little and have a trivial negative impact on my self-esteem, if any.

Critically, however, a domain-specific self-concept contributes not only to global self-worth to the extent it is valued, but also specifically to outcomes in that domain. Developmental psychologists have thus charted the consequences and trajectories of positive and negative self-concepts in various contexts (for reviews, see Bracken, 1996b; Damon & Hart, 1988; Harter, 1999; Marsh, 1990b, 1990c; Rosenberg, 1979; Wigfield et al., 1997). For example, a child who holds negative views of his or her abilities and attributes in a given academic domain receives lower grades on average than children with positive self-concepts in that domain, even after accounting for prior academic performance in that domain (Marsh, 1990a).

Most evidence suggests self-concepts in primary domains (like academics, athletics, physical appearance, behavioral conduct, and sociability) solidify during the transition from childhood to adolescence, even though evaluative self-knowledge may be differentiated to some degree as early as 5 years of age (Crain, 1996; Wigfield et al., 1997). Across multiple measures, average correlations between domain-specific self-concepts have been shown to decrease in both cross-sectional and longitudinal assessments during childhood and adolescence (Harter, Bresnick, Bouchey, & Whitesell, 1997; Marsh, 1990b, 1990c; Marsh & Ayotte, 2003; Marsh, Craven, & Debus, 1999). In other words, early views of the self are relatively dominated by valence rather than actual domain-specific content; children who believe they are smart are also highly likely to also believe they make friends easily, are good at sports, and so on. With increasing age, domain-specific self-concepts also become more stable and closely aligned with external indicators (e.g., higher correlations appear between academic self-concepts and grades, while correlations between other domain-specific self-concepts and academic outcomes drop, as do those between academic self-concepts and outcomes pursuant to other domains; Marsh, Craven, & Debus, 1998; Wigfield et al., 1997).

In any given domain, research has suggested the self-concept is populated by relevant trait descriptions that integrate many instances of behavior with perceived or actual evaluations made by others about our attributes and abilities. The cognitive ability to combine specific behavioral features of the self (I can run fast and throw far) into higher-order generalizations of characteristics that drive behavior (I am athletic) appears in middle childhood, approximately around age eight or nine (for reviews, see Damon & Hart, 1988; Harter, 1999; Rosenberg, 1986). Although a young child may use traits words to describe him or herself, they usually reflect single instances of behavior; there is no evidence that their use is based on abstractions of consistent qualities and recurring behaviors. On the other hand, a teenager might describe herself as popular because she gets invited to many parties, makes friends easily, believes her classmates think she is very well liked at school, and so on. An important point this highlights about self-evaluations is that, in addition to being complex internal cognitive representations, they may also be dependent on what we think others think of ourselves. This general theoretical perspective, also known as symbolic interactionism, proposes that self-concepts develop via the internalization of others' appraisals of us (Baldwin, 1895; Cooley, 1902; Mead, 1934). In this way, close others (and society at large) play a role in shaping our self-concepts, through their evaluations of our attributes and abilities. While family members typically hold the strongest influence over the developing self in childhood, peers occupy an increasingly important position during adolescence (Steinberg & Morris, 2001; Steinberg & Silverberg, 1986). Furthermore, while parents have an important role in fostering academic achievement and values throughout adolescence (Bouchey & Harter, 2005), peers tend to have more influence over social behaviors, views about interpersonal competence, and popularity (Gardner & Steinberg, 2005). Table 8.1 provides a summary description of these various components of self-evaluations, including definitions, examples, and the stage(s) during which a given component may be of particular importance in self-development.

In this chapter, we review developmental and social cognitive neuroscience evidence for the neural foundations of self-evaluations from three perspectives highlighted above. The first and most basic perspective examines the neural correlates of general evaluative self-knowledge. The second point of view explores how the neural representation

Table 8.1 Components of Self-Evaluations

Component	Definition	Example	Focal Periods
Self-esteem	Global sense of self-worth	"I like myself."	Early childhood
General Self-knowledge	Composed of traits that represent higher-order generalizations of recurring behaviors, attributes, or abilities	"I am friendly."	Middle childhood
Domain-specific Self-concepts	Organization of self-knowledge by various contexts	"I am good at science, but bad at sports."	Late childhood–early adolescence
Reflected Self-appraisals	Process of incorporating the perspectives of other individuals about the self, depending on the domain, various evaluative sources may have more influence	"Other kids at school think that I'm popular."	Adolescence

of domain-specific self-concepts may differ from that of general evaluative self-knowledge. The third and final perspective focuses on systems that are engaged in taking someone else's perspective on ourselves, both across and within domains. Figure 8.1 illustrates the brain structures mentioned throughout the chapter that provide critical support for various components of self-evaluations.

The proposed role of each region will be described in greater detail throughout the manuscript, but here we briefly summarize the general functions of each structure of interest. Anterior medial prefrontal cortex (MPFC; putative Brodmann's Area [BA] 10 and 32) has been strongly implicated in self-reflection. So has precuneus and posterior cingulate in medial posterior parietal cortex (MPPC; BA 7 and 31), in addition to its roles in episodic memory and mental imagery. Moving from MPFC up towards the apex of the head along the anterior cortical surface, dorsal medial prefrontal cortex (DMPFC; BA 9 [medially] and 8) is frequently engaged by more general social cognitive tasks including person perception and mentalizing. These functions have also been attributed to a key lateral parietal region known as the temporal parietal junction (TPJ; at the intersection of BA 22, 39, 40), which some have proposed is necessary for theory of mind and perspective-taking. Nearby, posterior superior temporal sulcus (pSTS) may feed primary

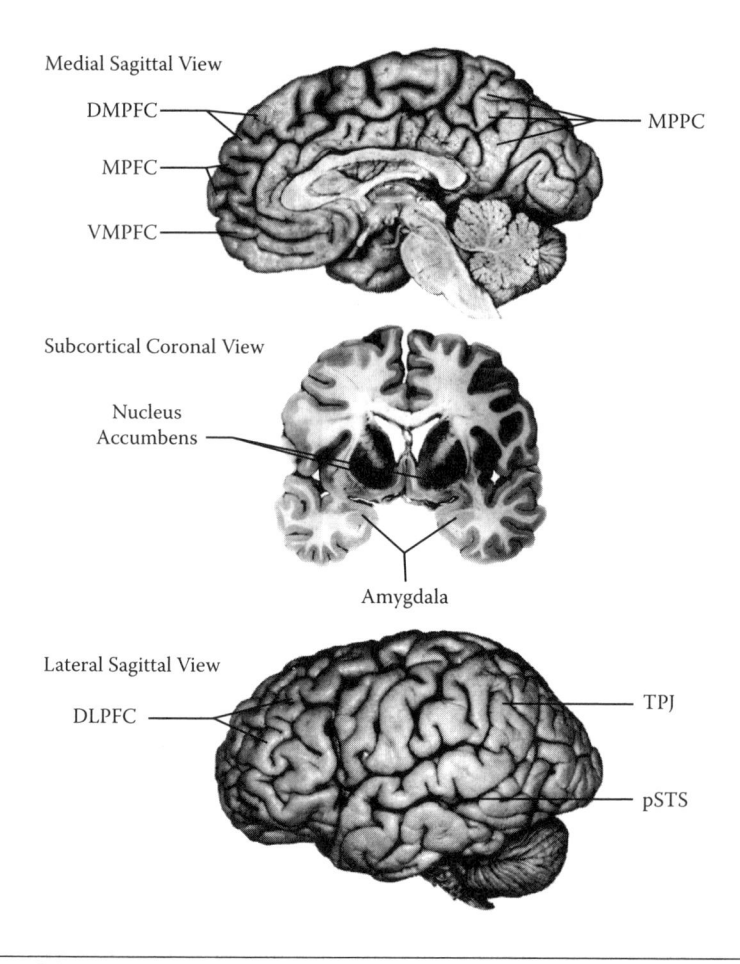

Figure 8.1 Brain regions centrally involved in various components of self-evaluations.

sensory information, particularly that which is relevant to the social domain, to higher-order processing regions. Dorsolateral prefrontal cortex (DLPFC; BA 46 and 9 [laterally]) and the hippocampus (a subcortical structure not pictured) subserve controlled processes of working memory and episodic memory storage. Other important subcortical structures include the amygdala, a site responsible for automatic affective associations and emotional learning, as well as the nucleus accumbens, a region involved in reward and approach motivation. Finally, similar affective, evaluative, and motivational functions are attributed to ventral medial prefrontal cortex (VMPFC; BA 10 and 11 [medially]), which is found by moving from MPFC down along the anterior cortical surface.

NEURAL CORRELATES OF GENERAL EVALUATIVE SELF-KNOWLEDGE

More than a dozen neuroimaging studies of general self-knowledge retrieval have been conducted in adult samples. Collectively, these studies provide a substantial consensus as to the neural systems that are likely to support mature self-evaluations. These studies typically ask adults to respond whether trait words across a variety of domains describe themselves, or assess their personal preferences. Two regions have consistently been associated with this manner of evaluative, self-referential processing: MPFC and MPPC. Such self-knowledge retrieval tasks typically produce relatively greater activity in MPFC and MPPC compared with other social or semantic processes, ranging from reporting on the personality and preferences of friends or famous individuals to making judgments about the visual appearance of words presented onscreen (D'Argembeau et al., 2005; D'Argembeau et al., 2007; Fink et al., 1996; Heatherton et al., 2006; Johnson et al., 2002; Kelley et al., 2002; Lieberman, Jarcho, & Satpute, 2004; Zysset, Huber, Ferstl, & von Cramon, 2002). Relatedly, accessing autobiographical memories and reflecting on them tends to engage putative BA 10 in MPFC (among other regions), while retrieving episodic memories does not reliably do so (for a review, see Gilboa, 2004). An additional fact of note is that MPFC and MPPC possess some of the highest resting metabolic rates in the brain. Activity in these regions tends to transiently decrease during complex, goal-directed tasks that focus participants on external factors, leading some to suggest this pair of cortical midline structures is thus responsible for processing that is internally directed or self-focused, and may provide us with an ongoing sense of one's "self" in relation to one's environment during rest (e.g., Gusnard, Akbudak, Shulman, & Raichle, 2001). Because these regions are often more active during rest than other cognitive tasks, they are often referred to as the "default network," alongside lateral parietal regions including TPJ.

However, several neuroimaging studies have not found that the activity in MPFC or MPPC is unique to self-evaluations (e.g., Craik et al., 1999; Kircher et al., 2002; Ochsner et al., 2005; Schmitz, Kawahara-Baccus, & Johnson, 2004; Vanderwal, Hunyadi, Grupe, Connors, & Schultz, 2008). In these instances, evaluative self-knowledge retrieval and control tasks (like retrieving knowledge about other social targets) typically engage MPFC and/or MPPC to a similar extent. Therefore, it is still debated whether the neural systems supporting self-knowledge processes differ from those supporting evaluations of other people. Growing evidence suggests that the most anterior subregion of MPFC (BA 10 rather than

BA 8 and/or BA 9, which we refer to as DMPFC) is more likely to be recruited to process information about individuals when they are seen as similar to ourselves (Mitchell, Banaji, & Macrae, 2005a; Mitchell, Macrae, & Banaji, 2006). A balanced viewpoint might thus be that although MPFC and MPPC are essential to evaluative self-knowledge, this medial fronto-parietal network may also support our understanding of other people, particularly those with whom we are close.

Building upon this foundation of research, we conducted the first study of the neural correlates of evaluative self-knowledge retrieval processes in a developmental sample (Pfeifer et al., 2007), which suggested that children exhibit both similarities to and differences from the adult patterns described above. We compared 9- and 10-year-old children and young adults completing a scanner task in which participants alternated between retrieving knowledge about themselves or a fictional, familiar other (Harry Potter). The major similarity between children and adults was that each age group engaged dorsal and/or anterior MPFC as well as MPPC during this social cognitive task of retrieving knowledge about oneself and another individual, regions previously shown to be involved in both processes.

Yet we also observed significant differences between children and adults in this study. Activity in anterior MPFC (BA 10 and BA 32) was significantly enhanced in children compared to adults; the BOLD signal was both stronger in amplitude and covered a larger spatial extent in the children. While this region was significantly more active in both children and adults when retrieving self-knowledge than knowledge about Harry Potter, in adults this manifested as relatively less deactivation compared to a resting baseline, but in children evaluative self-knowledge retrieval elicited activity in anterior MPFC above a resting baseline. This may indicate that children engage cortical midline structures less while resting than do adults—perhaps because children are less self-reflective than adults, or possibly because there are developmental changes in the tonic activation of these regions, independent of ongoing mental processes. Recently, a study compared activity in the default network across children and adults (aged 7–9 and 21–31 years, respectively), finding that while the regions involved in the default network were consistent across age groups, the functional connectivity between VMPFC, MPPC, and lateral parietal areas including TPJ is significantly stronger in adults than children (Fair et al., 2008). Nevertheless, this study did not report any significant differences in the absolute level of default network activity in children and adults. How much the development of the default network contributes to changes in the neural systems supporting general evaluative self-knowledge thus remains an open question for future research. Therefore, if changes in default network activity are not responsible, one alternative explanation for our findings of relatively greater activity in MPFC during self-knowledge retrieval in children than in adults is that self-reflection is qualitatively different across the two age groups.

Finally, in this study and contrary to other studies, both children and adults engaged MPPC more when thinking about Harry Potter's academic and social qualities than their own. Specifically, MPPC was less active during self-reflection than during social knowledge retrieval and a resting baseline. Furthermore, children used more anterior subregions affiliated with mental imagery and perspective taking, whereas adults used more posterior subregions associated with episodic memory retrieval (for a review of functional subdivisions within MPPC, see Cavanna & Trimble, 2006). We theorized this might be due to some particular quality of Harry Potter, a fictional character depicted extensively in movies and books, since this pattern has not been observed in any other studies contrasting self- and social knowledge retrieval (in which the other targets include friends, family members, or famous individuals like politicians). These results may also suggest that adults are less likely than children to retrieve episodic memories when reporting on themselves, perhaps relying on stored semantic knowledge of the traits they possess. Indeed, some evidence for this was observed in that one of the only regions activated more in adults than children during self-knowledge retrieval was lateral temporal cortex (BA 22), a region often implicated in semantic storage (see also Lieberman et al., 2004).

NEURAL FOUNDATIONS OF DOMAIN-SPECIFIC SELF-CONCEPTS

Exploring the neural systems supporting general evaluative self-knowledge, as in the studies described above, is a reasonable first step in developmental social cognitive neuroscience investigations of the self. To further advance this field, however, these inquiries eventually need to take into account the multidimensional nature of the self (Bracken, 1996a). As mentioned in the introduction, domain-specific self-concepts strongly contribute not only to a child's global self-image, but also specifically to outcomes in these domains (Harter, 1999). Strong correlations between self-concepts and relevant behaviors have been demonstrated in a variety of domains, including greater social competence and more peer acceptance as well as less aggression, loneliness, depression, and anxiety (Barry & Wigfield, 2002; Bellmore & Cillessen, 2003;

Berndt & Burgy, 1996); popularity and increased frequency of risky behaviors such as drug or alcohol use (Stein, Roeser, & Markus, 1998); negative body image and greater eating disorder pathology (Hargreaves & Tiggeman, 2002); as well as positive athletic self-concept and greater participation in school sports (Jacobs, Lanza, Osgood, Eccles, & Wigfield, 2002). What remains unclear from this line of behavioral research is how—via what mechanism(s)—domain-specific self-concepts exert such strong effects on behavior.

Most neuroimaging research on self-knowledge studies the self in a global sense, representing (and averaging across) a wide sampling of domains, expertise, and evaluations. To our knowledge there have been only two prior studies conducting direct inquiries of whether and how the neural systems involved in self processes differ across domains (Lieberman et al., 2004; Rameson & Lieberman, 2007). This work, unlike others, specifically examined retrieval of self-knowledge by domain. In the first study, adult participants had an abundance of experience in one domain (on average, 10 years) and considered their performance and participation in that domain a central or defining aspect of themselves, while in the other domain they had little experience and/or did not identify strongly with it. Results showed they activated ventral MPFC (VMPFC), nucleus accumbens, amygdala, lateral temporal cortex, and inferior parietal cortex more when retrieving knowledge in high- versus low-experience domains, whereas the reverse contrast led to more activity in dorsolateral PFC (DLPFC; Lieberman et al., 2004). Evidence of schematicity (exhibiting enhanced speed of processing for information in a given domain; Markus, 1977) was also associated with increased activity in MPPC and decreased activity in hippocampus and dorsal MPFC (DMPFC) during self-knowledge retrieval from the schematic domain, relative to the low-experience one. In the second study, adult participants demonstrating schematicity in an athletic domain exhibited more activity while viewing information from that domain versus an academic domain in the nucleus accumbens and the amygdala; furthermore, the response in both structures positively correlated with recall of information from the athletic domain in a surprise test (Rameson & Lieberman, 2007).

These findings support a characterization of the processes involved in low-experience, low-identification domains as "evidence-based" retrieval of self-knowledge, because the neural structures involved (in particular, DLPFC and the hippocampus) are implicated in episodic and working memory. In contrast, the processes involved in schematic, high-experience, high-identification domains may be characterized as "intuition-based" retrieval of self-knowledge, because the active network of structures is associated with motivation, emotional learning, impulsive behavior, automaticity, and affective reactions. This distinction between evidentiary and intuitive self-knowledge is supported by a series of behavioral and neuropsychological studies conducted by Klein and colleagues (Klein, Loftus, & Kihsltrom, 1996; Klein, Loftus, Trafton, & Fuhrman, 1992; Klein, Rozendal, & Cosmides, 2002) demonstrating that autobiographical evidence is not always necessary to report on one's self. For example, temporarily amnesic patients' judgments about the self-descriptiveness of traits are as accurate as normal controls.

When might schemas—indicating the possession of intuitive self-knowledge—emerge for particular self-concepts in the course of normal development? An intense level of identification with a domain (which actually constitutes a core defining feature of self-schemas; Markus, 1977) could reasonably be expected to appear during the transition between late childhood and early adolescence, as self-concepts in primary spheres like academics or sociality cohere and become more strongly related to subjective task values (Steinberg & Morris, 2001; Wigfield et al., 1997). Indeed, the earliest mentions of potential automatization of domain-specific self-concepts (or equivalently, the presence of self-schemas) surface at this developmental stage (Harter, 1999; Higgins, 1991; Marsh, 1990a; Siegler, 1991; Stein et al., 1998). This is also the time during which identities begin to surface around activities or abilities and "crowds" coalesce at school (e.g., jocks, nerds, and so forth [Brown, 2004]).

To begin examining the neural foundations of domain-specific self-concepts in children, we conducted a study including 52 typically developing 9- and 10 year-olds (20 boys, 32 girls) who completed the same scanner task described above, in which blocks alternated between retrieving knowledge about oneself and Harry Potter (Pfeifer, Dapretto, & Lieberman, in preparation). Critically, some blocks queried academic self-perceptions, and others tapped perceptions of social competence. Several hours after the scan, children completed the Self Perception Profile for Children (SPPC; Harter, 1985). The SPPC includes two subscales assessing academic and social competence, so mean scores on these subscales were calculated for each child and used as regressors.

The primary objective of this study was to see whether strongly positive perceptions of one's competence in a domain (a proxy for identification and experience) moderated patterns of brain activity supporting general evaluative self-knowledge. We found that activity in the amygdala—one of the structures especially strongly implicated in automatic affective associations, motivation, and emotional learning—was greater during self-knowledge retrieval from a high-competence domain relative to a low-competence domain, increasingly so to the extent children reported more positive self-images in the former domain than the latter ($r(50) = 0.48$, $t = 4.44$). A similar association was also observed in MPPC and DLPFC, two regions involved in episodic and working memory ($r(50)$s = 0.56 and 0.44, ts = 4.78 and 4.15, respectively). This combination of intuitive- and evidence-based systems suggests that children may have been transitioning from general processes of self-knowledge retrieval to expert processes, some of which may function in any high-experience domain, whereas others depend on the presence of self-schemas in a particular domain. Therefore, the overall pattern suggests shifts from evidentiary to intuitive self-knowledge retrieval processes are merely underway and not yet completed in 9- and 10-year-olds, even in basic domains like the ones used here.

Due to computer difficulties while collecting reaction time data, we have yet to determine whether any children also possessed self-schemas in these domains, but as activity in only one automatic and affective structure (the amygdala) was correlated with self-concepts, the neuroimaging data suggest that most children had not yet developed schemas about their relevant abilities or lack thereof. This leads to one practical implication of our findings. Self-schemas are thought to be particularly resistant to contrary information and difficult to change (Markus, 1977)—thus, if 9- and 10-year-old children have not yet developed a self-schema as a social outcast or academic failure, there may still be a good chance of modifying these negative self-perceptions (Brinthaupt & Lipka, 1994). In other words, while typically developing children clearly have the capability to reflect on their personal attributes and qualities at this age, they probably have not yet come to a particularly entrenched viewpoint on the self, even in such common domains as sociality and academics. However, children on average did possess moderately positive views of themselves in both domains, so it remains to be determined whether persistently negative self-concepts correlate with intuitive, automatic, and affective self-knowledge systems this early in development. Such a study would require targeted recruitment of children that possess these negative self-perceptions.

NEURAL CORRELATES OF REFLECTED SELF-APPRAISAL PROCESSES

In the previous sections, we discussed the neural systems supporting general evaluative self-knowledge retrieval, as well as domain-specific self-concepts. Our introduction pointed out not only how self-knowledge may be shaped by perceptions of how others view the self, but also how family members and peers have varied levels of influence in certain domains across development. In this final subsection, we explore how these facets of self development intersect in the brain via two questions: How do MPFC, MPPC, and other brain regions enable reflected self-appraisals (in other words, taking another individual's perspective on the self) across development, and is this moderated by whose perspective is being taken in a given domain?

To our knowledge, only two previous neuroimaging studies have directly examined the neural correlates of reflected self-appraisals in adults. Findings from these studies overlapped to some degree; both reported a high degree of similarity overall between direct and reflected self-appraisals, and both found that reflected self-appraisals may be associated with more activity in orbitalfrontal and insular cortex (D'Argembeau et al., 2007; Ochsner et al., 2005). Meanwhile, social cognitive neuroscience research more generally suggests four key regions that may also be involved in reflected appraisals. These areas have been emphasized by reviews as cornerstones of mentalizing and other unique aspects of human social cognition, and include temporal-parietal junction (TPJ), dorsal MPFC (DMPFC), posterior superior temporal sulcus (pSTS), and the temporal poles (Frith & Frith, 2006; Frith & Frith, 2003; Saxe, 2006). Third-person perspective-taking processes, such as reasoning about other people's mental contents or beliefs, are thought to rely on a region at the intersection of inferior parietal lobule and posterior superior temporal gyrus, frequently referred to as TPJ (Aichhorn, Perner, Kronbichler, Staffen, & Ladurner, 2006; Apperly, Samson, Chiavarino, & Humphreys, 2004; D'Argembeau et al., 2007; Ruby & Decety, 2003; Samson, Apperly, Chiavarino, & Humphreys, 2004; Saxe & Kanwisher, 2003; Saxe & Wexler, 2005). Mental state attribution and impression formation also typically engages DMPFC (Mitchell et al., 2005a; Mitchell, Macrae, & Banaji, 2005b, 2006). Extracting information about goals and intentions from biological motion within a social context is thought to be a primary function of pSTS (Pelphrey, Morris, & McCarthy 2004; Pelphrey, Viola, & McCarthy, 2004). Finally, the temporal poles may be responsible for storing social and personal semantic knowledge, and linking perceptions with emotions (Olson, Plotzker, & Ezzyat, 2007).

We recently conducted a study including early adolescents (aged 11–13 years) and adults designed to address the two questions opening this subsection. That is, we (a) compared the neural correlates of direct and reflected self-appraisals made by adolescents and adults, and (b) for adolescents in particular, explored the influence of parents and peers across domains on the networks supporting reflected appraisals (Pfeifer, Masten, Borofsky, Dapretto, Lieberman, & Fuligni, in press). The scanner task was nearly identical to that used in our previous studies, in that we asked about academic and social qualities, but instead of contrasting the self with Harry Potter we also asked participants to tell us what they believed their mother, best friend, and classmates thought about them. We found that in both age groups, making these reflected self-appraisals engaged regions associated with self-reflection (MPFC and MPPC) as well as social cognition and perspective taking (TPJ, DMPFC, and pSTS). Furthermore, adolescents exhibited more activity in MPFC and MPPC when making reflected appraisals in a domain that was consistent with an evaluative source's sphere of influence (i.e., taking a best friend's perspective on the social self, or a mother's perspective on the academic self). Perhaps the medial fronto-parietal network composed of MPFC and MPPC is most sensitive to processing information about ourselves in relation to others, rather than in a context-independent fashion. This suggests that reflected self-appraisals made in a domain where a given evaluative source possesses a high degree of influence may be flagged by our brains as being more self-relevant.

Finally, there was an additional and unexpected discovery made in this study: adolescents also recruited TPJ, DMPFC, and pSTS (in addition to MPFC and MPPC) during direct self-appraisals, whereas adults only engaged MPFC and MPPC. This suggested the possibility that direct self-reflection in teenagers incorporates aspects of reflected appraisal processes. Despite not being asked to consider others' perspectives on themselves, adolescents engaged components of the social perception network commonly associated with doing so (including TPJ, DMPFC, and pSTS), in addition to recruiting the cortical midline structures affiliated with self-reflection and self-knowledge retrieval (in MPFC and MPPC). There are several possible developmental explanations for why direct self-appraisals appeared to possess characteristics of reflected self-appraisals in early adolescence. Perhaps self-appraisals are simply more dependent on what individuals believe others think about the self specifically during this period, as compared with both adulthood (as shown in this study) and childhood (as suggested by our other work discussed above; Pfeifer et al., 2007). However, we did not ask children to make reflected appraisals in the scanner. Perhaps once a task design provides participants with the idea to consider others' perspectives on the self, children are also apt to do so during direct self-reflection, just like adolescents but unlike adults. Or more generally, younger samples may not follow directions as well as adults, and take outside perspectives on the self during direct self-appraisal conditions by accident. Future research should attempt to disentangle these possibilities.

DEVELOPMENTAL MODEL

One can draw at least two conclusions from this brief review of the relevant developmental and social cognitive neuroscience research addressing evaluative self-knowledge. First, although there is a reasonable amount of groundwork that has been laid, there is much work that needs to be done, and as we indicated throughout there are many opportunities for investigators to do so. Perhaps more important, this review demonstrates that typically developing children and adolescents do differ from adults in the neural systems supporting self-evaluations in meaningful ways.

These differences allow us to outline a tentative developmental model in which the brain provides a biological foundation for the self. In this model, children are likely to rely most heavily on the medial fronto-parietal network (MPFC and MPPC) to produce self-evaluations—doing the mental "work" involved in defining the self via traits that are abstracted from episodic memories of many instances of behaviors. This is particularly consistent with general reviews of the function of MPFC (BA 10), which suggest that this region is well positioned to support the integration of multiple, internally generated inputs (e.g., Christoff, Ream, Geddes, & Gabrieli, 2003; Dumontheil, Burgess, & Blakemore, 2008), as would be necessary to make higher-order generalizations about one's own abilities and attributes from past experiences. Furthermore, a recent longitudinal study of brain structure in 375 typically developing individuals found that MPFC, MPPC, DMPFC, and TPJ all follow cubic developmental trajectories of cortical thickness: increases in childhood followed by decreases in adolescence and eventual stability in young adulthood (Shaw et al., 2008). Peaks in grey matter density for these regions appear between the ages of 9 and 13 years (on average by 11 years of age), suggesting that after this period there may be a shift in the trajectory of functions associated with these brain regions, which includes self development. In high-competence domains, the brain seems to engage more strongly in working and episodic memory processing via enhanced recruitment of DLPFC and MPPC during the course of making self-evaluations, as well as attain a more affective, motivational orientation via activation of the amygdala.

Beginning in early adolescence, then, we hypothesize that individuals begin to habitually incorporate others' perspectives on the self during the process of making self-evaluations, as indicated by the involvement of TPJ in particular. On the one hand this may be seen as a change driven by external social factors, such as the growing importance of the peer group. We propose that it may also be driven by biological factors, such as the functional connectivity between MPFC, MPPC, and TPJ observed to increase throughout adolescence (Fair et al., 2008). Additionally during this time, when an evaluative source is perceived to be highly relevant in a domain (such as whether a peer thinks you are popular or not), the brain may tag this information as especially self-relevant, weighting its contribution towards that domain-specific self-concept.

Finally, by adulthood, the vast majority of general evaluative self-knowledge is more likely to be stored and retrieved (from lateral temporal cortex and temporal poles, sites maintaining general and personal or emotional semantic information) than generated on each occasion (see Lieberman et al., 2004; Pfeifer et al., 2007). For particular domains with which we identify strongly or in which we have vast amounts of experience, self-evaluations will be relatively more automatic and intuitive, and involve a different neural system that may include the amygdala and nucleus accumbens, among other regions. These structures facilitate speedy access to self-knowledge and are affiliated with affect and motivation, presumably supporting emotional learning and behavior patterns associated with that domain. Interestingly, because self-development continues throughout the lifespan, one may be able to observe the same neurodevelopmental trajectory in any new domain (for example, when an adult picks up a new hobby, changes careers, or becomes a parent).

FUTURE DIRECTIONS AND CONCLUSION

Throughout this chapter we have attempted to point out promising future directions for this subfield. These have included comparing the neural correlates of general self-knowledge retrieval with accessing information about other social targets besides Harry Potter; exploring the relationship between development of the default network and the functioning of the neural systems supporting evaluative self-knowledge, domain-specific self-concepts, and reflected self-appraisals; studying the neural systems engaged when children or adolescents make self-evaluations in domains for which they possess persistently negative self-concepts; and examining in greater detail how and when younger children and older adolescents recruit networks for social perspective-taking in service of self-perception, compared to early adolescents and adults. One particularly important avenue of research to pursue would be longitudinal investigations of the neural systems involved in general evaluative self-knowledge retrieval, domain-specific self-concepts, and reflected self-appraisals. For example, using such a strategy would allow us to potentially observe self-schemas emerge and the associated shift in the neural systems supporting self-knowledge in that domain from explicit, integrative, and evidentiary bases in MPFC and MPPC to automatic, affective and motivational bases in the amygdala and nucleus accumbens.

Another critical extension of research in this field would be to examine the functioning of these neural systems in children and adolescents with autism spectrum disorders (ASD), as compared to typically developing individuals. Research on autism has tended to emphasize dysfunctions in social perception, to the relative neglect of disordered self-perception. Recently, a study assessing various components of self-referential processing in adults with ASD demonstrated that memory for both the self and a similar, close other (best friend) was relatively impaired in adults with ASD, but not memory for information processed with reference to a dissimilar, nonclose other (Harry Potter) or to nonsocial features of the stimuli (Lombardo, Barnes, Wheelwright, & Baron-Cohen, 2007). Furthermore, there was a bidirectional interaction between abilities to think about the self and others. More advanced mentalizing skills led to better recall for self-relevant information, while self-focused attention and private self-consciousness was associated with enhanced mentalizing abilities. The authors concluded that being more self-focused was beneficial for individuals with ASD as it may support metacognition, self-reflection, and mentalizing—but that future research using neuroimaging techniques was needed to explore the degree of impairment in self-understanding experienced by these individuals, its relationship to the manner in which they think about close or distant others, and how these patterns develop.

In the past several years, two fMRI studies have found an abnormally hypoactive default network in adults with ASD (for a discussion see Iacoboni, 2006), including failures to deactivate MPFC during a cognitively demanding task (Kennedy, Redcay, & Courchesne, 2006), as well as weakened functional connectivity between anterior (MPFC) and posterior (MPPC) components of the network (Cherkassky, Kana, Keller, & Just, 2006; Just, Cherkassky, Keller, Kana, & Minshew, 2007). Most important, a very recent neuroimaging study of adults with ASD demonstrated that they did not engage cortical midline structures during self-knowledge retrieval (Moran, Qureshi, Singh, & Gabrieli, 2007). These three studies suggest that patterns of neural activity in these regions during self-referential processing should be studied in childhood and adolescence, as individuals with ASD are building a foundation of self-knowledge to support their future behavior and social interactions. A further, critical step in understanding atypical patterns of self-development in ASD will be not only to study the neural systems supporting self-concepts in a global fashion, but also those supporting the specific process of ascertaining others' perspectives on the self. This kind of social cognitive task—which combines social perspective-taking and self-reflection—includes components that are likely to pose special challenges for individuals with ASD, which may both amplify difficulties in interpersonal relationships and negatively impact their self development.

In conclusion, our understanding of the neural systems supporting self development (including general evaluative self-knowledge, domain-specific self-concepts, and reflected self-appraisals) is still in its early stages. Yet this field of developmental social cognitive neuroscience holds much promise, both applied and theoretical. We may be able to learn more about the systems that support typical self-development through the lens of children and adolescents with autism, who grapple with difficulties in self-perception, as well as create treatments that cultivate their abilities to amass evaluative self-knowledge and take other people's perspectives about them. The mechanisms that enable domain-specific self-concepts to motivate and guide behavior may be revealed, and this information may be used to design interventions to improve negative or affirm positive self-evaluations. A new perspective on the age-old debate about the self-centeredness of adolescents, and their susceptibility to external social influences, may be obtained. At the most philosophical level, a developmental social cognitive neuroscience approach to the self provides a window onto the biological foundations for the development, maintenance, and expression of our beliefs in the essential attributes and abilities that make us unique from any other individual in the world.

REFERENCES

Aichhorn, M., Perner, J., Kronbichler, M., Staffen, W., & Ladurner, G. (2006). Do visual perspective tasks need theory of mind? *NeuroImage, 30,* 1059.

Apperly, I. A., Samson, D., Chiavarino, C., & Humphreys, G. W. (2004). Frontal and temporo-parietal lobe contributions to theory of mind: Neuropsychological evidence from a false-belief task with reduced language and executive demands. *Journal of Cognitive Neuroscience, 16,* 1773–1784.

Baldwin, J. M. (1895). *Mental development of the child and the race: Methods and processes.* New York: Macmillan.

Barry, C. M., & Wigfield, A. (2002). Perceptions of friendship-making ability and perceptions of friends' deviant behavior: Childhood to adolescence. *Journal of Adolescence, 22,* 143–172.

Bellmore, A. D., & Cillessen, A. N. (2003). Children's meta-perceptions and meta-accuracy of acceptance and rejection by same-sex and other-sex peers. *Personal Relationships, 10,* 217–233.

Berndt, T. J., & Burgy, L. (1996). Social self-concept. In B. A. Bracken (Ed.), *Handbook of self-concept: Developmental, social, and clinical considerations* (pp. 171–209). New York: Wiley.

Bouchey, H. A., & Harter, S. (2005). Reflected appraisals, academic self-perceptions, and math/science performance during early adolescence. *Journal of Educational Psychology, 97,* 673–686.

Bracken, B. A. (1996a). Clinical applications of a context-dependent, multidimensional model of self-concept. In B. A. Bracken (Ed.), *Handbook of self-concept: Developmental, social, and clinical considerations.* New York: Wiley.

Bracken, B. A. (1996b). *Handbook of self-concept.* New York: Wiley.

Brinthaupt, T. M., & Lipka, R. P. (1994). *Changing the self: Philosophies, techniques, and experiences.* Albany, NY: State University of New York Press.

Brown, B. (2004). Adolescents' relationships with peers. In R. Lerner & L. Steinberg (Eds.), *Handbook of adolescent psychology* (pp. 363–394). New York: Wiley.

Cavanna, A. E., & Trimble, M. R. (2006). The precuneus: A review of its functional anatomy and behavioural correlates. *Brain, 129,* 564–583.

Christoff, K., Ream, J. M., Geddes, L. P. T., & Gabrieli, J. D. (2003). Evaluating self-generated information: Anterior prefrontal contributions to human cognition. *Behavioral Neuroscience, 117,* 1161–1168.

Cooley, C. H. (1902). *Human nature and the social order.* New York: Charles Scribner's Sons.

Craik, F. I. M., Moroz, T. M., Moscovitch, M., Stuss, D. T., Winocur, G., Tulving, E. et al. (1999). In search of the self: A positron emission tomography study. *Psychological Science, 10,* 26–34.

Crain, R. M. (1996). The influence of age, race, and gender on child and adolescent self-concept. In B. A. Bracken (Ed.), *Handbook of self-concept: Developmental, social, and clinical considerations* (pp. 395–420). New York: Wiley.

D'Argembeau, A., Collette, F., Van der Linden, M., Laureys, S., Del Fiore, G., Degueldre, C. et al. (2005). Self-referential reflective activity and its relationship with rest: A PET study. *NeuroImage, 25,* 616–624.

D'Argembeau, A., Ruby, P., Collette, F., Degueldre, C., Balteau, E., Luxen, A. et al. (2007). Distinct regions of the medial prefrontal cortex are associated with self-referential processing and perspective taking. *Journal of Cognitive Neuroscience, 19,* 935–944.

Damon, W., & Hart, D. (1988). *Self-understanding in childhood and adolescence.* New York, NY: Cambridge University Press.

Dumontheil, I., Burgess, P. W., & Blakemore, S. J. (2008). Development of rostral prefrontal cortex and cognitive and behavioural disorders. *Developmental Medicine and Child Neurology, 50,* 168–181.

Fair, D. A., Cohen, A. L., Dosenbach, N. U., Church, J. A., Miezin, F. M., Barch, D. M. et al. (2008). The maturing architecture of the brain's default network. *Proceedings of the National Academy of Sciences USA, 105,* 4028–4032.

Fink, G. R., Markowitsch, H. J., Reinkemeier, M., Bruckbauer, T., Kessler, J., & Heiss, W. D. (1996). Cerebral representation of one's own past: Neural networks involved in autobiographical memory. *Journal of Neuroscience, 16,* 4275–4282.

Frith, C. D., & Frith, U. (2006). The neural basis of mentalizing. *Neuron, 50,* 531–534.

Frith, U., & Frith, C. D. (2003). Development and neurophysiology of mentalizing. *Philosophical Transactions of the Royal Society of London: B Biological Sciences, 358,* 459–473.

Gardner, M., & Steinberg, L. (2005). Peer influence on risk taking, risk preference, and risky decision making in adolescence and adulthood: An experimental study. *Developmental Psychology, 41,* 625–635.

Gilboa, A. (2004). Autobiographical and episodic memory—one and the same? Evidence from prefrontal activation in neuroimaging studies. *Neuropsychologia, 42,* 1336.

Gusnard, D. A., Akbudak, E., Shulman, G. L., & Raichle, M. E. (2001). Medial prefrontal cortex and self-referential mental activity: Relation to a default mode of brain function. *Proceedings of the National Academy of Sciences USA, 98,* 4259–4264.

Hargreaves, D., & Tiggeman, M. (2002). The role of appearance schematicity in the development of adolescent body dissatisfaction. *Cognitive Therapy and Research, 26,* 691–700.

Harter, S. (1999). *The construction of the self: A developmental perspective.* New York, NY: Guilford Press.

Harter, S., Bresnick, S., Bouchey, H. A., & Whitesell, N. R. (1997). The development of multiple role-related selves during adolescence. *Development and Psychopathology, 9,* 835–854.

Heatherton, T. F., Wyland, C. L., Macrae, C. N., Demos, K. E., Denny, B. T., & Kelley, W. M. (2006). Medial prefrontal activity differentiates self from close others. *Social Cognitive and Affective Neuroscience, 1,* 18–25.

Higgins, E. T. (1991). Development of self-regulatory and self-evaluative processes: Costs, benefits, and tradeoffs. In M. R. Gunnar & L. A. Sroufe (Eds.), *Self processes and development: The Minnesota Symposia on child development* (Vol. 23, pp. 125–166). Hillsdale, NJ: Erlbaum.

Jacobs, J. E., Lanza, S., Osgood, D. W., Eccles, J., & Wigfield, A. (2002). Changes in children's self-competence and values: Gender and domain differences across grades one through twelve. *Child Development, 73,* 509–527.

James, W. (1890). *Principles of psychology.* Chicago: Encyclopedia Britannica.

Johnson, S. C., Baxter, L. C., Wilder, L. S., Pipe, J. G., Heiserman, J. E., & Prigatano, G. P. (2002). Neural correlates of self-reflection. *Brain, 125*, 1808–1814.

Kelley, W. M., Macrae, C. N., Wyland, C. L., Caglar, S., Inati, S., & Heatherton, T. F. (2002). Finding the self? An event-related fMRI study. *Journal of Cognitive Neuroscience, 14*, 785–794.

Kircher, T. T., Brammer, M., Bullmore, E., Simmons, A., Bartels, M., & David, A. S. (2002). The neural correlates of intentional and incidental self processing. *Neuropsychologia, 40*, 683–692.

Klein, S. B., Loftus, J., & Kihsltrom, J. F. (1996). Self-knowledge of an amnesic patient: Toward a neuropsychology of personality and social psychology. *Journal of Experimental Psychology: General, 125*, 250–260.

Klein, S. B., Loftus, J., Trafton, J. G., & Fuhrman, R. W. (1992). Use of exemplars and abstractions in trait judgments: A model of trait knowledge about the self and others. *Journal of Personality and Social Psychology, 63*, 739–753.

Klein, S. B., Rozendal, K., & Cosmides, L. (2002). A social-cognitive neuroscience analysis of the self. *Social Cognition, 20*, 105–135.

Lieberman, M. D. (2007). Social cognitive neuroscience: A review of core processes. *Annual Review of Psychology, 58*, 259–289.

Lieberman, M. D., Jarcho, J. M., & Satpute, A. B. (2004). Evidence-based and intuition-based self-knowledge: An fMRI study. *Journal of Personality and Social Psychology, 87*, 421–435.

Lombardo, M. V., Barnes, J. L., Wheelwright, S. J., & Baron-Cohen, S. (2007). Self-referential cognition and empathy in autism. *PLoS ONE, 2*(9), e883.

Markus, H. R. (1977). Self-schemata and processing information about the self. *Journal of Personality and Social Psychology, 35*, 63–78.

Marsh, H. W. (1990a). Causal ordering of academic self-concept and academic achievement: A multiwave, longitudinal panel analysis. *Journal of Educational Psychology, 82*, 646–656.

Marsh, H. W. (1990b). *Self description questionnaire (SDQ) I: A theoretical and empirical basis for the measurement of multiple dimensions of preadolescent self-concept: A test manual and research monograph.* Macarthur, NSW Australia: Faculty of Education, University of Western Sydney.

Marsh, H. W. (1990c). *Self description questionnaire (SDQ) II: A theoretical and empirical basis for the measurement of multiple dimensions of adolescent self-concept: A test manual and research monograph.* Macarthur, NSW Australia: Faculty of Education, University of Western Sydney.

Marsh, H. W., & Ayotte, V. (2003). Does self-concept become more differentiated with age? The differential distinctiveness hypothesis. *Journal of Educational Psychology, 95*, 687–706.

Marsh, H. W., Craven, R., & Debus, R. (1998). Structure, stability, and development of young children's self-concepts: A multicohort-multioccasion study. *Child Development, 69*, 1030–1053.

Marsh, H. W., Craven, R., & Debus, R. (1999). Separation of competency and affect components of multiple dimensions of academic self-concept: A developmental perspective. *Merrill-Palmer Quarterly, 45*, 567–601.

Mead, G. H. (1934). *Mind, self and society from the standpoint of a social behaviorist.* Chicago, IL: University of Chicago Press.

Meltzoff, A. N., & Moore, M. K. (1977). Imitation of facial and manual gestures by human neonates. *Science, 198,* 74–78.

Mitchell, J. P., Banaji, M. R., & Macrae, C. N. (2005a). The link between social cognition and self-referential thought in the medial prefrontal cortex. *Journal of Cognitive Neuroscience, 17,* 1306–1315.

Mitchell, J. P., Macrae, C. N., & Banaji, M. R. (2005b). Forming impressions of people versus inanimate objects: Social-cognitive processing in the medial prefrontal cortex. *NeuroImage, 26,* 251–257.

Mitchell, J. P., Macrae, C. N., & Banaji, M. R. (2006). Dissociable medial prefrontal contributions to judgments of similar and dissimilar others. *Neuron, 50,* 655–663.

Ochsner, K. N., Beer, J. S., Robertson, E. R., Cooper, J. C., Gabrieli, J. D., Kihsltrom, J. F. et al. (2005). The neural correlates of direct and reflected self-knowledge. *NeuroImage, 28,* 797–814.

Olson, I. R., Plotzker, A., & Ezzyat, Y. (2007). The enigmatic temporal pole: A review of findings on social and emotional processing. *Brain, 130,* 1718–1731.

Pelphrey, K. A., Morris, J. P., & McCarthy, G. (2004). Grasping the intentions of others: The perceived intentionality of an action influences activity in the superior temporal sulcus during social perception. *Journal of Cognitive Neuroscience, 16,* 1706–1716.

Pelphrey, K. A., Viola, R. J., & McCarthy, G. (2004). When strangers pass: Processing of mutual and averted social gaze in the superior temporal sulcus. *Psychological Science, 15,* 598–603.

Pfeifer, J. H., Dapretto, M., & Lieberman, M. D. (in preparation). It's important to me: Factors influencing self-relevant information processing in the pre-teen brain.

Pfeifer, J. H., Lieberman, M. D., & Dapretto, M. (2007). "I know you are but what am I?" Neural bases of self- and social knowledge retrieval in children and adults. *Journal of Cognitive Neuroscience, 19,* 1323–1337.

Pfeifer, J. H., Masten, C. L., Borofsky, L., Dapretto, M., Lieberman, M. D., & Fuligni, A. J. (in press). Neural correlates of direct and reflected self-appraisals in adolescents and adults: When social perspective-taking informs self-perception. *Child Development.*

Rameson, L., & Lieberman, M. D. (2007). Thinking about the self from a social cognitive neuroscience perspective. *Psychological Inquiry, 18,* 117–122.

Rosenberg, M. (1979). *Conceiving the self.* New York, NY: Basic Books.

Rosenberg, M. (1986). Self-concept from middle childhood through adolescence. In J. Suls & A. G. Greenwald (Eds.), *Psychological perspectives on the self* (Vol. 3, pp. 107–135). Hillsdale, NJ: Erlbaum.

Ruby, P., & Decety, J. (2003). What you believe versus what you think they believe: A neuroimaging study of conceptual perspective-taking. *European Journal of Neuroscience, 17,* 2475–2480.

Samson, D., Apperly, I. A., Chiavarino, C., & Humphreys, G. W. (2004). Left temporoparietal junction is necessary for representing someone else's belief. *Nature Neuroscience, 7,* 499.

Saxe, R. (2006). Uniquely human social cognition. *Current Opinion in Neurobiology, 16,* 235.

Saxe, R., & Kanwisher, N. (2003). People thinking about thinking people: The role of the temporo-parietal junction in "theory of mind." *NeuroImage, 19,* 1835–1842.

Saxe, R., & Wexler, A. (2005). Making sense of another mind: The role of the right temporo-parietal junction. *Neuropsychologia, 43,* 1391–1399.

Schmitz, T. W., Kawahara-Baccus, T. N., & Johnson, S. C. (2004). Metacognitive evaluation, self-relevance, and the right prefrontal cortex. *NeuroImage, 22,* 941–947.

Shaw, P., Kabani, N. J., Lerch, J. P., Eckstrand, K., Lenroot, R., Gogtay, N. et al. (2008). Neurodevelopmental trajectories of the human cerebral cortex. *Journal of Neuroscience, 28,* 3586–3594.

Siegler, R. S. (1991). *Children's thinking.* Englewood Cliffs, NJ: Prentice Hall.

Stein, K. F., Roeser, R., & Markus, H. R. (1998). Self-schemas and possible selves as predictors and outcomes of risky behaviors in adolescents. *Nursing Research, 47,* 96–106.

Steinberg, L., & Morris, A. S. (2001). Adolescent development. *Annual Review of Psychology, 52,* 83–110.

Steinberg, L., & Silverberg, S. B. (1986). The vicissitudes of autonomy in early adolescence. *Child Development, 57,* 841–851.

Vanderwal, T., Hunyadi, E., Grupe, D. W., Connors, C. M., & Schultz, R. T. (2008). Self, mother and abstract other: An fMRI study of reflective social processing. *NeuroImage, 41,* 1437–1446.

Vartanian, L. R. (2000). Revisiting the imaginary audience and personal fable constructs of adolescent egocentrism: A conceptual review. *Adolescence, 35,* 639–661.

Wigfield, A., Eccles, J., Yoon, K. S., Harold, R. D., Arbreton, A. J. A., & Blumenfeld, P. C. (1997). Changes in children's competence beliefs and subjective task values across the elementary school years: A three-year study. *Journal of Educational Psychology, 89,* 451–469.

Zysset, S., Huber, O., Ferstl, E., & von Cramon, D. Y. (2002). The anterior frontomedian cortex and evaluative judgment: An fMRI study. *NeuroImage, 15,* 983.

9

NEURODEVELOPMENT UNDERLYING ADOLESCENT BEHAVIOR
A Neurobiological Model

Monique Ernst
Michael Hardin
*Neurodevelopment of Reward Systems/Mood and
Anxiety Programs/NIMH/NIH/DHHS*

This review addresses the unique behavioral changes occurring during adolescence, and their neural correlates. It is organized into four sections. First, we identify four cardinal behavioral changes that characterize adolescence. Second, we describe the general course of brain development, with an emphasis on the unique aspects pertaining to adolescence. Third, we restrict our focus to three neurobehavioral systems that, together, can account for the typical adolescent behavioral pattern. This triadic representation of behavioral determinants is crystallized in the Triadic Model. The Triadic Model is particularly helpful to understand how the orchestration of neural changes with time can modulate behavior in typically developing individuals, but also in psychopathology. Finally, we provide a review of the few functional neuroimaging studies of reward-related behaviors in adolescents that can inform hypotheses generated by the Triadic Model.

ADOLESCENT BEHAVIOR

Adolescence is characterized by unique changes in motivated behaviors that reflect the typical development of cognitive, affective, and social processes. Among these changes, (1) cognitive impulsivity, (2) risk seeking, (3) affective lability and intensity, and (4) social peer primacy are considered to be critical in shaping adolescent behavior (Dahl, 2004; Ernst & Paulus, 2005; Ernst, Pine, & Hardin, 2006). Each of these behavioral characteristics is briefly addressed below.

Cognitive (as opposed to motor) impulsivity is a recurrent theme in descriptions of adolescents' behavior. Cognitive impulsivity refers to the difficulty of waiting for outcomes of actions, particularly for those with potential rewards. Delay aversion represents a form of cognitive impulsivity. Tasks designed to examine choice preference between stimuli signaling immediate rewards of low value, and stimuli signaling delayed rewards of high value are frequently used in studies of cognitive impulsivity (e.g., Glimcher, Kable, & Louie, 2007; Hariri et al., 2006; Roesch, Calu, Burke, & Schoenbaum, 2007). Studies of the ability to delay reward outcomes with age have shown improvement of this skill with age (Scheres & Sanfey, 2006). Scheres and Sanfey (2006) found that children (6- to 11-year-olds) discounted temporal delay more steeply than adolescents (12- to 17-year-olds). In addition, their findings suggested that steeper discounting in young children was driven by reward immediacy rather than by delay aversion, whereas adolescents showed the opposite pattern. The paucity of such studies in typically developing adolescents makes conclusions only tentative.

Risk-taking behavior is another typical presentation of adolescents' deportment. Risky decisions and actions assume uncertainty of the outcomes. Such uncertainty is usually characterized by potentially favorable outcomes (e.g., reward) balanced against potentially unfavorable outcomes (e.g., loss). Three core psychological features are subsumed under risk: reward attraction, loss aversion, and intolerance of uncertainty. Each of these characteristics can be studied independently. Notwithstanding, such a line of research, based on the careful fragmentation of risk taking, has not yet been applied specifically to the neurodevelopment of adolescence.

Many real-world risk-taking behaviors, such as drug use, unprotected sex, and reckless driving, but also mountain climbing, diving, or parachute jumping, are recognized to emerge, rise, and eventually peak in adolescence (Boyer, 2006; Steinberg, 2004). Most developmental studies of risk taking measure real-life risk-taking rates based on observational or self-report sources. Less work has been devoted to empirical assessment of risk-taking using laboratory-based paradigms. The development of such paradigms (e.g., Crone & van der Molen, 2004; Ernst et al., 2006; Lejuez, Aklin, Zvolensky, & Pedulla, 2003; Lejuez et al., 2002; Overman et al., 2004) is challenging because of the difficulty of ensuring ecological validity and of the complexity of the construct of decision making. Based on a cognitive neuroscience approach, these tasks are designed to probe discrete aspects of risk-taking, which may not reflect the actual everyday-life propensity for risk-taking behavior. With this caveat in mind, empirical data tend to show greater risk-taking in adolescence compared to adulthood (Deakin, Aitken, Robbins, & Sahakian, 2004; Hooper, Luciana, Conklin, & Yarger, 2004), but still greater risk-taking in childhood than in adolescence (e.g., Crone & van der Molen, 2004; Overman, 2004). Yet again, a recent study found no age-related changes in risk-taking in participants 8 to 19 years old (Van Leijenhorst, Westenberg, & Crone, 2008). The discrepancy between real-life and empirical risk taking developmental findings is a growing area of research.

The emotional life of adolescents is often described as peaking in lability and intensity. Adolescents appear prone to rapid mood swings. For example, it would not be uncharacteristic for an adolescent to feel euphoric one minute, desperate the next instant, then suddenly outraged, then quickly elated again. A popular interpretation attributes these frequent mood switches to "the raging hormones of adolescence." However, this explanation is not clearly supported by neurobehavioral studies. On the other hand, affective intensity refers to extreme emotions. This characteristic is well illustrated by the countless examples in the literature of idealistic strives and romantic passions of adolescent protagonists throughout history (Dahl, 2004). Affective lability has not yet been captured in the laboratory. A few psychopathology studies have focused on frustration tolerance, which may tap some unique aspects of mood lability (e.g., Rich et al., 2007). Much more research needs to be done to first define this concept of affective lability and intensity from a cognitive neuroscience framework, and then to identify simple components that can be studied in controlled experiments.

Finally, the adolescent social landscape is changing dramatically from being family centered to being peer centered (Steinberg, 2005). Adolescents are found to spend on average 75% of their time interacting with peers, but only 8% interacting with adults (Brown & Tapert, 2004; Spear, 2000). The peak social saliency in adolescence impacts decision making. For example, risk-taking behavior is amplified in a social context relative to a nonsocial context (Gardner & Steinberg, 2005). The high saliency of social stimuli is a booster not only of reward seeking and risk-taking, but also of anxiety and avoidance, as reflected by the adolescent's greater sensitivity to peers' opinions (e.g., Cohen & Prinstein, 2006). Evolutionary theories associate this transition with the need of individuals to acquire independence and join new social circles for successful reproduction. Here, again, to our knowledge, no cognitive neuroscience studies have investigated in a systematic way the developmental changes during adolescence of the various parameters accounting for the increased social salience in adolescence. This line of research, however, is beginning to be formulated and a few studies have been published (Guyer, McClure-Tone, Shiffrin, Pine, & Nelson, in press; McClure et al., 2007).

The emergence of these four dimensions of behavior seems to be biologically determined as it appears highly conserved across species. It obeys a genetically based internal clock and is shaped by the interaction of biological and environmental factors. Three major neurobiological systems are proposed as primary candidates for playing a fundamental role in this adolescent metamorphosis. They include (1) the amygdala and related circuits for their role in emotion, threat, and social information processing; (2) the striatum and related circuits for their role in reward, approach behavior, and habitual behavior; and (3) the prefrontal cortex and related circuits for their role in modulating affective and cognitive processes. The development of these structures across adolescence should provide some clues to the neurobiology underlying the behavioral changes that mark this transition period.

BRAIN DEVELOPMENT

Two critical factors have contributed to the recent interest in the neurobiological substrates of adolescent behavior. On the one hand, the advent of noninvasive functional neuroimaging techniques has provided the tools to examine this question, and on the other hand, the recognition that adolescence is a critical period of vulnerability for the onset of psychopathology has provided the scientific basis to this quest (Rumsey & Ernst, 2009).

Since the early 1990s when child and adolescent development were first studied with structural and functional neuroimaging techniques, a number of studies have emphasized the importance of the adolescent period. The first set of these early studies culminated in two fundamental findings.

First, adolescence was recognized as a period of critical structural and functional neural changes. These changes include cell/dendrite/synapse proliferation followed by elimination (see Rubia et al., 2006;

Toga, Thompson, & Sowell, 2006). Parallel developmental studies of gray compared to white matter measurements have indicated ongoing processes of myelination (Giedd, 2004). Complementary to these works in humans, animal studies have revealed a host of specific neurochemical modifications, such as enhanced inhibitory GABAergic prefrontal modulation (Lewis, Cruz, Eggan, & Erickson, 2004), and alterations in dopamine responses to stimulants (Laviola, Pascucci, & Pieretti, 2001). However, the scope of these functional changes, at the cellular and molecular level, is still far from being fully identified or understood.

The second set of findings indicated that different brain structures and regions undergo changes along unique trajectories. This point is clearly illustrated in work conducted by Giedd and colleagues (1999) showing a scatter of inflection points along developmental volumetric curves of various brain regions across childhood and adolescence (Figure 9.1). These inflection points represent a change in the direction of volumetric modifications, with increases occurring to the left of the inflection points and decreases to the right. This overall pattern suggests asynchrony in the developmental trajectories of brain regions, and their associated functional systems. While Giedd's early work included mostly cortical regions, similar developmental heterogeneity could be expected in the development of subcortical regions such as that amygdala or striatum.

While much less is known about ontogenic changes in subcortical regions than in cortical areas, a number of comparative animal studies suggest unique anatomical and functional changes. For example, the amygdala develops strong connections with prefrontal cortex during adolescence, and these connections prevail on the GABAergic system (Cunningham, Bhattacharyya, & Benes, 2002). Additionally, the striatum responds in a unique way to stimulant challenges in adolescence compared to other age groups by showing a peak activity as evidenced by c-fos activation (Andersen, LeBlanc, & Lyss, 2001).

Much like cortical regions, subcortical structures are anatomically and functionally heterogeneous, as they are composed of discrete subunits with distinct projections and functions. For example, the amygdala comprises several nuclei, including the basolateral nucleus (BLA), the central nucleus (CeA) and the medial nucleus (MeA). Likewise, the striatum is composed of the caudate nucleus, putamen, and nucleus accumbens. As seen with cortical regions, it is conceivable that the different subunits of these subcortical structures also show asynchronous development.

In fact, this regional asynchrony in the ontogeny of the central nervous system is pivotal to the developmental neurobiological model of behavioral patterns, and in particular, decision-making patterns.

Brain Development

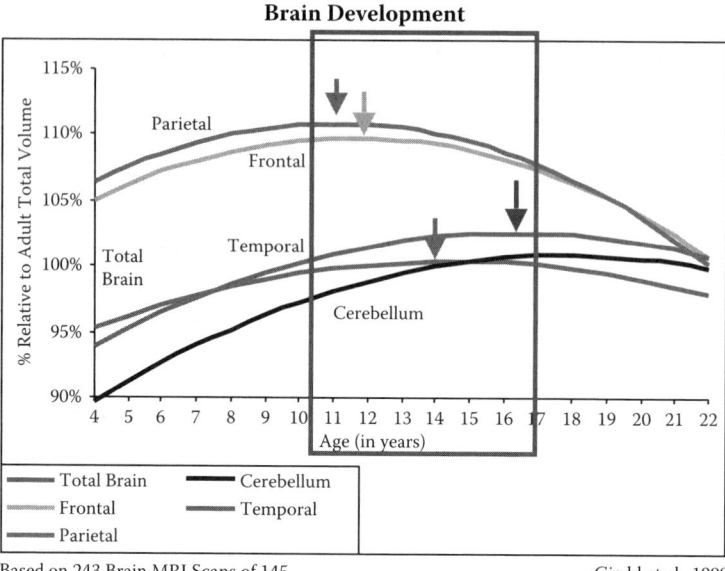

Figure 9.1 Developmental changes of regional brain percentage of total brain volumes between the ages of 4 and 22 years. This work is based on a combination of cross-sectional and longitudinal data collection. These data result from the analysis of 243 MRI scans from 145 children and adolescents. The adolescent period is highlighted by the red rectangle that comprises ages between 10 and 17 years. During this period, unique changes occur, including processes of dendritic multiplications followed by dendritic eliminations, and myelination. Three points are apparent in this graph: (a) Substantial structural ontogenic changes occur throughout development. (b) During the adolescence period, the nature of these changes shows qualitative uniqueness as evidenced by the presence of inflection points (arrows) for each curve, indicating that during this transition period, the direction of changes is reversing. (c) These inflection points are distributed across the whole adolescence period, denoting an asynchrony across brain regions of their qualitative alterations, and suggesting unique functional consequences. (This figure has been modified from Giedd et al., 1999. *Nature Neuroscience, 2*, 861–863.)

Triadic Model of Decision Making

The anatomical, functional and maturational heterogeneity of neural networks (e.g., Table 9.1) poses a challenge for the formulation of a multisystem model that can guide research on the biological substrates of decision making. To facilitate formulating such a model, we first organized the complex construct of decision making into three basic behavioral processes: (a) approach, (b) avoidance, and (c) cognitive regulation. These three fundamental processes provide a framework for the study of the four behavioral changes described before.

Table 9.1 Anatomical and Functional Heterogeneity of the Triadic Nodes

Amygdala	Striatum	Medial PFC
Anatomy		
Basal nucleus	Caudate nucleus	Frontal pole
Lateral nucleus	Putamen	Orbitofrontal cortex
Central nucleus	Globus pallidum	Anterior cingulate
	Nucleus accumbens	Supplementary motor area
Function		
Attention orienting	Motor responses	Self-assessment
Threat avoidance	Habits	Conflict monitoring
Current affective value	Motivation	Action planning
Conditioning	Conditioning	Conditioning
Reward processing	Reward processing	Affective value
Dominant Role		
Avoidance	Approach	Modulation

Cognitive impulsivity can be understood as poor regulation of the cost of delaying reward, a cost that is translated as a decreased reward value and, in turn, a diminished approach response. Risk seeking can result from either a dominant approach system, a weak avoidance system, a biased regulatory tone towards approach responses, or any combinations of the above. Affective lability and intensity evokes poor regulatory capacity, which would affect the balance between approach (response to positively valenced stimuli) and avoidance (response to negatively valenced stimuli). Finally, social peer primacy, as already mentioned, significantly affects the equilibrium of the Triadic Model. This equilibrium can be tilted either towards approach or towards avoidance, depending on the nature of the social context, and on individual sensitivity of the network responding to social cues. For example, behaviorally inhibited individuals characterized by severe shyness will tend to activate preferentially the avoidance system in social contexts, compared to individuals with an exuberant temperament who will tend to activate preferentially the approach system in social contexts. The interaction between the social network (e.g., Blakemore, 2008; Nelson, Leibenluft, McClure, & Pine, 2005) and the Triadic Model is beyond the scope of this review. However, the large overlap between the structures involved in the social network and those in the Triadic Model is worth noting (see Ernst, Romeo, & Andersen, 2008).

Three-System Model of Decision Making

The processes involved in approach behavior, avoidance behavior, and the cognitive regulation of both behavioral polarities form the backbone of the Triadic Model (Ernst, Pine, & Hardin, 2006). Concepts of approach and avoidance behavior have been used to characterize personalities and styles of behavioral response (Davidson, Jackson, & Kalin, 2000; Gray, 1994). The main tenet of these conceptualizations is that motivated behavior manifests as two primary expressions: approach toward a stimulus or avoidance of a stimulus. Stimulus approach is associated with rewarding (i.e., appetitive, positively valenced) outcomes, while avoidance is associated with punishing (i.e., aversive, negatively valenced) outcomes. This behavioral schism is also widely used in the conditioning literature, which provides neural mechanisms underlying the coding of these responses (e.g., Martin-Soelch, Linthicum, & Ernst, 2007). In the Triadic Model, a third system is added that serves to regulate and balance the approach and avoidance behavior systems. This balance is unique to each individual, and varies as a function of development. It can be viewed as a homeostatic set point that can be explored accordingly.

The neural networks underlying these three systems are distinct, yet overlapping. Controversies continue to dispute the functional boundaries of these systems. Here, we adopt the following schemes, based on the large literature on processes of reward (e.g., Di Chiara & Bassareo, 2007; Wise, 2004), fear/threat (e.g., Aggleton, 2000; Davis, 2006; LeDoux, 2000), and executive function (e.g., Aron, Robbins, & Poldrack, 2004; Miller, 2000; Miller & Cohen, 2001; Ridderinkhof, van den Wildenberg, Segalowitz, & Carter, 2004). Accordingly, the neural circuits manifesting dominant responses to appetitive stimuli include striatal structures and orbitofrontal cortex. The circuits particularly sensitive to aversive stimuli include amygdala, hippocampus, and insula. Finally, the circuits associated with regulatory function include regions of the prefrontal cortex, particularly medial prefrontal areas (Figure 9.2).

Anatomical and Functional Boundaries

Amygdala, striatum, and prefrontal cortex each supports a selected range of functions (e.g., Table 9.1). These functions overlap across systems. For example, the amygdala is involved in the coding of affective values of both positive and negative stimuli (Baxter & Murray, 2002). Through direct projections, the amygdala sends this information to the orbitofrontal cortex, which stores the representation of these values (Baxter & Murray, 2002; Holland & Gallagher, 2004). However,

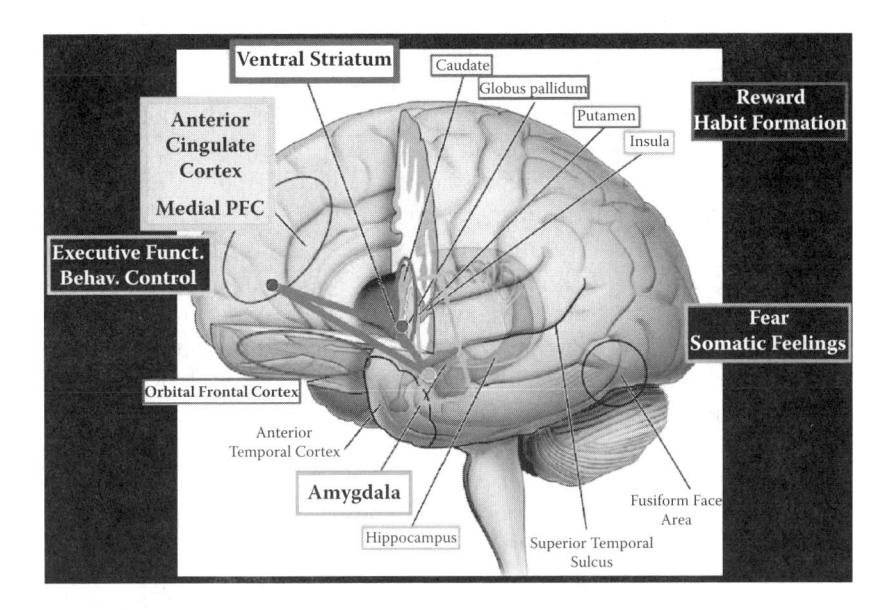

Figure 9.2 (See color insert following p. 242) Schematic representation of a sagittal view of the brain, with the lateral slab of the prefrontal region being removed, exposing its medial section. Regions associated with the three functional systems of the Triadic Model are highlighted. The approach system includes striatum and orbitofrontal cortex. The avoidance system includes the amygdala, hippocampus, and insula. The modulatory system includes the medial prefrontal cortex, including the anterior cingulate cortex. The triangle delineates the Triadic Model. Of note, this formulation is a simplification of the function of these regions, as discussed in the text.

dissociation in the sensitivity of these structures to the valence of the stimuli, or the behavior generated in response to these stimuli, can be detected based on functional neuroimaging. Activation of the amygdala is more often and more strongly associated with exposure to negative stimuli and to behaviors of avoidance (Rauch, Shin, & Wright, 2003), whereas activation of the orbitofrontal cortex shows stronger affinity with positive stimuli and behaviors of approach (Kringelbach, 2005).

Similarly, the striatum is also engaged in the processing of both appetitive and aversive stimuli (e.g., Jensen et al., 2003). However, the striatum is more tightly linked to actions (i.e., motor output) in response to stimuli, whereas the amygdala activation is not as strongly tied to actions per se or their modulation. The striatum integrates information from midbrain dopamine neurons (ventral tegmental area and substantia nigra), the amygdala, and the prefrontal cortex. This integrated information is then funneled to the globus pallidum, back to thalamus, and eventually to the motor cortex for the execution of behavior (see Ernst & Fudge, 2009). In contrast to the amygdala, the striatum appears to be more responsive to appetitive stimuli than to aversive stimuli. Furthermore, based on the functional organization scheme proposed by Haber and colleagues (Haber, Fudge, & McFarland, 2000; Haber, Kim, Mailly, & Calzavara, 2006), the ventral striatum, including the nucleus accumbens (NAc), represents the affective/motivation coding part of the striatum and is more closely connected to medial than to lateral prefrontal cortical regions. It is also the striatal region that receives direct amygdala inputs (Fudge, Breitbart, & McClain, 2004).

Overall, the ultimate test for assessing the dominant role of a neural structure is to study the behavioral consequences of activation or inactivation of this structure. With regard to the amygdala, lesion studies in both humans and animals converge on the expression of decreased fear responses, and enhanced approach-type behaviors, such as the hyperorality or hypersexuality seen in the bilateral medial temporal lobe lesions of Kluver-Bucy syndrome (Hayman, Rexer, Pavol, Strite, & Meyers, 1998). With regard to the striatum, inactivation of the striatum has been associated with reduction in motivation, locomotion and positive affect (e.g., Nestler & Carlezon, 2006). These observations support a dominant role of amygdala in facilitating avoidance and striatum in facilitating approach behavior.

The medial prefrontal cortex, including dorsal and ventral anterior cingulate cortex (ACC), and the medial part of the frontal pole, serves a number of functions, including affective processing for the ventral ACC (see Bush, Luu, & Posner, 2000), conflict monitoring and error processing for the dorsal ACC (Carter & van Veen, 2007), and self-monitoring for the medial frontal pole (e.g., Amodio & Frith, 2006). These different functions can also contribute to modulating behavior, in part through feedback to amygdala and striatum (Figure 9.3).

Fractal Triadic Model in Adolescents

The representative structures of each behavioral system (amygdala, striatum, and medial prefrontal cortex) are themselves coordinating information about appetitive and aversive stimuli. Each structure integrates these two flows of information in a way that influences behavioral output in unique directions. This is represented schematically in Figure 9.4 (the Fractal Triadic Model). While details are not presented, the functional subunits of these structures are likely to subserve different roles within the Triadic Model, based on identified functional subspecializations at this level (see Ernst & Fudge, 2009).

The ultimate utility of a model is to provide a matrix for generating hypotheses about mechanisms that can be experimentally tested. As described above, adolescent decision making is characterized by (a)

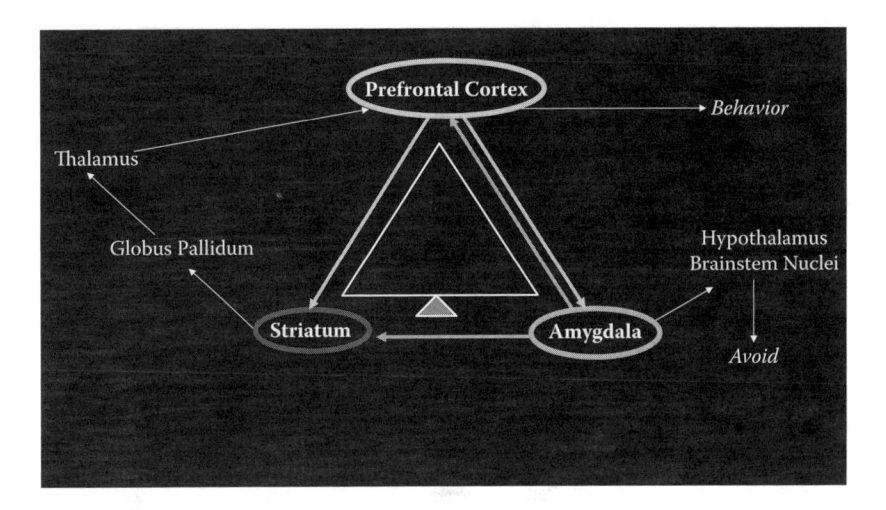

Neurodevelopment Underlying Adolescent Behavior • 175

Figure 9.3 Schematic representation of the connectivity among the triadic nodes. The amygdala receives and processes information from cortical somatosensory areas, and then relays this processed information to hypothalamus and brain stem nuclei for immediate/automatic action, and to prefrontal cortex and striatum for more elaborated responses. The prefrontal cortex sends information back to the amygdala and striatum, modulating their output. The striatum integrates the activity originating from the amygdala and prefrontal cortex, and funnels it to the globus pallidum, which then transmits it back to prefrontal cortex, including motor areas to generate behavioral responses. There are no direct projections from the striatum onto the amygdala. This is a schematic representation which, for purposes of simplicity, is incomplete and ignores other structures and projections involved with these systems.

an exacerbation of approach behaviors that are consistent with novelty and risk seeking, (b) a reduction in avoidance behaviors and/or sensitivity to punishment, and (c) difficulty modulating decision making in highly emotional or socially related situations. These characteristics could arise from the prominent engagement of the approach system (i.e., striatum circuits), a lesser contribution of the avoidance system (i.e., amygdala circuits), and/or poor regulatory function (i.e., prefrontal circuits) (See Figure 9.4). Though only a few have been conducted, functional neuroimaging studies can be used to test the hypotheses generated by this model.

FUNCTIONAL NEUROIMAGING

The study of decision making and influence of rewards on cognitive performance permits us to examine motivated behavior in a systematic way. Here, we selected works that focused on decision-making paradigms administered to adolescents. At present, six different paradigms

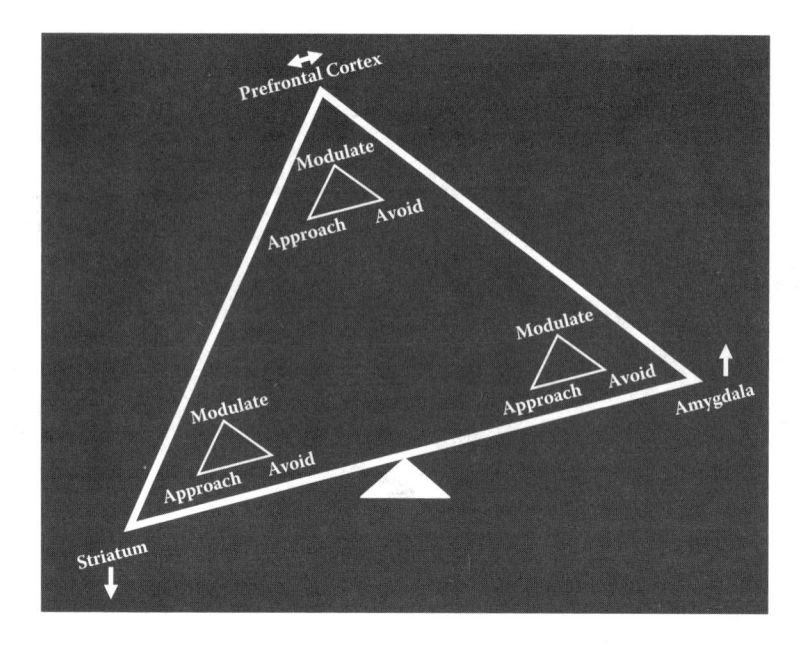

Figure 9.4 Fractal Triadic Model from a developmental perspective. This figure represents the balance of the Triadic Model among three systems that contribute to motivated behavior (i.e., decision-making), and comprise systems serving approach, avoidance, and regulation of these two systems. This balance is tilted towards approach behavior, and away from avoidance behavior in adolescents, compared to the pattern in adults. This balance could be conceptualized as a homeostatic set point that is biologically determined and varies as a function of age, sex, and psychopathology.

have been reported in studies of reward-related processes in adolescents (Table 9.2). Four of these paradigms have used a decision-making scheme, and two have manipulated the influence of rewards on cognitive performance, without overt decision making.

Decision Making

fMRI Paradigms The four decision paradigms used so far all involved a dichotomous selection (Bjork, Smith, Danube, & Hommer, 2007; Ernst et al., 2005; Eshel, Nelson, Blair, Pine, & Ernst, 2007; May et al., 2004; Van Leijenhorst, Crone, & Bunge, 2006). However, they differ on a number of characteristics.

May et al. (2004) used a simple guessing task providing a 50% chance of a positive outcome. The outcome was either a $1.00 gain or a $0.50 loss. Van Leijenhorst et al. (2006) used a task that gave probabilistic information on the option most likely to have a favorable outcome. Thus, this task always had an optimal answer (the most probable option). The outcome was in points, +1 for a positive feedback, and −1 for a negative

Table 9.2 Description of the Paradigms Used in the Developmental Studies of Decision-Making and Reward-Related Processes

Study	Paradigm			
May et al., 2004	Cue	Guess	Number	Feedback
		if guess is greater than '5'		

Guessing card game
Is the next card >5? (50%) probability
2.5s cue; 0.5s delay; 1.0s feedback; 12s ITI
Total trial 16s
Feedback:
Correct: + $1.00
Incorrect: − $0.50
Neutral (5): 0.00
Every subject won $27.00
REWARD condition + $1.00

Galvan et al., 2006	Cue (1 sec)	Response (2 sec)	Reward (1 sec)	

Pirate task
Delayed response two-choice task
3 cues (high, medium, low reward value)
Press button on the side where the cue was presented
1s cue; 2s delay (fix); 2s target (treasure chest); 2s delay (fix); 1s feedback; 12s ITI
Total trial 20s
Feedback:
3 reward values
End of task, received $25 Amount of $ not specified at the beginning of task or throughout the task

Van Leijenhorst et al., 2006				
500 ms.	2000 ms.	1500 ms.	200 ms.	

Cake task
Which flavor (chocolate or strawberry) would someone pick with their eyes closed?
Varies probability, always a correct answer
0.5s fix; 3.5s cue; 2.0s feedback; ITI jittered between 2s and 8s
Total trial 6s
Feedback:
Points: Correct + 1, Incorrect −1

Table 9.2 Description of the Paradigms Used in the Developmental Studies of Decision-Making and Reward-Related Processes (Continued)

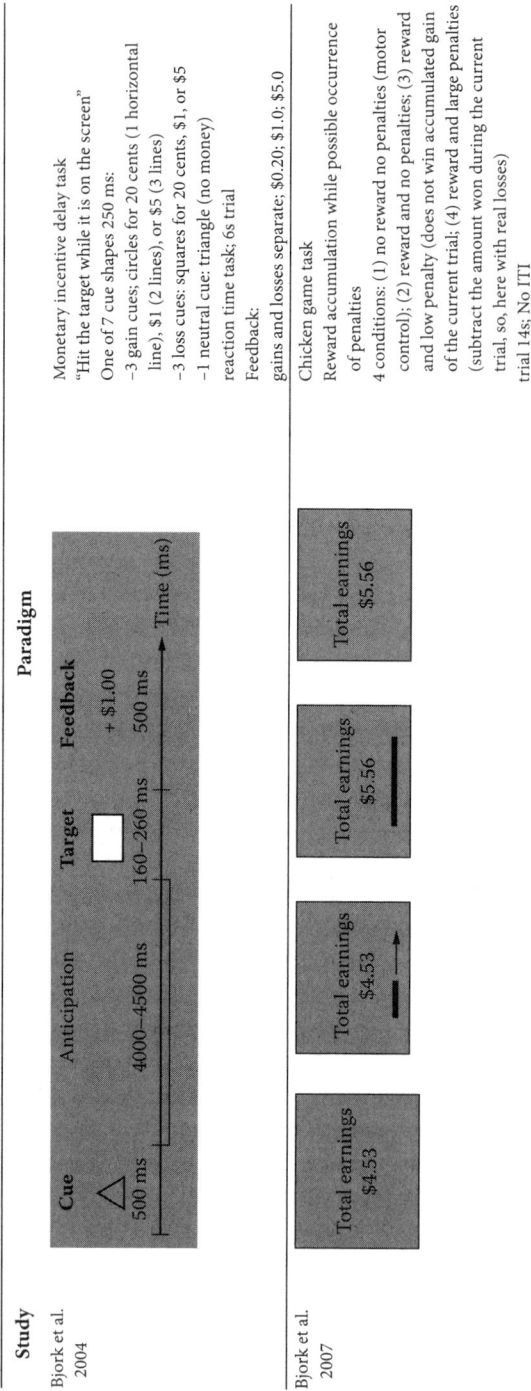

Study	Paradigm	
Bjork et al. 2004		Monetary incentive delay task "Hit the target while it is on the screen" One of 7 cue shapes 250 ms: –3 gain cues; circles for 20 cents (1 horizontal line), $1 (2 lines), or $5 (3 lines) –3 loss cues: squares for 20 cents, $1, or $5 –1 neutral cue: triangle (no money) reaction time task; 6s trial Feedback: gains and losses separate; $0.20; $1.0; $5.0
Bjork et al. 2007		Chicken game task Reward accumulation while possible occurrence of penalties 4 conditions: (1) no reward no penalties (motor control); (2) reward and no penalties; (3) reward and low penalty (does not win accumulated gain of the current trial; (4) reward and large penalties (subtract the amount won during the current trial, so, here with real losses) trial 14s; No ITI

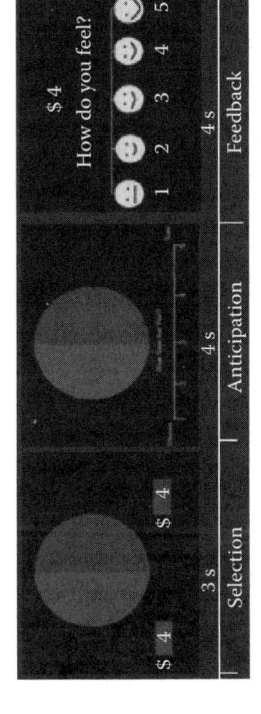

Ernst et al. 2005
Eshel et al. 2007

Wheel of fortune task
4 conditions: $ amount and probability vary:
- 10/90 wheel: 10% chance of winning $4 (10%–$4) vs. no win, and 90% chance of winning $0.50 (90%–$0.5) vs. no win
- 30/70: 30%–$2, 70%–$1
- 50/50: 50%–$4
- 50/50: 50%–$0.5
Ernst et al. 2005
Only 2 conditions included: (1) 50/50 with high reward ($4.00) and (2) 50/50 with low reward ($0.50)
Assess magnitude of reward and not probability of reward
Eshel et al. 2007
Only 2 conditions included: (1) 10/90: highly risky vs. highly safe (2) 30/70: moderately risky vs. moderately safe

Note: ITI = inter-trial interval.

feedback. Ernst et al. (2005) and Eshel et al. (2007) used a task in which both probability and magnitude of gains were manipulated, in order to evoke risk-taking behavior. Subjects were asked to choose between an option of high likelihood but of small reward (e.g., 90% chance of a $4 gain), and an option of low likelihood but high reward (e.g., 10% chance of a $.50 gain). Finally, Bjork et al. (2007) used a task that manipulated penalties. In addition, this task pitted the possibility of higher monetary accumulation with time against the possibility of penalties that could occur at any time during the duration of monetary accumulation. The choice was between taking more risk and letting rewards accumulate, or stopping the task before a penalty occurred.

In addition to the differences in parameters of the decision-making tasks, the stage of decision making that was analyzed in these studies was different. Decision making can be decomposed temporally into three stages, including (1) assessment and formation of preference, (2) selection and execution of the preference, and (3) feedback and learning (Ernst & Paulus, 2005). Anticipation is a state that precedes feedback, and can be difficult to isolate from the processes occurring during the first two stages. Three studies examined the first stage, i.e., assessment and preference formation (Bjork et al., 2007; Eshel et al., 2007; van Leijenhorst et al., 2006). Two studies analyzed responses to feedback (Ernst et al., 2005; Van Leijenhorst et al., 2006). One study did not distinguish between stages and analyzed the pattern of the fMRI signal across the whole trial (May et al., 2004).

Findings During Selection Findings showed that, overall, similar neural circuits were recruited when adolescents (May et al., 2004) and adults (Delgado, Nystrom, Fissell, Noll, & Fiez, 2000) performed the same task. These structures included striatum, OFC, and medial PFC. Although no head-to-head comparison was possible because of slight differences in the acquisition of data, the amygdala was activated in the adult study and not in the adolescent study, possibly reflecting a stronger involvement of this structure in adults than in adolescents.

During the first stage of decision making (i.e., formation of preference), one study reported no significant activation differences between adolescents and adults when "certain" selections were compared to "uncertain" selections (Van Leijenhorst et al., 2006). However, the other study reported greater activation of the ventrolateral PFC and dorsal ACC in adults than in adolescents when risky choices were compared to safe choices (Eshel et al., 2007). A critical difference between these two studies was that one study modulated reward magnitude and risk (Eshel et al., 2007), whereas the other study manipulated only probability (Van

Leijenhorst et al., 2006). In one study, there was no best choices, only preferred choices (Eshel et al., 2007), whereas in the other study there was always a best choice (Van Leijenhorst et al., 2006). The discrepancy in findings suggests that these factors are important to detect a deficit in the engagement of prefrontal cortex in adolescents compared to adults. This deficit in prefrontal cortical activation may underlie the predicted maturational delay in the behavioral regulatory system of the Triadic model.

The fourth study (Bjork et al., 2007) focused on the initial assessment stage of the decision-making process, even before having to elicit a preference and make a decision ("stop" or "let it go"). Findings revealed greater deactivation in medial PFC in adults relative to adolescents when subjects assessed the possibility of incurring penalties compared to when there was no possibility of penalties. This finding had two potential implications. First, consistent with Eshel et al. (2007), adolescents seemed to be less able to modulate prefrontal circuits when risk was present. Second, the possibility of penalties might have been less meaningful for adolescents than for adults, and thus less likely to mobilize cognitive resources to control behavior. This interpretation is consistent with the notion of lower sensitivity to punishment in adolescents (Spear, 2000).

Findings During Feedback Two of the decision-making studies also reported on adolescent responses to outcome (Ernst et al., 2005; van Leijenhorst et al., 2006). Van Leijenhorst et al. (2006) found greater lateral OFC activation in adolescents compared to adults for losses (−1 point) when compared to gains (+1 point). This finding is surprising, as this structure was hypothesized to be more sensitive to positive stimuli. The negative outcome in this study was not translated into personal losses, and was more likely to occur when the outcome was most uncertain. Although the authors tried to control for the factor of certainty, this potential confound could not be ruled out. In contrast, Ernst et al. (2005) compared brain activation when a large monetary gain was obtained to brain activation when a large monetary gain was missed. They reported greater activation of the dorsolateral PFC and amygdala in adults than in adolescents, but greater activation of the NAc in adolescents than in adults. These age differences were consistent with predictions based on the Triadic Model.

Task Incentives

fMRI Paradigms Two studies used paradigms that examined the effects of incentives on the performance of an independent cognitive task, a reaction time task (Bjork et al., 2004), and a delay spatial matching task (Galvan et al., 2006) (Table 2). These tasks did not involve decision making per se. In both tasks, subjects were first presented with a cue that indicated the value of the trial. In Bjork et al. (2004), subjects faced trials during which they could win money, lose money or have no monetary sanction. Subjects were told that they would take home the cumulative amount of money won at the end of the task. In Galvan et al.'s study (2006), subjects were presented trials that depicted a large value, a moderate value, or a small value. A direct translation into potential monetary gains was not given to the subjects.

Findings Galvan et al. (2006) analyzed the fMRI signal across the entire trial. Of note, this study included three age groups—children, adolescents, and adults. Findings revealed that, overall, the level of peak activation in ventral striatum was greater for adolescents compared to both adults and children. In contrast, the OFC showed an age-related group effect with the greatest activity occurring in children compared to adults and adolescents. The OFC seemed to show a similar extent of activation in adolescents and adults, whereas the NAc exhibited a larger extent of activation in adolescents than in adults. The authors concluded that the OFC and NAc differed in maturational profiles. Consistent with earlier findings and with predictions based on the Triadic Model (Ernst et al., 2005), this study suggested a stronger reward system in adolescents than in adults, and also children.

In their reaction time task study, Bjork et al. (2004) focused on two different periods of the task: anticipation/preparation to act and feedback. The analysis of the delay period prior to pressing the button when the target appeared was understood as reflecting anticipation of outcome. Adults showed higher ventral striatal activation than adolescents during anticipation of gain versus during no gain involved (neutral cue). The authors proposed that adolescents evidenced a hypofunctional reward system. Such a weak reward system would then require extra stimulation to be maintained at a homeostatic level of activity, thereby contributing to high levels of risk-seeking behavior. This interpretation was akin to the homeostatic model of addiction proposed by Koob (2003), arguing for the role of addictive substances in resetting the homeostatic state of the reward system to a level requiring repeated exposure to these rewarding addictive substances. An alternative interpretation leads to the opposite conclusion. Because performance level was similar in adults and in adolescents, it is conceivable that the reward system of adolescents might be more efficient than in adults, i.e., less activation necessary to energize actions to a similar level.

Finally, Bjork et al. (2004) also analyzed the feedback phase, when subjects received performance outcome. Contrary to other studies (Ernst et al., 2005; Van Leijenhorst et al., 2006), Bjork et al. (2004) failed to detect any activation differences between adolescents and adults in any of the feedback contrasts.

CONCLUSIONS

The Fractal Triadic Model offers a platform for studying maturational changes in the functional neural systems that can explain at least partly the typical manifestations of adolescent behaviors, i.e., risk taking, cognitive impulsivity, poor emotional regulation, and hypersensitivity to social factors. Although neural systems models of decision-making research to date have focused principally on mechanisms that account for risk taking, particularly in the context of addiction (e.g., Bechara, 2005; Berridge, 1996; Jentsch & Taylor, 1999; Volkow & Fowler, 2000) and in individuals with brain lesions (e.g., Bechara & Van Der Linden, 2005; Fellows & Farah, 2005; Seitz, Nickel, & Azari, 2006), the extension of this research to ontogenic changes and other psychopathologies (e.g., anxiety, depression) is fairly recent. Of note, the integration of the proposed social information processing network by Nelson et al. (2005) with the matrix of the Triadic Model is particularly relevant for understanding the unique modulation of decision making by social context in adolescence (Ernst & Fudge, 2009).

The parameters that influence the behavioral output of the triadic system are multiple, and can be organized along characteristics of stimuli, actors (e.g., age, psychopathology), and environment. The few studies reviewed above begin to give a flavor of the critical need to control for these factors in future work. Factors mentioned in this review include stimulus features (e.g., probabilistic outcomes, reward magnitudes), and type of cognitive performance (decision making per se versus incentives during an independent task performance). The impact of social context is probably the factor most unique to the adolescent period (Gardner & Steinberg, 2005).

Indeed, for adolescents, the most salient aspect of the environment is how much social involvement is present. Interest in this direction of research is emerging, and studies will soon provide information on the biological basis of such critical developmental aspect of adolescent behavior. Preliminary work in young adults (Nawa, Nelson, Pine, & Ernst, 2008) seems to indicate that the presence of a social other, who plays a betting task alongside a player who is being scanned, significantly affects the neural systems involved in the Triadic Model. When the social other was present, amygdala and medial prefrontal cortex were activated. When the social other was absent, ventral striatum was activated.

These preliminary findings suggested that the presence of a neutral social other may have triggered an alarm signal within the amygdala that warned the individuals of a potential danger. In contrast, in the absence of a social other, the brain might have functioned under a reward mode to perform the reward-related task. The next step is to use this same paradigm in adolescents. We might expect greater involvement of the reward system during the social context in adolescents compared to adults because of the expected higher rewarding effect of the social presence of a peer for adolescents than for adults.

This review highlights the need for a systematic research program to help understand typically developing adolescents. This step is pivotal to the understanding of predictors of deviant behaviors or ongoing psychopathology.

REFERENCES

Aggleton, J. P. (2000). *The amygdala: A functional analysis* (2nd ed.). Oxford: Oxford University Press.

Andersen, S. L., LeBlanc, C. J., & Lyss, P. J. (2001). Maturational increases in c-fos expression in the ascending dopamine systems. *Synapse, 41*, 345–350.

Baxter, M. G., & Murray, E. A. (2002). The amygdala and reward. *Nature Reviews Neuroscience, 3*, 563–573.

Bechara, A. (2005). Decision making, impulse control and loss of willpower to resist drugs: A neurocognitive perspective. *Nature Neuroscience, 8*, 1458–1463.

Bechara, A., & Van Der Linden, M. (2005). Decision-making and impulse control after frontal lobe injuries. *Current Opinion in Neurology, 18*, 734–739.

Berridge, K. C. (1996). Food reward: Brain substrates of wanting and liking. *Neuroscience Biobehavioral Review, 20*, 1–25.

Bjork, J. M., Knutson, B., Fong, G. W., Caggiano, D. M., Bennett, S. M., & Hommer, D. W. (2004). Incentive-elicited brain activation in adolescents: Similarities and differences from young adults. *Journal of Neuroscience, 24*, 1793–1802.

Bjork, J. M., Smith, A. R., Danube, C. L., & Hommer, D. W. (2007). Developmental differences in posterior mesofrontal cortex recruitment by risky rewards. *Journal of Neuroscience, 27*, 4839–4849.

Blakemore, S. J. (2008). The social brain in adolescence. *Nature Reviews Neuroscience, 9*, 267–277.

Boyer, T. W. (2006). The development of risk taking: A multi-perspective review. *Developmental Review, 26*, 291–345.

Brown, S. A., & Tapert, S. F. (2004). Adolescence and the trajectory of alcohol use: Basic to clinical studies. *Annuals of the New York Academy of Science, 1021*, 234–244.

Bush, G., Luu, P., & Posner, M. I. (2000). Cognitive and emotional influences in anterior cingulate cortex. *Trends in Cognitive Science, 4*, 215–222.

Carter, C. S., & van Veen, V. (2007). Anterior cingulate cortex and conflict detection: An update of theory and data. *Cognitive Affective Behavioral Neuroscience, 7*, 367–379.

Cohen, G. L., & Prinstein, M. J. (2006). Peer contagion of aggression and health risk behavior among adolescent males: An experimental investigation of effects on public conduct and private attitudes. *Child Development, 77*, 967–983.

Crone, E. A., & van der Molen, M. W. (2004). Developmental changes in real life decision making: Performance on a gambling task previously shown to depend on the ventromedial prefrontal cortex. *Developmental Neuropsychology, 25*, 251–279.

Cunningham, M. G., Bhattacharyya, S., & Benes, F. M. (2002). Amygdalo-cortical sprouting continues into early adulthood: Implications for the development of normal and abnormal function during adolescence. *Journal of Computational Neurology, 453*, 116–130.

Dahl, R. E. (2004). Adolescent brain development: A period of vulnerabilities and opportunities. Keynote address. *Annals of the New York Academy of Science, 1021*, 1–22.

Davidson, R. J., Jackson, D. C., & Kalin, N. H. (2000). Emotion, plasticity, context, and regulation: Perspectives from affective neuroscience. *Psychological Bulletin, 126*, 890–909.

Davis, M. (2006). Neural systems involved in fear and anxiety measured with fear-potentiated startle. *American Psychologist, 61*, 741–756.

Deakin, J., Aitken, M., Robbins, T., & Sahakian, B. J. (2004). Risk taking during decision-making in normal volunteers changes with age. *Journal of the International Neuropsychological Society, 10*, 590–598.

Delgado, M. R., Nystrom, L. E., Fissell, C., Noll, D. C., & Fiez, J. A. (2000). Tracking the hemodynamic responses to reward and punishment in the striatum. *Journal of Neurophysiology, 84*, 3072–3077.

Di Chiara, G., & Bassareo, V. (2007). Reward system and addiction: What dopamine does and doesn't do. *Current Opinion in Pharmacology, 7*, 69–76.

Ernst, M., & Fudge, J. L. (2009). A developmental neurobiological model of motivated behavior: Anatomy, connectivity and ontogeny of the triadic nodes. *Neuroscience and Biobehavioral Reviews, 33*(3), 367–382.

Ernst, M., Nelson, E. E., Jazbec, S., McClure, E. B., Monk, C. S., Leibenluft, E. et al. (2005). Amygdala and nucleus accumbens in responses to receipt and omission of gains in adults and adolescents. *NeuroImage, 25*, 1279–1291.

Ernst, M., Pine, D. S., & Hardin, M. (2006). Triadic model of the neurobiology of motivated behavior in adolescence. *Psychological Medicine, 36*, 299–312.

Ernst, M., Romeo, R. D., & Andersen, S. L. (in press). Neurobiology of the development of motivated behaviors in adolescence: A window into a neural systems model. *Pharmacology, Biochemistry, and Behavior.*

Eshel, N., Nelson, E. E., Blair, R. J., Pine, D. S., & Ernst, M. (2007). Neural substrates of choice selection in adults and adolescents: Development of the ventrolateral prefrontal and anterior cingulate cortices. *Neuropsychologia, 45,* 1270–1279.

Fellows, L. K., & Farah, M. J. (2005). Different underlying impairments in decision-making following ventromedial and dorsolateral frontal lobe damage in humans. *Cerebral Cortex, 15,* 58–63.

Fudge, J. L., Breitbart, M. A., & McClain, C. (2004). Amygdaloid inputs define a caudal component of the ventral striatum in primates. *Journal of Computational Neurology, 476,* 330–347.

Galvan, A., Hare, T. A., Parra, C. E., Penn, J., Voss, H., Glover, G. et al. (2006). Earlier development of the accumbens relative to orbitofrontal cortex might underlie risk-taking behavior in adolescents. *Journal of Neuroscience, 26,* 6885–6892.

Gardner, M., & Steinberg, L. (2005). Peer influence on risk taking, risk preference, and risky decision making in adolescence and adulthood: An experimental study. *Developmental Psychology, 41,* 625–635.

Giedd, J. N. (2004). Structural magnetic resonance imaging of the adolescent brain. *Ann Annuls of the New York Academy of Science, 1021,* 77–85.

Giedd, J. N., Blumenthal, J., Jeffries, N. O., Castellanos, F. X., Liu, H., Zijdenbos, A. et al. (1999). Brain development during childhood and adolescence: A longitudinal MRI study. *Nature Neuroscience, 2,* 861–863.

Glimcher, P. W., Kable, J. W., & Louie, K. (2007). Neuroeconomic studies of impulsivity: Now or just as soon as possible? *American Economic Review, 97,* 142–147.

Gray, J. A. (1994). Three fundamental emotion systems. In P. Ekman & R. J. Davidson (Eds.), *The nature of emotion: Fundamental questions* (pp. 243–247). New York: Oxford University Press.

Guyer, A. E., McClure-Tone, E. B., Shiffrin, N. D., Pine, D. S., & Nelson, E. E. (in press). Neural correlates of anticipated peer evaluation in adolescence. *Child Development.*

Haber, S. N., Fudge, J. L., & McFarland, N. R. (2000). Striatonigrostriatal pathways in primates form an ascending spiral from the shell to the dorsolateral striatum. *Journal of Neuroscience, 20,* 2369–2382.

Haber, S. N., Kim, K. S., Mailly, P., & Calzavara, R. (2006). Reward-related cortical inputs define a large striatal region in primates that interface with associative cortical connections, providing a substrate for incentive-based learning. *Journal of Neuroscience, 26,* 8368–8376.

Hariri, A. R., Brown, S. M., Williamson, D. E., Flory, J. D., de Wit, H., & Manuck, S. B. (2006). Preference for immediate over delayed rewards is associated with magnitude of ventral striatal activity. *Journal of Neuroscience, 26,* 13213–13217.

Holland, P. C., & Gallagher, M. (2004). Amygdala-frontal interactions and reward expectancy. *Current Opinion in Neurobiology, 14*, 148–155.

Hooper, C. J., Luciana, M., Conklin, H. M., & Yarger, R. S. (2004). Adolescents' performance on the Iowa Gambling Task: Implications for the development of decision making and ventromedial prefrontal cortex. *Developmental Psychology, 40*, 1148–1158.

Jensen, J., McIntosh, A. R., Crawley, A. P., Mikulis, D. J., Remington, G., & Kapur, S. (2003). Direct activation of the ventral striatum in anticipation of aversive stimuli. *Neuron, 40*, 1251–1257.

Jentsch, J. D., & Taylor, J. R. (1999). Impulsivity resulting from frontostriatal dysfunction in drug abuse: Implications for the control of behavior by reward-related stimuli. *Psychopharmacology, 146*, 373–390.

Koob, G. F. (2003). Alcoholism: Allostasis and beyond. *Alcoholism: Clinical and Experimental Research, 27*, 232–243.

Kringelbach, M. L. (2005). The human orbitofrontal cortex: Linking reward to hedonic experience. *Nature Reviews Neuroscience, 6*, 691–702.

Laviola, G., Pascucci, T., & Pieretti, S. (2001). Striatal dopamine sensitization to D-amphetamine in periadolescent but not in adult rats. *Pharmacology and Biochemical Behavior, 68*, 115–124.

LeDoux, J. E. (2000). Emotion circuits in the brain. *Annual Review of Neuroscience, 23*, 155–184.

Lejuez, C. W., Aklin, W. M., Zvolensky, M. J., & Pedulla, C. M. (2003). Evaluation of the balloon analogue risk task (BART) as a predictor of adolescent real-world risk-taking behaviours. *Journal of Adolescence, 26*, 475–479.

Lejuez, C. W., Read, J. P., Kahler, C. W., Richards, J. B., Ramsey, S. E., Stuart, G. L. et al. (2002). Evaluation of a behavioral measure of risk taking: The balloon analogue risk task (BART). *Journal of Experimental Psychology: Applied, 8*, 75–84.

Lewis, D. A., Cruz, D., Eggan, S., & Erickson, S. (2004). Postnatal development of prefrontal inhibitory circuits and the pathophysiology of cognitive dysfunction in schizophrenia. *Annuls of the New York Academy of Science, 1021*, 64–76.

Martin-Soelch, C., Linthicum, J., & Ernst, M. (2007). Appetitive conditioning: Neural bases and implications for psychopathology. *Neuroscience Biobehavioral Review, 31*, 426–440.

May, J. C., Delgado, M. R., Dahl, R. E., Stenger, V. A., Ryan, N. D., Fiez, J. A. et al. (2004). Event-related functional magnetic resonance imaging of reward-related brain circuitry in children and adolescents. *Biological Psychiatry, 55*(4), 359–366.

McClure, E. B., Parrish, J. M., Nelson, E. E., Easter, J., Thorne, J. F., Rilling, J. K. et al. (2007). Responses to conflict and cooperation in adolescents with anxiety and mood disorders. *Journal of Abnormal Child Psychology, 35*, 567–577.

Nawa, N. E., Nelson, E. E., Pine, D. S., & Ernst, M. (2008). Do you make a difference? Social context in a betting task. *Social Cognitive and Affective Neuroscience, 3*(4), 367–376.

Nelson, E. E., Leibenluft, E., McClure, E. B., & Pine, D. S. (2005). The social re-orientation of adolescence: A neuroscience perspective on the process and its relation to psychopathology. *Psychological Medicine, 35,* 163–174.

Nestler, E. J., & Carlezon, W. A., Jr. (2006). The mesolimbic dopamine reward circuit in depression. *Biological Psychiatry, 59,* 1151–1159.

Overman, W. H. (2004). Sex differences in early childhood, adolescence, and adulthood on cognitive tasks that rely on orbital prefrontal cortex. *Brain and Cognition, 55,* 134–147.

Overman, W. H., Frassrand, K., Ansel, S., Trawalter, S., Bies, B., & Redmond, A. (2004). Performance on the Iowa card task by adolescents and adults. *Neuropsychologia, 42,* 1838–1851.

Rauch, S. L., Shin, L. M., & Wright, C. I. (2003). Neuroimaging studies of amygdala function in anxiety disorders. *Annuals of the New York Academy of Science, 985,* 389–410.

Rich, B. A., Schmajuk, M., Perez-Edgar, K. E., Fox, N. A., Pine, D. S., & Leibenluft, E. (2007). Different psychophysiological and behavioral responses elicited by frustration in pediatric bipolar disorder and severe mood dysregulation. *American Journal of Psychiatry, 164,* 309–317.

Roesch, M. R., Calu, D. J., Burke, K. A., & Schoenbaum, G. (2007). Should I stay or should I go? Transformation of time-discounted rewards in orbitofrontal cortex and associated brain circuits. *Annuals of the New York Academy of Science, 1104,* 21–34.

Rumsey, J., & Ernst, M. (Eds.) (2009). *Neuroimaging in developmental clinical neuroscience.* Cambridge: Cambridge University Press.

Scheres, A., & Sanfey, A. G. (2006). Individual differences in decision making: Drive and Reward Responsiveness affect strategic bargaining in economic games. *Behavioral and Brain Functions, 2,* 35.

Seitz, R. J., Nickel, J., & Azari, N. P. (2006). Functional modularity of the medial prefrontal cortex: Involvement in human empathy. *Neuropsychology, 20,* 743–751.

Spear, L. P. (2000). The adolescent brain and age-related behavioral manifestations. *Neuroscience Biobehavioral Reviews, 24,* 417–463.

Steinberg, L. (2004). Risk taking in adolescence: What changes, and why? *Annals of the New York Academy of Science, 1021,* 51–58.

Steinberg, L. (2005). Cognitive and affective development in adolescence. *Trends in Cognitive Science, 9,* 69–74.

Van Leijenhorst, L., Crone, E. A., & Bunge, S. A. (2006). Neural correlates of developmental differences in risk estimation and feedback processing. *Neuropsychologia, 44,* 2158–2170.

Van Leijenhorst, L., Westenberg, P. M., & Crone, E. A. (2008). A developmental study of risky decisions on the cake gambling task: Age and gender analyses of probability estimation and reward evaluation. *Developmental Neuropsychology, 33,* 179–196.

Volkow, N. D., & Fowler, J. S. (2000). Addiction, a disease of compulsion and drive: Involvement of the orbitofrontal cortex. *Cerebral Cortex, 10,* 318–325.

Wise, R. A. (2004). Dopamine, learning and motivation. *Nature Reviews Neuroscience, 5,* 483–494.

10

THE TERRIBLE TWELVES

Abigail A. Baird
Vassar College

We are born, so to speak, twice over; born into existence, and born into life; born a human being, and born a man.

—Jean-Jacques Rousseau

Collectively, the scientific literature has described the function of human development as enabling the individual to pass their genetic material to the next generation. Because of this predisposition, development is an extremely plastic process, enabling human infants to follow an infinite number of developmental trajectories. Ideally, the developmental path is constrained by environmental demands that not only complement the individual's biology, but also forge behavioral outcomes that result in the individual's eventual reproductive success (see Waddington, 1975). Certain processes within human development build upon themselves. These recursive processes can be seen in both neural and behavioral development (see Gould, 1977). Other processes develop in such a way that each period constitutes a unique time with individual environmental demands and idiosyncratic milestones (see Bjorklund, 1997 for a review). It is the complex interaction of these two developmental courses (and likely others yet to be discovered) that give rise to observable human development. Arguably, the most fascinating instantiations of this occur during both the toddler and adolescent years, when preprogrammed biology and environmental demand interact to produce periods of rapid development. In order to cope with the inordinate number and magnitude of physical and mental changes that take place during these periods, there is a sudden need to acquire and assimilate an overabundance of crucial information. To accomplish this, evolution has provided humans with sensitive periods, periods of time during development when the individual is able to acquire and assimilate information with a speed and precision not seen at other points in development. This enables children to rapidly acquire new skills and abilities, often with little or no conscious awareness.

A great deal has been written about the first two years of life, the first sensitive period in human development—a period most frequently recognized for the emergence of spoken language, and the onset of the "terrible twos." It can be argued that humans undergo a second sensitive period during the sociocultural expression of puberty, known as adolescence. As in the first sensitive period, the adolescent is designed to be exceptionally attune to, and capable of learning about, stimuli that are critical to survival. This essay will offer support for the idea that there are two sensitive periods during human development, the first of which occurs close to the second year and the second close to the twelfth.

While the directives of these developmental periods are superficially unique, a closer examination reveals that they may reflect initial and secondary refinements to the same larger systems. At their centers, both periods involve major changes to the cognitive processes that support abstract thinking. According to Piagetian theory, symbolic representation is the central accomplishment of the preoperational period, which extends from 2 to 7 years. The most widely recognized manifestation of this is the massive improvements in the child's ability to generate and comprehend meaningful language. A strikingly similar period of rapid improvement in communication takes place during adolescence, when the individual achieves fluency in social and emotional communication.

Among children, abstract thought is initially expressed through mental representations of objects that are no longer in the child's presence, as well as through imaginary play, both of which require abstract thought. Abstraction again takes center stage during the formal operational stage, between the ages of 11 and 15. During the formal operational stage, young adolescents are able to move beyond concrete thought and imagine things that have not yet happened (Piaget & Inhelder, 1969). During both stages, maturational change enables the internal representation of the thoughts and feelings of others. Initially, the child is focused on understanding standards imparted by parents or caretakers, while the adolescent is focused on acquiring peer-based standards. The ability to hold multiple points of view in mind also enables great strides to be made in self-awareness. The 2-year-old comes to realize that their internal world is unique, and of their own making. This ability is revised in adolescence, when individuals become increasingly aware of their own beliefs, aspirations, and identities. In summary, the major developmental advances that take place during the second year are, first, the emergence of language; second, a considerable leap in abstract thought and reasoning; and third, significant growth in self-awareness. Each of these competencies will be detailed below, and compared to similar achievements during the early adolescent years.

COMMUNICATION

Sometime between the first and third birthdays, virtually all children acquire the ability to communicate via spoken language. The way children "acquire language" parallels the acquisition of social communication in adolescence. A parsimonious description of how and why language emerges continues to elude theorists; however, all agree that both the maturation of the brain and social experience play critical roles. The consistency and ubiquity with which language develops is staggering; despite this, theorists have struggled to describe such development, seemingly unable to divine the formula of neural and environmental factors that produce it. Noam Chomsky (1972) considered language acquisition to be an "automatic human endeavor." Chomsky described this acquisition as something naturally befalling children placed in an appropriate environment. Steven Pinker (1994) asserted that language is a human instinct based on genetic instructions and on the maturing of language centers in the child's brain. What is clear is that infants are born with the capacity for acquiring language, not the language itself. Human infants possess biology primed for environmental input and uniquely adept at learning language during the first two years. During this sensitive period in development, simultaneously universal (the capacity itself) and individual (one's specific language and idioms) processes exist. Although the precise means by which humans acquire language remains elusive, two factors are very clear: infants are able to learn multiple languages simultaneously, and learning language is an unconscious process (Pinker, 1994). It has been estimated that during this prime time for language development, infants learn eight to ten words a day; these words can be from one language or three, depending on the child's environmental demands (see Karmiloff & Karmiloff-Smith, 2001, for a review of these ideas). One obvious consequence of this developmental progression manifests in difficulty learning a new language later in life. An adult may acquire proficiency in a new language, but will never sound like a native speaker.

Similarly, it can be hypothesized that during adolescence, a teen must learn the social language of adulthood, or forever relate to others with a "socio-cultural accent." It is also quite plausible that the learning that takes place during adolescence, as in infancy, occurs within a "sensitive time," meaning adolescents' biology is uniquely sensitive to socially salient information, and they are able to acquire this information at an astounding rate. This sudden increase in attention to social salience may help explain why typical teenagers are able to recall every detail about their favorite band, but have absolutely no idea where South Dakota is located, despite an entire year in geography class. In terms of peer relationships, knowledge about current music is social currency; the precise location of some place you have never been is not.

While the ultimate objective of human development is always survival, this goal takes unique forms during different periods of development. During infancy, survival is contingent upon the infant's ability to elicit the support of his or her caretakers, whereas during the second infancy, survival rests on the ability to join adult society, with all of the associated rights and privileges, including access to potential mates. Human beings' remarkable developmental plasticity may very well have set the evolutionary stage for special sensitivity to variable social contexts (e.g., Adolphs, 2003; Carruthers, 2002), and is undoubtedly one of the great accomplishments of human evolution. A number of researchers have described the human brain as a "social tool" (see Flinn, 2004, for an excellent review of this idea). Given that human beings are not hardwired for specific social behaviors, their acquisition is reliant on a myriad of developmental processes. The primary goal of adolescence is to learn enough to make the transition to, and survive in, the adult world, a place where individuals will spend roughly five-sixths of their living years. Simply, adolescence is the sensitive period for social development. During adolescence, the genetic and hormonal influences of puberty prime the brain for learning how to understand, and in turn navigate, social relationships. During this time, individuals are bombarded with socio-cognitive information and are capable of processing it with unprecedented speed and efficiency, rivaled only by that which takes place during infancy. Finally, here is a highly functional reason for delaying social maturation until adolescence. Simply, different stages of development call for very different social skills. Infants learn to speak the language of their place of birth, and in doing so ensure adults' willingness to care for them. Adolescents learn to speak the language of their culture (peers), and in doing so acquire access to adult society as well as to potential mates, thereby securing the opportunity to reproduce.

The vast complexity of social relationships first becomes apparent during adolescence. For example, teens often feel the need to get home as quickly as possible in order to rehash the school day that just ended, talking for hours with the people they have just spend the day with, whether it be over instant messenger or the telephone. This underscores both the intricate and elusive nature of these critical relationships. Adolescent relationships appear highly dynamic, if not altogether unstable. This dynamism is a necessary component of relationships during this time, as the participants are inexperienced in the newfound complexity and nuance of adolescent social interaction. The very nature of social relationships, and the information they convey, generates an infinite number of possible scenarios. As a result, it is left to the individual to extract heuristics or rules from his or her experience. The adolescent social world requires the ability to quickly modify behavioral strategies in real time with rapidly changing demand contingencies. Those who fare best during adolescence will be those who are able to integrate their abstract knowledge, past experiences, and current situational demands. The dynamic nature of adolescent culture creates an organic hierarchy, precipitating a "survival of the coolest" contest for gaining access to friends and potential mates.

While the neural underpinnings of language acquisition are not precisely understood, most agree with the Wernicke-Lichtheim-Geshwind model that describes three main regions, and their interactions, as being critical for language development. The superior temporal gyrus (Wernicke's area), the angular gyrus, and the inferior frontal gyrus (Broca's area) are all connected by the arcuate fasciculus, and are critical for language acquisition (see Damasio, Tranel, Grabowski, Adolphs, & Damasio, 2004 for a review). There is a well-documented growth in layer III of the cortex during the second year. Layer III is comprised of largely pyramidal neurons and is responsible for communication between cortical regions. This growth is accompanied by peak levels of glutamate binding, as well as a spurt in GABA activity in inhibitory interneurons of layer III during the second year of life (Huttenlocher & Dabholkar, 1997; Mrzljak, Uylings, van Eden, & Judas 1990; Slater, McConnell, D'Souza et al., 1992).

These same regions undergo significant change again during puberty, as social communication takes center stage. Interestingly, the density of gray matter in both the frontal and parietal lobes peaks at approximately 12 years, followed by a steady decline into adulthood (Geidd et al., 1999; Gotay et al., 2004; Sowell et al., 2003). Similar to many other measures of pubertal development, these gray matter changes peak at an earlier point for girls than for boys. Toga and colleagues have described a slower and ongoing developmental trajectory in the temporal lobe, including the superior temporal sulcus (Toga et al., 2006). Increases in the white matter, and hence connectivity, of the left arcuate fasciculus have been documented during puberty (Paus et al., 1999). It has been repeatedly demonstrated that increased myelination has a direct impact on the speed and efficiency of neural processing (Baird, Colvin, VanHorn, Inati, S., & Gazzaniga, 2005). At the level of the neuron, increased myelination leads to increased action potential propagation speed and reduced signal attenuation. At a macroscopic level this type of maturation facilitates synchrony and coordination, both regionally and across the whole brain. While it is possible that these developments do facilitate improvements in adolescents' spoken communication, it is also quite possible that these changes enable improvements in social cognition, well documented to also be associated with these regions. In particular the superior temporal sulcus (and the neighboring cortex) has been shown to play a critical role in the perception of social cues such as eye, face, and body movements in both humans and monkeys (see Adolphs, 2001; Allison et al., 2000, for reviews). Therefore, there is good reason to propose that the pubertal changes in these regions relate closely to the myriad of modifications in social communication observed during adolescence.

IMPROVEMENTS TO ABSTRACT THOUGHT AND REASONING

According to Piaget, the second year marks the transition from the sensorimotor period to the preoperational period. The cognitive hallmark of the preoperational period is the ability to create and maintain symbolic representations (Piaget & Inhelder, 1969). The utility of this function is clear with regard to language acquisition, but it is also a fundamental aspect of cognition, upon which many higher-order functions will be built. There are a number of ways in which this ability manifests itself in the 2-year-old. Perhaps easiest to observe is deferred imitation where children will imitate or "role play" behavior they have observed in the nonimmediate past. Another key function that symbolic representation enables is the internalization of rules, or "standards" as described by Jerome Kagan. The ability to maintain an internal representation of parents' rules and regulations is critical to survival within the family context. Additional evidence for internal abstract standards can be seen in 2-year-olds upset at broken toys, soiled clothes, and prohibited actions (Kagan, 1981). The ability to create and manipulate abstract ideas undergoes a major transformation during adolescence.

As described by Piaget (1954) adolescent thought becomes more abstract, logical, and idealistic. Adolescents are more capable of examining their own thoughts, others' thoughts, and what others are thinking about them; further, they are more likely than younger children to interpret and monitor the world around them. For the first time in their development, adolescents are no longer limited to actual, concrete experiences as anchors for thought. In fact, early adolescent thought often disregards the constraints of reality (Broughton, 1978). They can cognitively generate make-believe situations, events that are entirely hypothetical possibilities, or strictly abstract propositions. In doing so, adolescents are able to make predictions about future outcomes and possible consequences, as well as how they might feel about them. During the later phases, the adolescent learns to better regulate their thoughts, primarily by measuring the products of her reasoning against experience and imposing monitoring or inhibitory cognitions when appropriate. The primary gain in adolescent cognition is that in addition to being able to generate abstract thought, they are able to reason about the products of their cognition. This "thinking about thinking" forms the foundation for both metacognition and introspection. However, it is clear that the emergence of these abilities depends in great part on experience, and therefore does not appear across all situational domains simultaneously. Simply, mature thought is more likely to be used in areas where adolescents have the most experience and knowledge (Carey, 1988). During development, adolescents acquire elaborate knowledge through extensive experience and practice in multiple settings (e.g., home, school, work). Greater experience, together with an improved system for organizing and retrieving memories of that experience, enable the adolescent to recall and apply a greater number of experiences to new situations. The changes just described in thinking and feeling largely result from the synergistic maturation of working memory capacity, selective attention, error detection, and inhibition, all of which have been shown to improve with maturational changes in brain structure and function. Perhaps the most consistently reported finding associated with adolescent brain development is the maturation of the prefrontal cortex.

The prefrontal cortex is of paramount interest in human development largely because of its well-understood function with regard to cognitive, social, and emotional processes in adulthood. Converging evidence of prolonged development and organization throughout childhood and adolescence (Chugani et al., 1987; Diamond, 1988, 1996; Huttenlocher, 1979) underscores the important parallel between prefrontal development and cognitive development. Perhaps the most consistently reported finding associated with adolescent brain development is the decrease of gray matter and the increase of white matter throughout the cortex, but most significantly within frontal cortex (see Giedd et al., 1999; Sowell et al., 1999, for reviews). The decline of gray matter in prefrontal cortex in adolescence has been taken to be a marker of neural maturation as a result of synaptic pruning (Bourgeois, Goldman-Rakic, & Rakic, 1994; Paus et al., 2001). The delayed maturation of this brain region allows neural and behavioral plasticity, which in turn enables the individual to adapt to the particular demands of their unique environment.

The prolonged development and organizational connectivity of prefrontal cortex throughout childhood and adolescence provides the neural scaffolding for the observed changes in both cognitive and emotional processes. When an adolescent rapidly gains six inches in height and thirty pounds in weight, as is typical in a year, the effects on coordination can be striking. References to the physically awkward teen abound in our culture; these same attributions should be applied to the growth of abstract reasoning and executive control. Akin to the way that both infants and adolescents gradually gain control over the changing entity that is their body and brain, the building blocks of abstract reasoning are assembled, toppled, and reassembled until a structure fitting the individual's needs is created. And it is through the ever-evolving architecture of this structure that adolescents not only make some of the memorable "errors of coordination" i.e., bad choices, for which they are well known, but also arrive at young adulthood with the important life lessons that ensure their position in adult society.

GROWTH IN SELF-AWARENESS

The readily discernable changes that take place in self-awareness at both 2 and 12, are largely the synergistic product of the developments described above; namely, improvements in communication and cognitive ability. By 2 years of age, the individual is well aware of their physical existence, as demonstrated by empirical findings such as the rouge test (see Lewis & Brooks-Gunn, 1979). However, as a result of improvements in cognition and language, the 2-year-old begins to speak about him/herself using the pronoun "I," and can communicate knowledge of their own feelings, intentions, actions, and competencies (Kagan, 1981). This newfound self-awareness appears to also manifest in toddlers' egocentrism. Within the developmental realm, egocentrism refers to an incomplete differentiation between the self and the world (Piaget, 1954). This means that young children have a very hard time "putting themselves in other people's shoes," and often believe that the people around them are experiencing everything that the toddler experiences (not limited to thoughts, egocentrism extends to perceptual experiences as well, see Piaget, 1954, for a review). Despite these gains in cognition, 2-year-olds are not very skilled at integrating their own thoughts with those of others, especially when they do not agree. For example, it is very difficult for the average 2-year-old to understand why they cannot have a cookie when they want one. In the mind of a 2-year-old, they really want the cookie, and because they are feeling this way, surely their caretaker is feeling this way; and as a result it makes absolutely no sense when the caretaker refuses the request. As a result of these interactions (which appear in a variety of domains), where the child is aware of their own wants, but is unable to reconcile that these wants may differ from those of their caretakers, there is a often a significant increase in parent-child conflict during the second year of the child's life (Laible & Thompson, 2002). As difficult as this time may be for parents and children, it is part of a critical developmental progression. In order to become a mature member of society, the individual must learn to recognize their individual wants and needs and integrate those with the wants and needs of those around them; the "terrible two's" is an important milestone in this process.

In many ways, adults can be conceptualized as "external frontal lobes" for their children, helping to interpret environmental demands, and construct and execute appropriate responses. Given the behavioral consequences of having an immature (uncoordinated) frontal system, adults subsume a number of frontal functions by instructing children in the absence of their own abstract reasoning. The next significant advance in development takes place during adolescence when abstract thought enables an individual to envision and anticipate situations that they have not directly experienced, and forever changes the relationship between the adolescent and the adults around them. The adolescent has the capacity to discern future feelings and to make subtle distinctions regarding expressed emotion. Moreover, affective states become integrated with formal thought operations. During adolescence emotion is translocated from the self to self-in-relationship, a domain where the complex unfolding of emotional states in both the self and other can be mutually recognized. Much like their 2-year-old counterparts, however, during the emergence of these abilities adolescents experience another period of egocentrism, and just as in the first period of egocentrism, increased conflict with parents ensues (see D'Angelo & Omar, 2003 for a review).

The theory of adolescent egocentrism delineates two separate but related ideation patterns: the imaginary audience and the personal fable. The imaginary audience refers to adolescents' tendency to believe that others are always watching and evaluating them; the personal fable refers to the belief that the self is unique, invulnerable, and omnipotent (Elkind, 1976). The cognitions reflected by both constructs seem to parsimoniously describe the reasons for feelings and behaviors typically associated with early adolescence, such as self-consciousness, conformity to peer group norms, and risk taking (Vartanian, 2000).

The imaginary audience and personal fable are creations of the important improvements in adolescent cognition. They appear to be byproducts of increases in abstract thought and self-awareness. Much in the way that a child learning a new rule of language commits errors of application (cars, dolls, apples, sheeps), adolescents make errors in applying their new reasoning skills. The adolescent errs in over-applying their new ability to recognize the thoughts and feelings of others, and in this instance cannot differentiate their own feelings from those of others (imagined audience) and at the other extreme fails to realize the commonality of their own thoughts and feelings and feels overly unique (personal fable). The imagined audience is thought to be constantly scrutinizing the adolescent, and there is an assumption on the part of the adolescent that the evaluations of the imagined audience will match their own (Elkind, 1976). The personal fable reflects the mistaken belief that one's feelings and experiences are uniquely different from those of others (Elkind, 1976). The adolescent may therefore come to believe that "others cannot understand what I'm going through," "that won't happen to me," and "I can handle anything." The imaginary audience and personal fable seem to describe what have been viewed as typical facets of adolescent behavior. For example, self-consciousness and conformity to the peer group with regard to appearance can be understood as stemming from the belief that others are always watching and judging. Feelings of isolation and risk-taking behavior can be viewed as products of a personal fable (Vartanian, 2000).

At the core of the imagined audience is a significant improvement in perspective taking, which is fueled by the emergence of the ability to think increasingly abstractly that accompanies early adolescence. Perspective taking is the ultimate integration of emotion and cognition, and it relies on the perception of self and other, as well as a cognitive appreciation of emotional states. Thus, perspective taking requires both cognitive and emotional perspective taking, as well as self-regulation. Perspective taking is believed to undergo a series of stage-like developmental progressions (Selman, 1980). The stages of perspective taking reflect the transition from egocentric to sociocentric functioning and an eventual understanding of the internal and external states of others and their social context. Depending on situational demands, an appropriate response could require emotional perspective taking (empathy), cognitive perspective taking (theory of mind), or the integration of both. Selman's theory, which charts the developmental course by which children come to be aware of and coordinate the cognitive, social, and emotional perspectives of self and others, provides a better backdrop for understanding both the imagined audience and personal fable.

Lapsley and Murphy (1985) have proposed that Selman's theory of social perspective taking and interpersonal understanding provides an excellent explanation for the emergence, and eventual minimization, of the imagined audience and personal fable. It has been suggested that both the imaginary audience and personal fable might be products of what Selman described as Level 3 social perspective-taking ability. Level 3 corresponds to the age period during which both thoughts related to the imaginary audience and personal fable peak, most often between 10 and 15 years old. This period is typified by the ability to consider self and other perspectives simultaneously from a third-party, or "observing ego" perspective. This differs substantially from the abilities of the Level 2 child, who is limited to considering one perspective at a time, sequentially such that self and other are never able to co-occupy the same cognitive space.

The new abilities related to the observing ego perspective also enable adolescents to view themselves as both the agent and an object in social interaction; a lack of coordination or skill with this new ability is believed to underlie the imaginary audience ideation, due to the enhanced self-consciousness inherent in this cognitive development (Damon & Hart, 1982; Lapsley & Murphy, 1985). The acquisition of Level 4 social perspective-taking ability is believed to decrease imaginary audience and personal fable ideations. Upon reaching this final stage of development, the older adolescent is capable of considering and coordinating multiple third-party perspectives that form a "generalized social perspective" (Selman, 1980). This perspective alleviates self-consciousness, as the adolescent can better see the self within the context of the "larger matrix of social perspectives" (Lapsley & Murphy, 1985, p. 214). This is fundamentally an issue of increasing capacity, and in this way this last stage is akin to meta social cognition in that the Level

4 appreciation of unconscious mental processes helps to appropriately scale back the imaginary audience and personal fable ideation.

Hoffman (1991) has added to this, suggesting that by late childhood and early adolescence, coinciding with perspective taking and self-concept development, adolescents can empathize with a generalized group of others and their life situation. This newfound ability to empathize with a group of needy others might predict relatively sophisticated forms of moral behaviors—behaviors that involve groups of people. Thus, this transition may be important in the development of empathy and may help to explain relatively sophisticated social behaviors in adolescence and adulthood. Somewhere between the influence of the imagined audience and empathy for your social group exists the powerful import of the peer group.

During a typical weekend, adolescents spend more than twice as much time with peers as they do with their parents (Condry, 1987). In a regular school day, teens have an average of 299 interactions with peers (Barker & Wright, 1951). During the limited time teens spend away from their friends, if they are not conversing on the phone or computer, they are likely thinking about their peer groups. Based on the empirical and theoretical data describing the importance of social self-perceptions during the teenage years (Jacobs, Vernon, & Eccles, 2004), it is perfectly reasonable for teenagers to place great importance on their friendships. Adolescents turn to peer groups for emotional support and perceive group approval as an indication of social acceptability (Brown, Mounts, Lamborn, & Steinberg, 1993).

During adolescence, the neural hardware supporting cognitive development becomes increasingly coordinated with the hardware that enables emotion. One specific frontal region that may be responsible for the coordination of cognition and emotion is the anterior cingulate cortex, an area known for its prominent role in the mediation and control of emotional, attentional, motivational, social, and cognitive behaviors (Vogt, Finch, & Olson, 1992). A significant positive relationship between age and total anterior cingulate volume (which has been attributed to increases in white matter) has been well documented (Casey et al., 1997). It is thought that this relationship may reflect improved cortical-cortical and cortical-subcortical coordination. The observed projections from both cortical and subcortical regions to the cingulate in adult subjects are known to contribute to the coordination and regulation of cognitive and emotional processes. This has been elegantly demonstrated by the work of Ochsner and colleagues, who have described a process of "reappraisal" whereby individuals use cognitive control strategies to understand, contextualize, and minimize the negative impact of negative stimuli (see Ochsner & Gross, 2005 for a review). Activity of the dorsal portion of the anterior cingulate cortex has been shown to play a crucial role in autonomic control and the conscious interpretation of somatic state. For example, Shin and colleagues (2000) have reported significant increases in blood flow to cingulate cortex during a guilt-related script-driven imagery study, underscoring the idea that the medial cingulate is critical for interpreting the somatic components of emotions like guilt. Importantly, the examples above demonstrate that with the help of the cingulate, abstract thought is capable of generating an emotional state, which can in turn be interpreted by the individual. This is one means by which abstract cognition may contribute to abstract emotion (i.e., "how would I feel if …"). A critical question with regard to human development has been the exact maturational course of these projections. Maturation of the dorsal anterior cingulate cortex has been consistently related to self-control and behavioral inhibition (see Isomura & Takada, 2004 for a recent review). The dorsal anterior cingulate is thought to be an important center for the creation of second-order representations of body state. Second-order representations are the product of integrating first-order sensory information from insular and somatosensory cortices with cognitive and contextual information available to the cingulate. Critchley et al. (2001) found that right anterior cingulate and left posterior cingulate were key areas for the creation of second-order representations. Further, the authors suggested that right anterior cingulate plays a specific executive role in the integration of autonomic responses with behavioral effort (Critchley, Wiens, Rotshtein, Ohman, & Dolan, 2004).

The processes reviewed above have discernable and reliable representation in both behavioral and neural development. Refinements in the structure and function of the prefrontal cortex enables adolescents to think abstractly about idealized versions of themselves, and formulate ways in which they will acquire the experiences and skills needed to develop their desired identity. Developmental changes in a number of brain regions, in particular the anterior cingulate cortex, are providing the additional "hardware" for teens' increasing understand of how their visceral sense relates to their emotional and cognitive processes.

SUMMARY

This essay has highlighted a number of important points of commonality between the changes that occur during the 2nd and 12th years of life. In Piagetian terms, the beginning of preoperational thought and the beginning of formal operational thought both herald significant improvements to communication, abstract thought and feeling, and self-awareness. These changes are also likely reliant upon similar networks within the brain; however, more detailed work in young children and adolescents will have to be done before this is known for certain. Perhaps most important, however, is the idea that what we witness at the onset of adolescence may in fact not be a crucible to be endured by adults, but instead a highly functional recapitulation of the critical milestones of early childhood. Together, the ideas above suggest that while adolescents are arguably less cuddly than toddlers, they may deserve just as much understanding, patience, and support as their "cuter" counterparts.

ACKNOWLEDGMENTS

I wish to thank Debra M. Zeifman for her valuable insights and conversations during the preparation of the ideas in this manuscript. I am also extremely grateful to Shari Silver, who was essential to this manuscript's preparation.

REFERENCES

Adolphs, R. (2001). The neurobiology of social cognition. *Current Opinion in Neurobiology. Special Issue: Cognitive Neuroscience, 11*(2), 231–239.

Adolphs, R. (2003). Cognitive neuroscience of human social behaviour. *Nature Reviews Neuroscience, 4*(3), 165–178.

Baird, A. A., Colvin, M. K., VanHorn, J. D., Inati, S., & Gazzaniga, M. S. (2005). Functional connectivity: Integrating behavioral, diffusion tensor imaging, and functional magnetic resonance imaging data sets. *Journal of Cognitive Neuroscience, 17*(4), 687–693.

Barker, R. G., & Wright, H. F. (1951). *One boy's day: A specimen record of behavior.* New York: Harper & Brothers.

Bjorklund, D. F. (1997). The role of immaturity in human development. *Psychological Bulletin, 122*(2), 153–169.

Bourgeois, J. P., Goldman-Rakic, P. S., & Rakic, P. (1994). Synaptogenesis in the prefrontal cortex of rhesus monkeys. *Cerebral Cortex, 4*(1): 78–96.

Broughton, J. M. (1978). Criticism of the developmental approach to morality. *Catalog of Selected Documents in Psychology, 8,* MS. 1756, p. 82

Brown, B. B., Mounts, N., Lamborn, S. D., & Steinberg, L. (1993). Parenting practices and peer group affiliation in adolescence. *Child Development, 64,* 467–482.

Carey, S. (1988). Are children fundamentally different kinds of thinkers and learners than adults? In K. Richardson & S. Sheldon (Eds.), *Cognitive development to adolescence.* Hillsdale, NJ: Erlbaum.

Carruthers, P. (2002). The cognitive functions of language. *Behavioral and Brain Sciences, 25,* 657–726.

Casey, B. J., Trainor, R., Giedd, J. N., Vauss, Y., Vaituzis, C. K., Hamburger, S. D., Kozuch, P., & Rapoport, J. L. (1997). The role of the anterior cingulate in automatic and controlled processes: A developmental neuroanatomical study. *Developmental Psychobiology, 30*, 61–69.

Chomsky, N. (1972). Psychology and ideology. *Cognition, 1*(1), 11–46.

Chugani, H. T., Phelps, M. E., & Mazziotta, J. C. (1987). Positron emission tomography study of human brain functional development. *Annuls of Neurology, 22*, 487–497.

Condry, J. (1987). Enhancing motivation: A social developmental perspective. In M. L. Maehr & D. A. Kleiber (Eds.), *Advances in motivation and achievement, Vol. 5: Enhancing motivation*. Greenwich, CT: JAI Press.

Critchley H. D., Mathias C. J., & Dolan R. J. (2001). Neuroanatomical basis for first- and second-order representations of bodily states. *Nature Neuroscience, 4*(2), 207–212.

Critchley H. D., Wiens S., Rotshtein P., Ohman A., & Dolan R. J. (2004). Neural systems supporting interoceptive awareness. *Nature Neuroscience, 7*(2), 189–195.

D'Angelo, S. L., & Omar, H. A. (2003). Parenting adolescents. *International Journal of Adolescent Mental Health, 15*(1), 11–19.

Damasio, H., Tranel, D., Grabowski, T., Adolphs, R., & Damasio, A. (2004). Neural systems behind word and concept retrieval. *Cognition, 92*, 179–229.

Damon, W., & Hart, D. (1982). The development of self-understanding from infancy through adolescence. *Child Development, 53*, 841–864.

Diamond, A. (1988). Abilities and neural mechanisms underlying AB performance. *Child Development, 59*(2), 523–527.

Diamond, A. (1996). Impaired sensitivity to visual contrast in children treated early and continuously for phenylketonuria. *Brain, 119*(Pt. 2), 523–538.

Elkind, D. (1976). Egocentrism in adolescence. *Child Development, 38*, 1025–1034.

Flinn, M. V. (2004). Culture and developmental plasticity: Evolution of the social brain. In K. MacDonald, & R. L. Burgess (Eds.), *Evolutionary perspectives on child development* (pp. 73–98). Thousand Oaks, CA: Sage.

Giedd, J. N., Blumenthal, J., Jeffries, N. O., Castellanos, F. X., Liu, H., Zijdenbos, A., Paus, T., Evans, A. C., & Rapapport, J. L. (1999). Brain development during childhood and adolescence. *Nature Neuroscience, 2*, 861–863.

Gould, S. J. (1977). *Ontogeny and phylogeny*. Harvard University Press, Cambridge.

Hoffman, M. L. (1991). Empathy, social cognition, and moral action. In W. M. Kurtines, & J. L. Gewirtz (Eds.), *Handbook of moral behavior and development, vol. I: Theory*, Hillsdale, NJ: Lawrence Erlbaum.

Huttenlocher, P. R. (1979). Synaptic density in human frontal cortex—developmental changes and effects of aging. *Brain Res, 163*, 195–205.

Huttenlocher, P. R., & Dabholkar, A. S. (1997). Regional differences in synaptogenesis in human cerebral cortex. *Journal of Comparative Neurology, 387*, 167–178.

Isomura Y., & Takada M. (2004). Neural mechanisms of versatile functions in primate anterior cingulate cortex. *Reviews in the Neurosciences, 15*, 279–291.

Jacobs, J. E., Vernon, M. K., & Eccles, J. S. (2004). Relations between social self-perceptions, time use, and prosocial or problem behaviors during adolescence. *Journal of Adolescent Research, 19*, 45–62.

Kagan, J., 1981. *The second year*. Cambridge, MA: Harvard University Press.

Karmiloff, K., & Karmiloff-Smith, A., (2001) *Pathways to language: From fetus to adolescent*. Cambridge, MA: Harvard University Press.

Laible, D. J., & Thompson, R. A. (2002). Mother–child conflict in the toddler years: Lessons in emotion, morality, and relationships. *Child Development, 73*(4), 1187–1203.

Lapsley, D. K., & Murphy, M. N. (1985). Another look at the theoretical assumptions of adolescent egocentrism. *Developmental Review, 5*, 201–217.

Lewis, M., & Brooks-Gunn, J. (1979). *Social cognition and the acquisition of self*. Plenum Press: New York.

Mrzljak, L., Uylings, H. B., van Eden, C. G., & Judas, M., 1990. Neuronal development in human prefrontal cortex in prenatal and postnatal stages. *Progress in Brain Research, 85*, 185–222.

Ochsner, K., & Gross, J., (2005) The cognitive control of emotion, *Trends in Cognitive Neuroscience, 9*(5), 242–9.

Paus, T., Zijdenbos, A., Worsley, K., Collins, D. L., Blumenthal, J., Giedd, J. N. et al. (1999). Structural maturation of neural pathways in children and adolescents: In vivo study. *Science, 283*(5409), 1908–1911.

Paus, T., Collins, D. L., Evans, A. C., Leonard, G., Pike, B., & Zijdenbos, A. (2001). Maturation of white matter in the human brain: A review of magnetic resonance studies. *Brain Research Bulletin, 54*(3), 255–266.

Piaget, J. (1954). *The construction of reality in the child*. New York: Basic Books.

Piaget, J., & Inhelder, B. (1969). *The psychology of the child*. Basic Books: New York.

Pinker, S. (1994). *The language instinct*. New York: William Morrow & Co.

Selman, R. L. (1980). *The growth of interpersonal understanding: Developmental and clinical analyses*. New York: Academic Press.

Slater, P., McConnell, S., D'Souza, S. W., Barson, A. J., Simpson, M. D., & Gilchrist, A. C. (1992). Age related changes in binding to excitatory amino acid uptake site in temporal cortex of the human brain. *Brain Research: Developmental Brain Research, 65*, 157–160.

Sowell, E. R., Thompson, P. M., Holmes, C. J., Batth, R., Jernigan, T. L., & Toga, A. W. (1999). Localizing age-related changes in brain structure between childhood and adolescence using statistical parametric mapping. *NeuroImage, 9*(6 Pt. 1), 587–597.

Sowell, E. R., Peterson, B. S., Thompson, P. M., Welcome, S. E., Henkenius, A. L., & Toga, A. W. (2003). Mapping cortical change across the human life span. *Nature Neuroscience, 6*, 309–315.

Vartanian, L. R. (2000). Revisiting the imaginary audience and personal fable constructs of adolescent egocentrism: A conceptual review. *Adolescence, 35*, 639–662.

Vogt, B.A., Finch, D.M., & Olson, C.R. (1992). Functional heterogeneity in cingulate cortex: The anterior executive and posterior evaluative regions. *Cerebral Cortex, 2*, 435–443.

Waddington, C. H. (1975). *The evolution of an evolutionist*. Edinburgh: Edinburgh University Press.

11

PARADOXES IN ADOLESCENT RISK TAKING

Linda Van Leijenhorst
Eveline A. Crone
Leiden University

In this chapter we will review two influential hypotheses about risk taking in adolescence. On the one hand, the literature on adolescent development provides us with a considerable number of experiments that predict a decrease in risk-taking behavior with age as a consequence of increased cognitive control. On the other hand, many studies report an increase in risk taking in adolescence. We will give a short review of this literature and the literature on the neurobiological underpinnings of adolescent risk taking. In addition, we will propose a new working model that can account for these seemingly contradictory reports. This model describes risk taking under neutral circumstances (such as in the laboratory) in terms of increasing cognitive control abilities, but under emotional circumstances (such as when large rewards are at stake, or in the presence of peers) as a competition between emotion-related and control-related neural networks.

THE RISK-SEEKING ADOLESCENT

Adolescence is a remarkable period in life. In a relatively short period of time, roughly between 10 and 20 years of age, children undergo a fascinating transformation and turn into adults. At the beginning of the 20th century, G. S. Hall, who is often regarded as the founder of adolescent psychology, described adolescence as a period of storm and stress, a time characterized by conflict with parents, mood swings, and an increase in sensation seeking and risky behavior (Hall, 1904). This view of adolescents is still present these days; the stereotypical adolescent is portrayed as rebellious, impulsive, and risk taking.

Even though the development of risk-taking behavior has been studied from different perspectives, and using different methods (Boyer, 2006), there is no clear scientific support for the stereotype of adolescents being risk-takers. On the one hand, self-report studies provide evidence for this stereotypic view by showing that adolescents, compared to children and adults, take more risks in daily life. These measures show an increase in the number of traffic accidents, crime, the use of illegal drugs, tobacco and alcohol, and unsafe sex (Furby & Beyth-Marom, 1992; Steinberg, 2004). In addition, using self-report measures, sensation seeking has been shown to be the highest in adolescence (Arnett, 1996; Zuckerman, 1994). In contrast, the results of experimental studies are less clear; there is almost no evidence of an increase in risk-taking behavior in adolescence on laboratory tasks.

Studies using laboratory tasks have reported a general decrease in risk taking from childhood to adulthood, or no developmental differences in risk taking in adolescence at all. This pattern has been reported in at least three different risk-taking paradigms. First, the ability to weigh short-term against long-term benefits has been shown to improve throughout adolescence (Crone & Van der Molen, 2004; Hooper et al., 2004) in studies in which participants were asked to complete age-appropriate versions of the Iowa Gambling Task (IGT). The IGT is a widely used neuropsychological task that simulates real-life decision making in the way rewards, punishments, and future consequences of decisions need to be considered. Young children's behavior is demonstrated to be mostly driven by the magnitude of immediate rewards, while with age, participants learn to focus on the behavior that is most advantageous in the long run. Second, studies have reported that impulsivity decreases from childhood on (Eigsti et al., 2006; Mischel, Shoda, & Rodriguez, 1989). For example, with age, children become better at delaying gratification and these changes have been reported until puberty (Scheres et al., 2006). Third, Reyna and colleagues (Reyna & Ellis, 1994; Rivers, Reyna, & Mills, in press) report an age-related increase in intuitive decision making. According to their theory, children are less sensitive to framing effects, whereas adults are more likely to apply abstract principles in new situations and rely on gist-based processing. Further, many cognitive skills that underlie risk taking, such as the ability to judge probabilities, have been shown to develop well before puberty (Schlottmann, 2001; Van Leijenhorst, Westenberg, & Crone, 2008). Together, the behavioral literature to date shows that in straightforward risky situations children as young as 5 years of age can accurately estimate risks, but the ability to judge more complex risky situations increases gradually over the course of childhood and adolescence. These results are consistent with Piaget and Inhelder's (1974) conclusion that a full understanding of risks (including the notion that long-term outcomes may outweigh short-term benefits) continues to develop until the developmental stage of formal operations is reached around age 12. However, there is no behavioral evidence for a decline in decision-making performance in adolescence that could account for adolescent risky or irrational behavior. In addition, even though the consequences of adolescent risk taking can be grave (e.g., resulting in drug or alcohol addiction, traffic accidents, crime, or even death), most children get through adolescence relatively calmly (Arnett, 1999; Dahl, 2004; Masten et al., 1999). In sum, several findings support the stereotype of the risk-taking adolescent, whereas additional findings challenge this stereotype, suggesting that many questions about the precise nature of developmental changes that characterize adolescence are still to be answered.

A promising approach to explaining these seemingly contradictory findings is the study of brain maturation. In recent years the development of neuroimaging techniques such as magnetic resonance imaging (MRI) has transformed both our understanding of the neurological changes that occur during adolescence and the way in which we think about adolescent development. The benefit of using brain-based measures such as MRI, is that behavioral changes can be studied in terms of collaborating or competing brain networks, which underlie the outcome of a behavioral action. The results from a large-scale longitudinal study on the structural development of normal developing brains in which 145 healthy children and adolescents ranging in age from 4 to 22 years participated has revealed a more protracted development than was previously thought, with important changes taking place throughout adolescence (Giedd et al., 1999; Gogtay et al., 2004; Sowell et al., 2004). Even though the overall size of the brain of a 9-year-old is comparable to the size of an adult's brain, its structure differs. Recent MRI studies have shown that gray matter volume, or the total amount of neurons and connections between neurons, follows an inverted U-shaped developmental trajectory (Giedd et al., 1999). The number of neurons and connections increases from birth on and reaches a peak at the beginning of adolescence; from this point on the amount of gray matter will decrease. The adolescent brain begins to change and becomes more efficient, neurons and connections that are not necessary disappear, and important connections are strengthened. In contrast, white matter volume shows a linear increase which continues into adulthood (Giedd et al., 1999). The rate at which the brain matures differs between different brain regions, and regions in the prefrontal cortex (PFC) and parietal cortex are among the last regions in which gray matter volume reaches its peak (Casey, Tottenham, Liston, & Durston, 2005; Huttenlocher, 1979; Sowell et al., 2004). These regions are especially important for the control of emotions and behavior (Fuster, 2002), and the slow maturation of these brain regions parallels the slow maturation of the functions associated with it (Casey, Giedd, & Thomas, 2000).

In addition to these measures of structural brain maturation, the emergence of functional MRI (fMRI) has allowed us to examine the brain in action. Since this is a noninvasive technique that is safe to use in healthy children and adolescents, it enables us to examine the development of the neural substrates of risk taking behavior more directly. Differences in brain activation patterns between children and adolescents from different ages may provide insight into the seemingly conflicting behavioral findings. In this chapter we will discuss two important views on the development of adolescent risk-taking behavior, and we attempt to provide testable hypotheses and suggestions for future studies.

THE INCREASING COGNITIVE CONTROL HYPOTHESIS

One influential view in the literature on risk taking suggests that cognitive control abilities increase over the course of childhood and adolescence, and that risk taking decreases as a consequence. Cognitive control refers to the cognitive processes that enable us to control our behavior and perform goal-directed actions. Cognitive control processes such as working memory and inhibition have been shown to reach mature levels during adolescence (Davidson, Amso, Anderson, & Diamond, 2006; Diamond, 2002; Huizinga, Dolan, & Van der Molen, 2006). Adolescents' immature cognitive control abilities could be the reason for their impulsivity, risky behavior, and sometimes seemingly irrational decisions. In a cognitive psychology context, decision making is defined as choosing between competing courses of action, and this is seen as a fundamental component of human cognition. When decisions are associated with possible negative outcomes, they involve taking a risk. Decision making requires evaluating the possible outcome of actions and estimating the probability with which these occur. These abilities have been shown to change with development throughout adolescence. For example, the ability to reject an immediate reward in favor of a larger but delayed reward develops slowly. This ability begins to develop in preschool-aged children (Mischel et al., 1989); most of these children are unable to wait for a delayed reward. The ability to delay gratification at age 4 appears to be predictive of inhibition abilities in adolescence (Eigsti et al., 2006). Several studies have shown that the ability to resist the need for immediate reward improves throughout adolescence (Crone & Van der Molen, 2004; Hooper et al., 2004; Overman et al., 2004). A preference for immediate rewards may still be present in adults, but unlike children and adolescents, adults have the cognitive control skills that allow them to focus on the long-term consequences of their decisions and act rationally, most of the time.

Neuroscientific support for this theory comes from fMRI studies on the development of cognitive control abilities such as those that show that activation in areas related to these abilities still changes in adolescence. For example, prior studies have suggested increased activation in prefrontal and parietal cortices with age during a working memory task (Kwon, Reiss, & Menon, 2002) and during the resolution of response interference (Adleman et al., 2002). However, to date, there are very few studies that have examined the role of these regions in adolescent risk taking directly. Two recent fMRI studies that examined this topic found differences in brain activation patterns between children and adolescents. In a first study, 9- to 12-year-old children and young adults performed a simple probability estimation task (Van Leijenhorst, Crone, & Bunge, 2006). In this task, participants were asked to try to gain as many points as possible by making decisions in situations in which the probability of winning a reward was high (low-risk gambles) or low (high-risk gambles). Both age groups selected the choice with the highest probability of winning on most of the trials, although there was a trend of children to make more errors in the high-risk conditions. We examined brain activation patterns at the moment that participants made their decision and at the moment they saw the outcome of their choice. Overall, children and adults recruited similar brain regions when performing this task, but there were differences in the amount of activation between the age groups. At the moment of the decision, the anterior cingulate cortex (ACC) was more active for high-risk gambles than for low-risk gambles, and this difference was larger for 9–12 year olds than for adults. This suggests that decision making in children was associated with more response conflict (Carter, Botvinick, & Cohen, 1999). Activation in two other regions that have been linked to cognitive control and decision making, the dorsolateral PFC and the orbitofrontal cortex (OFC) were also more active during high-risk relative to low-risk choices, but these regions were not differentially activated for children and adults. When the outcome of gambles was presented, in children, relative to adults, the lateral OFC was more active for losses relative to wins. This difference was taken to suggest that children experienced losses as more aversive than adults.

A second study by Eshel, Nelson, Blair, Pine, & Ernst (2007) used a similar paradigm, but in this study both the probability with which rewards could be won, and the magnitude of the rewards was manipulated, making the task more difficult compared to the task used by Van Leijenhorst et al. In particular, this task allowed for a comparison of risky versus nonrisky choices, where a risky choice was defined as a gamble for a low-probability option, which could result in high reward, whereas a nonrisky choice was defined as a choice of a high-probability option that would result in a smaller reward. Adolescents (mean age: 13.3 years) and adults (mean age: 26.7 years) participated in this study. Adolescents and adults recruited similar brain regions in this study, but the difference in brain activation related to risky choices compared to nonrisky choices revealed more activation in areas related to cognitive control (ventrolateral PFC and ACC) in adults relative to adolescents. The authors argued that adolescents recruit prefrontal control areas to a lesser extent than adults do when they make risky decisions. Thus, when the task requires only probability estimation, 9- to 12-year-old children already perform at adult levels despite experiencing more response conflict for the high-risk judgments (Van Leijenhorst et al., 2006). In contrast, when the task also requires weighing of rewards, 9- to 17-year-old adolescents recruit fewer cognitive control areas than adults, suggesting that they allocate less cognitive control (Eshel et al., 2007).

One problem with this increase of cognitive control theory is that it would predict that children with the least mature reasoning skills and cognitive control abilities should show even more risk-taking behavior than adolescents. This seems to be in contrast to the self-report data which shows an increase in risk taking starting at the onset of adolescence. Therefore, the increasing cognitive control hypothesis can only account for the change in behavior that occurs with the transition from adolescence to adulthood, but cannot explain how risk taking can increase from childhood to adolescence. An alternative theory therefore suggests that cognitive control abilities compete with regions that are important for reward and arousal.

THE INCREASING EMOTION AND AROUSAL HYPOTHESIS

A second theory suggests that adolescence is characterized by an increase in the search for risk. According to this theory, with the emergence of puberty, adolescents begin to show risk-seeking behavior, and this need for thrills and adventure is thought to decrease again in adulthood. According to this theory, risk-taking behavior follows an inverted U-shaped developmental pattern with a peak in adolescence. Biological and physiological changes that start at the onset of puberty are thought to underlie this behavioral pattern. Adolescence can roughly be subdivided in two phases; the beginning of adolescence is marked by the onset of puberty around 10 years of age and lasts until about 15 years of age. While development during the whole period of adolescence (approximately between 10 and 20 years of age) is characterized by the maturation of psychological and psychosocial abilities (Steinberg, 2005), during puberty the physical transformation from child to adult occurs under the influence of hormones; children undergo growth spurts and the secondary sex characteristics develop (Spear, 2000).

A possible neurobiological explanation for the increased risk-seeking hypothesis argues that at the onset of puberty gonadal hormones influence the brain, especially neurotransmitter systems in limbic brain areas, leading to a higher sensitivity to appetitive stimuli (Nelson, Leibenluft, McClure, & Pine, 2005; Spear, 2000). In fact, recent imaging studies suggest that there are fundamental differences between adolescents and adults in the way rewards are experienced. A study on reward sensitivity that directly compared sensitivity to rewards in children, adolescents and adults by (Galvan et al., 2006) provides support for the hypothesis that adolescence is characterized by a peak in reward sensitivity. Participants ranging in age from 7 to 29 years participated in an fMRI experiment and performed a task in which they could win small, medium size, or large rewards by accurately responding to a cue. The nucleus accumbens, a region that is thought to play an important role in processing rewards in adults (Huettel, 2006; Knutson, Adams, Fong, & Hommer, 2001; McClure, Berns, & Montague, 2003), was more active in adolescents compared to children and adults. Interestingly, activation in PFC regions related to cognitive control showed an immature pattern in adolescents. OFC activity in adolescents was comparable to that in children, suggesting that reward-related regions in the brain are disproportionately active in this task relative to control systems that follow a more protracted developmental trajectory. The authors suggest that this imbalance between appetitive and control regions in the brain biases adolescents towards taking risks.

216 • Developmental Social Cognitive Neuroscience

Recent data from our own laboratory (Van Leijenhorst et al., 2009), in which we examined developmental changes in the neural correlates of the anticipation and processing of rewards in 10–12-, 14–15-, and 18–23-year-olds, supports the reward-sensitivity hypothesis. Participants were asked to activate slot machines that consecutively presented three pictures once they were activated. Participants only won money when these three pictures were the same. This design allowed for two comparisons: (a) anticipation of winning (two first pictures the same and still a chance of winning, versus two first pictures different and no chance of winning), and (b) outcome processing (three pictures the same, which shows winning versus a different last picture that shows not winning). The results demonstrate differences in the anticipation and processing of rewards between adolescents and young adults. Rewards and anticipation of rewards resulted in activation in limbic and arousal areas, including the nucleus accumbens and the insula, and elicited the most pronounced activation in the adolescent brain. In contrast, in adults we found control regions to be most active; the OFC was responsive to the omission of rewards in this age group, but not in adolescents (see Figure 11.1).

It should be noted that not all fMRI studies support the overactive emotion network hypothesis. For example, Bjork et al. (2004) reported

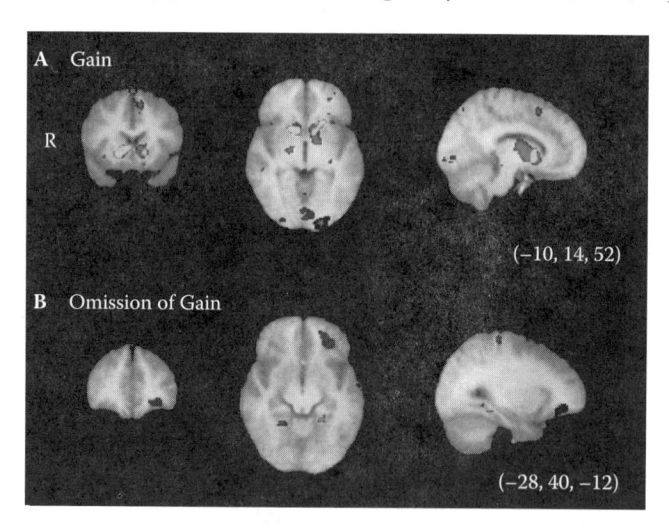

Figure 11.1 (See color insert following p. 242) Brain activation patterns related to the processing of (a) rewards, and (b) the omission of expected rewards. Activation for 10- to 12-year-olds in green, 14- to 15-year-olds in blue, and 18- to 25-year-olds in red. Winning was associated with activation of the bilateral striatum in 10- to 12-year-olds and 14- to 15-year-olds, whereas processing of not winning was associated with left lateral OFC in 18- to 25-year-olds.

that adolescents show underactivation in the nucleus accumbens when anticipating rewards, but an adult response when receiving rewards. It is currently unclear how these results can be integrated with other studies on risk taking in adolescents. One possibility is that differences in the behavioral requirements between the tasks used in these studies account for differences in the observed patterns of brain activation. In all studies reported above, it is possible that adolescents' risk-taking behavior is influenced by differences in the strategies used by participants from different ages when they approach a risky situation. For example, the task used by Bjork and colleagues was more difficult and therefore may have resulted in a failure to anticipate outcomes in adolescents because they simply gave up and waited until the final outcomes. In contrast, the tasks used by Galvan et al. and Van Leijenhorst et al. were very simple, and therefore it may be that, especially in these tasks, adolescents, driven by their active appetitive circuitry, focused on the possible rewards whereas adults, driven by a more active control circuitry, focused on the possible negative consequences.

A COMBINED PERSPECTIVE: THE COGNITION–EMOTION BALANCE HYPOTHESIS

The two theories presented above provide different predictions with respect to the development of risk taking. How can these perspectives be integrated to account for the differential findings in the literature? Here we present the cognition–emotion balance hypothesis, which is based on a brain-behavior perspective and which may provide a framework for understanding the differential developmental trajectories.

Neuroscientific studies have proved useful in showing that most of our decision making is based on a competition between two brain systems,* one that is particularly sensitive to reward and arousal, and a second system which is sensitive to long-term goals (Ernst et al., 2005; Nelson et al., 2005). The first, evolutionarily older system, builds on subcortical structures that have been linked to the processing of emotion, such as the amygdala and the nucleus accumbens, whereas a second

---

* Even though here we describe the emotion- and cognition-related neural networks as competing, it is also often argued that these networks work together in order to guide decision making. For example, the somatic marker hypothesis argues that emotions are necessary for decision making (Damasio, 1994) and more recent theories suggest that developmental changes in risk taking are associated with an increased integration of cognitive and emotional systems (see Casey, Getz, & Galvan, 2008; Rivers, Reyna, & Mills, 2008; Steinberg, 2008). We agree with these theories; therefore, the current ideas can also be seen as a better integration of cognition and emotion systems.

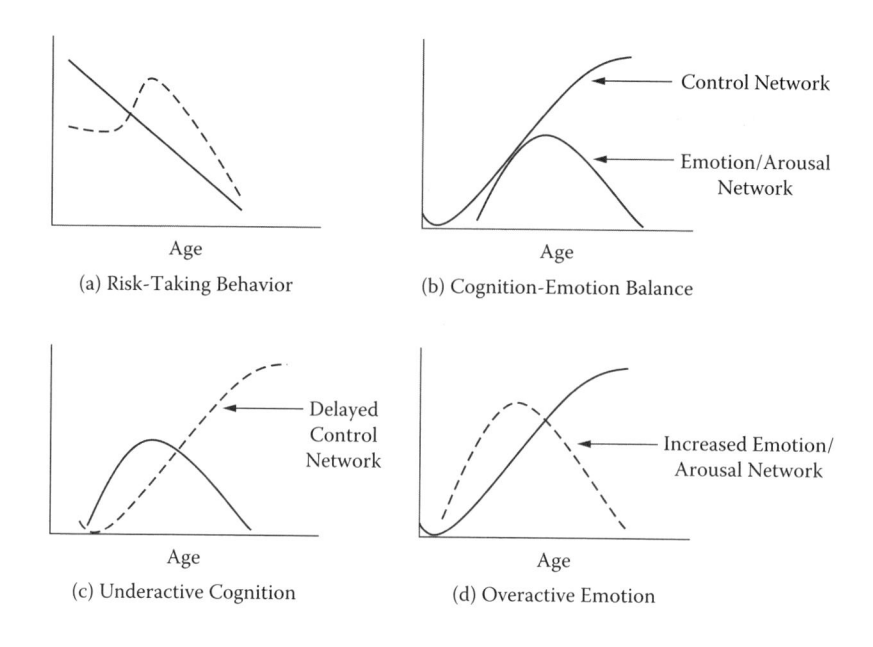

218 • Developmental Social Cognitive Neuroscience

Figure 11.2 (a) Showing the predicted behavioral pattern for both hypotheses (solid line = linearly decreasing risk taking, dotted line = increasing risk taking in adolescence), (b) showing pattern of brain maturation of control- and emotion-related areas under neutral conditions (c) showing pattern of brain maturation of control- and emotion-related areas under delayed cognitive control conditions, Figure (d) showing pattern of brain maturation of control- and emotion-related areas under increased emotion/arousal conditions.

evolutionary younger system builds on cortical brain areas, including the PFC and the parietal cortex (Adolphs, 2003). The cognition–emotion balance hypothesis predicts that both systems contribute to decision making, but that the outcome is dependent on the relative strength of each system in a given situation. As can be seen in Figure 11.2a, behavioral changes in risk taking across development are sometimes described in terms of a linear decrease (Crone & Van der Molen, 2004; Reyna & Ellis, 1994) and sometimes in terms of an increase that peaks in adolescence (Arnett, 1992; Steinberg, 2004). Figure 11.2b shows the pattern of behavioral changes with age in a neutral situation (such as often seen in laboratory tasks), and predicts a linear decrease in risk taking, concurrent with increased cognitive control. Figure 11.2c,d show two situations in which the emotion-arousal network becomes overactive relative to the control network, which results in a peak in risk taking in adolescence (such as often seen in real-life situations; see Figure 11.3 for a schematic representation of the emotion–cognition balance and implicated brain regions). This difference between the situation in a

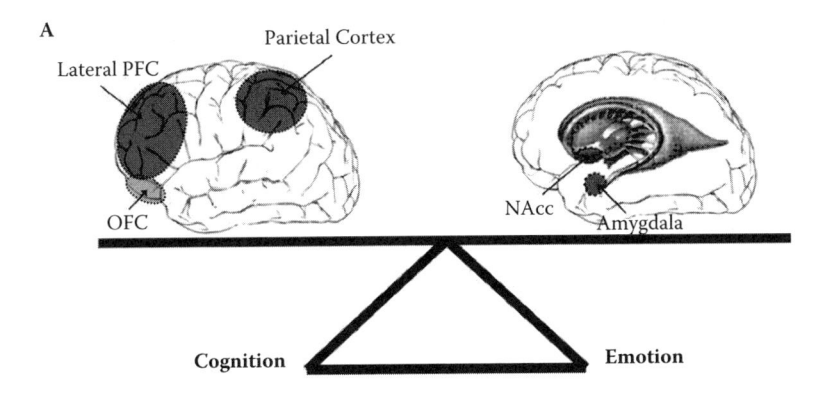

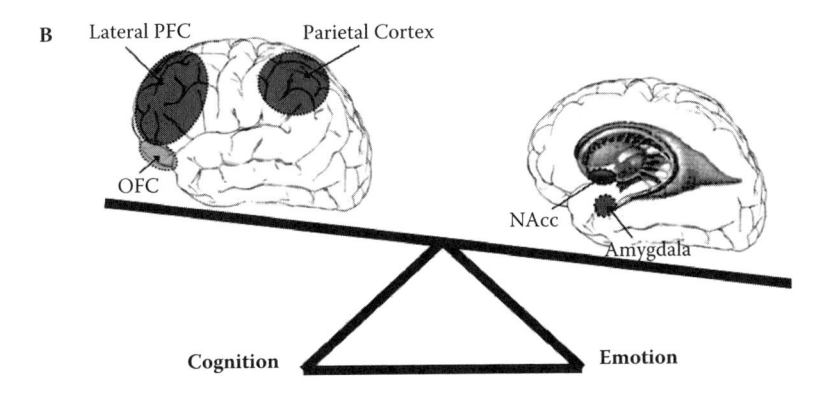

Figure 11.3 (a) A schematic representation of the emotion-cognition balance and associated brain regions (ACC and insula are not shown), in a neutral situation (a), such as in a laboratory, and in a situation with a relatively overactive emotion system (b), such as in real life.

laboratory and in real life could contribute to the discrepancy between self-report and observational, relative to laboratory data on adolescent risk taking. One recent study supports this suggestion.

During adolescence friendships change and peers become more and more important. For example, during adolescence as compared to childhood, more and more time is spent in the presence of peers than in the presence of parents. It has been suggested that the opinions of peers become more important as well (Harris, 1995). In an experimental study on the influence of peers on adolescent risk-taking, adolescent, (13–16 years), young adult (18–22 years), and adult (24 years), participants played a risk-taking game in the presence of peers or while they were alone (Gardner & Steinberg, 2005). This study showed a disproportionate increase in the number of risky decisions in the presence of peers in adolescents. It could be that the increased significance of peers in adolescence influenced the emotion and arousal brain network to such an extent that it led to adolescent risk taking. The need to fit in might outweigh the risks involved, and a situation that would seem risky to adults could be perceived differently by adolescents (e.g., as fun) because of the presence of their peers.

CONCLUSIONS AND FUTURE DIRECTIONS

This chapter described two existing possible hypotheses for adolescent changes in risk taking: increased cognitive control versus increased emotion and arousal seeking. To date, the neurobiological evidence for these hypotheses is not conclusive. Our literature review on the neurobiological underpinnings provides support for both accounts, and we suggest that cognitive and emotional development should be studied in combination to explain seemingly inconsistent patterns in the behavioral literature. Therefore, we have proposed a new working model, the cognition–emotion balance hypothesis, which describes risk seeking under neutral circumstances (such as in the laboratory) in terms of increasing cognitive control abilities, but under emotional circumstances (such as when large rewards are at stake, or in the presence of peers) as a competition between emotion-related and control-related neural networks (Figure 11.3).

The literature on adolescent risk-taking behavior provides important building blocks for theory development, but also lacks in specificity. We propose three specific suggestions that should be taken into account in future studies: (a) Adolescence is a unique developmental period that can be characterized by different types of developmental stages. For example, from a pubertal perspective, one can refer to teens who are prepubertal, pubertal, and postpubertal. From a cognitive and social perspective, one can refer to teens who are in early, middle, or late adolescence. These stages should be studied separately in order to enable the dissociation between the effects of hormones and brain maturation. (b) Some risk-taking behavior seems characteristic for adolescents but other types of risk-taking behavior can be maladaptive and continue in adulthood. Many types of risky behavior have been studied ranging from controlled experimental risks on laboratory tasks to real-life risks. Real-life risks can be problematic, such as smoking cigarettes, drinking alcohol, gambling, or displaying aggression, or they can be more accepted, such as listening to loud music and engaging in extreme sports. Future studies should dissociate normal developmental changes in risk taking from personality-related differences in sensation seeking. (c) Individual differences in the rate of maturation are particularly large in adolescence, and for this reason it will be important in future studies to include maturational indices in addition to age, such as detailed measures of pubertal development, and the level of psychosocial maturation (Cohn & Westenberg, 2004; Westenberg, Hauser, & Cohn, 2004), and to take gender differences into account.

We suggest that developmental models of risk taking can be articulated further on the basis of a brain maturation model, and by differentiation of anatomically segregated regions that contribute to risk taking. One of the major virtues of decomposing brain functions is that this method allows assessment of each function in terms of its developmental time course and psychophysiological manifestation. The model described in this chapter illustrates this approach. Previous studies examining the development of risk taking have used different tasks and methods, and the integration of these methods (including laboratory and real-life assessments, and cognitive, emotional, and social task manipulations) is necessary for a full understanding of this poorly understood developmental period. The task for the future is to gain converging evidence from behavioral, physiological, and brain-imaging studies that may provide a working model for functional and regional specification, and that may account for developmental changes and individual differences in risk taking.

REFERENCES

Adleman, N. E., Menon, V., Blasey, C. M., White, C. D., Warsofsky, I. S., Glover, G. H. et al. (2002). A developmental fMRI study of the Stroop color-word task. *NeuroImage, 16*, 61–75.

Adolphs, R. (2003). Cognitive neuroscience of human social behaviour. *Nature Reviews Neuroscience, 4*(3), 165-178.

Arnett, J. (1992). Reckless behavior in adolescence: A developmental perspective. *Developmental Review, 12*, 391–409.

Arnett, J. J. (1996). Sensation seeking, aggressiveness, and adolescent reckless behavior. *Personality and Individual Differences, 20*(6), 693–702.

Arnett, J. J. (1999). Adolescent storm and stress, reconsidered. *American Psychologist, 54*(5), 317–326.

Bjork, J. M., Knutson, B., Fong, G. W., Caggiano, D. M., Bennett, S. M., & Hommer, D. W. (2004). Incentive-elicited brain activation in adolescents: Similarities and differences from young adults. *Journal of Neuroscience, 24*, 1793–1802.

Boyer, T. W. (2006). The development of risk taking: A multi-perspective review. *Developmental Review, 26*(3), 291–345.

Carter, C. S., Botvinick, M. M., & Cohen, J. D. (1999). The contribution of the anterior cingulate cortex to executive processes in cognition. *Review of Neuroscience, 10*(1), 49–57.

Casey, B. J., Getz, S., & Galvan, A. (2008). The adolescent brain. *Developmental Review, 28*, 62–77.

Casey, B. J., Giedd, J. N., & Thomas, K. M. (2000). Structural and functional brain development and its relation to cognitive development. *Biological Psychology, 54*(1–3), 241–257.

Casey, B. J., Tottenham, N., Liston, C., & Durston, S. (2005). Imaging the developing brain: What have we learned about cognitive development? *Trends in Cognitive Science, 9*(3), 104–110.

Cohn, L. D., & Westenberg, P. M. (2004). Intelligence and maturity: Meta-analytic evidence for the incremental and discriminant validity of Loevinger's measure of ego development. *Journal of Personality and Social Psychology, 86*(5).

Crone, E. A., & Van der Molen, M. W. (2004). Developmental changes in real-life decision making: Performance on a gambling task previously shown to depend on the ventromedial prefrontal cortex. *Developmental Neuropsychology, 25*(3), 251–279.

Dahl, R. E. (2004). Adolescent brain development: A period of vulnerabilities and opportunities. Keynote address. *Annals of the New York Academy of Sciences, 1021*(1), 1–22.

Damasio, A. R. (1994). *Descartes' error*. New York: Grosset/Putnam.

Davidson, M. C., Amso, D., Anderson, L. C., & Diamond, A. (2006). Development of cognitive control and executive functions from 4 to 13 years: Evidence from manipulations of memory, inhibition, and task switching. *Neuropsychologia, 44*(11), 2037–2078.

Diamond, A. (2002). Normal development of prefrontal cortex from birth to young adulthood: Cognitive functions, anatomy and biochemistry. In D. T. Stuss & R. T. Knight (Eds.), *Principles of frontal lobe function* (pp. 466–503). London: Oxford University Press.

Eigsti, I., Zayas, V., Mischel, W., Shoda, Y., Ayduk, O., Dadlani, M. B. et al. (2006). Predicting cognitive control from preschool to late adolescence and young adulthood. *Psychological Science, 17*(6), 478–484.

Ernst, M., Nelson, E. E., Jazbec, S., McClure, E. B., Monk, C. S., Leibenluft, E. et al. (2005). Amygdala and nucleus accumbens in responses to receipt and omission of gains in adults and adolescents. *NeuroImage, 25*(4), 1279–1291.

Eshel, N., Nelson, E. E., Blair, J. R., Pine, D. S., & Ernst, M. (2007). Neural substrates of choice selection in adults and adolescents: Development of the ventrolateral prefrontal and anterior cingulate cortices. *Neuropsychologia, 45*, 1270–1279.

Furby, L., & Beyth-Marom, R. (1992). Risk taking in adolescence: A decision-making perspective. *Developmental Review, 12*, 1–44.

Fuster, J. M. (2002). Frontal lobe and cognitive development. *Journal of Neurocytology, 31*, 373–385.

Galvan, A., Hare, T. A., Parra, C. E., Penn, J., Voss, H., Glover, G. et al. (2006). Earlier development of the accumbens relative to orbitofrontal cortex might underlie risk taking behavior in adolescents. *Journal of Neuroscience, 26*(25), 6885–6892.

Gardner, M., & Steinberg, L. (2005). Peer influence on risk taking, risk preference, and risky decision making in adolescence and adulthood: An experimental study. *Developmental Psychology, 41*(4), 625–635.

Giedd, J. N., Blumenthal, J., Jeffries, N. O., Castellanos, F. X., Liu, H., Zijdenbos, A. et al. (1999). Brain development during childhood and adolescence: A longitudinal MRI study. *Nature Neuroscience, 2*, 861–863.

Gogtay, N., Giedd, J. N., Lusk, L., Hayashi, K. M., Greenstein, D., Vaituzis, A. C. et al. (2004). Dynamic mapping of human cortical development during childhood through early adulthood. *Proceedings of the National Academy of Science (U.S.A), 101*(21), 8174–8179.

Hall, G. S. (1904). *Adolescence: Its psychology and its relations to physiology, anthropology, sociology, sex, crime, religion, and education (Vols. I & II).* New York: D. Appleton & Co.

Harris, J. R. (1995). Where is the child's environment? A groups socialization theory of development. *Psychological Review, 102*(3), 458–489.

Hooper, C. J., Luciana, M., Conklin, H. M., & Yarger, R. S. (2004). Adolescents' performance on the Iowa Gambling Task: Implications for the development of decision making and ventromedial prefrontal cortex. *Developmental Psychology, 40*(6), 1148–1158.

Huettel, S. A. (2006). Behavioral, but not reward, risk modulates activation of prefrontal, parietal and insular cortices. *Cognitive, Affective, & Behavioral Neuroscience, 6*, 141–151.

Huizinga, M., Dolan, C. V., & Van der Molen, M. W. (2006). Age-related change in executive function: Developmental trends and a latent variable analysis. *Neuropsychologia, 44*(11), 2017–2036.

Huttenlocher, P. R. (1979). Synaptic density in human frontal cortex—developmental changes and effects of aging. *Brain Research, 163*, 195–205.

Knutson, B., Adams, C. M., Fong, G. W., & Hommer, D. (2001). Anticipation of increasing monetary reward selectively recruits nucleus accumbens. *Journal of Neuroscience, 21*, 1–5.

Kwon, H., Reiss, A. L., & Menon, V. (2002). Neural basis of protracted developmental changes in visuo-spatial working memory. *PNAS, 99*(20), 13336–13341.

Masten, A. S., Hubbard, J. J., Gest, S. D., Tellegen, A., Garmezy, N., & Ramirez, M. (1999). Competence in the context of adversity: Pathways to resilience and maladaptation from childhood to late adolescence. *Development and Psychopathology, 11*, 143–169.

McClure, S. M., Berns, G. S., & Montague, P. R. (2003). Temporal prediction errors in a passive learning task activate human striatum. *Neuron, 38*, 339–346.

Mischel, W., Shoda, Y., & Rodriguez, M. (1989). Delay of gratification in children. *Science, 244*, 933–938.

Nelson, E. E., Leibenluft, E., McClure, E. B., & Pine, D. S. (2005). The social re-orientation of adolescence: A neuroscience perspective on the process and its relation to psychopathology. *Psychological Medicine, 35*, 163–174.

Overman, W. H., Frassrand, K., Ansel, S., Trawalter, S., Bies, B., & Redmond, A. (2004). Performance on the Iowa card task by adolescents and adults. *Neuropsychologia, 42*(13), 1838–1851.

Reyna, V. F., & Ellis, S. C. (1994). Fuzzy-trace theory and framing effects in children's risky decision making. *Psychological Science, 5*, 275–279.

Rivers, S. E., Reyna, V. F., & Mills, B. (2008). Risk taking under the influence: A fuzzy-trace theory of emotion in adolescence. *Developmental Review, 28*, 107–144.

Rivers, S. E., Reyna, V. F., & Mills, B. (in press). Risk taking under the influence: A fuzzy-trace theory of emotion in adolescence. *Developmental Review*.

Scheres, A., Dijkstra, M., Ainslie, E., Balkan, J., Reynolds, B., Sonuga-Barke, E. et al. (2006). Temporal and probabilistic discounting of rewards in children and adolescents: Effects of age and symptoms. *Neuropsychologia, 44*, 2092–2103.

Schlottmann, A. (2001). Children's probability intuitions: Understanding the expected value of complex gambles. *Child Development, 72*(1), 103–122.

Sowell, E. R., Thompson, P. M., Leonard, C. M., Welcome, S. E., Kan, E., & Toga, A. W. (2004). Longitudinal mapping of cortical thickness and brain growth in normal children. *Journal of Neuroscience, 24*(38), 8223–8231.

Spear, L. P. (2000). The adolescent brain and age-related behavioral manifestations. *Neuroscience and Biobehavioral Reviews, 24*, 417– 463.

Steinberg, L. (2004). Risk taking in adolescence: What changes and why? *Annals of the New York Academy of Sciences, 1021*, 51–58.

Steinberg, L. (2005). Cognitive and affective development in adolescence. *Trends in Cognitive Sciences, 9*, 69–74.

Steinberg, L. (2008). A social neuroscience perspective on adolescent risk taking. *Developmental Review, 28*, 78–106.

Van Leijenhorst, L., Crone, E. A., & Bunge, S. A. (2006). Neural correlates of developmental differences in risk estimation and feedback processing. *Neuropsychologia, 44*(11), 2158–2170.

Van Leijenhorst, L., Westenberg, P. M., & Crone, E. A. (2008). A developmental study of risky decisions on the cake gambling task: Age and gender analyses of probability estimation and reward evaluation. *Developmental Neuropsychology, 33*, 179–196.

Van Leijenhorst, L., Zanolie, K., Van Meel, C. S., Westenberg, P. M., Rombouts, S. A. R. B., & Crone, E. A. (2009). What motivates the adolescent? Brain regions explaining reward sensitivity across adolescence. *Cerebral Cortex*.

Westenberg, P. M., Hauser, S. T., & Cohn, L. D. (2004). Sentence completion measurement of personality development. In M. J. Hilsenroth & D. Segal (Eds.), *Objective and projective assessment of personality and psychopathology: Vol. 2. Comprehensive handbook of psychological assessment* (pp. 595–616). New York: Wiley.

Zuckerman, M. (1994). *Behavioral expressions and biosocial bases of sensation seeking*. New York: Cambridge University Press.

12

BETWEEN NEURONS AND NEIGHBORHOODS
Innovative Methods to Assess the Development and Depth of Adolescent Social Awareness

Robert L. Selman
Luba Falk Feigenberg
Harvard University

Prologue: An angry girl in the school and neighborhood speaks her mind—

"When someone talks about you behind your back, or turns your friends against you, or says there is some boy you like but you don't, you gotta fight back, protect yourself, you know, let them have it. There's nothing else to do that would work. 'Cause, you know, girls have always said things that get under other girls' skin ... And, then, we're black in a racist society, so what choice do we really have? Put those together out there and that's why this is never going to end."

SOCIAL RELATIONS AND SOCIAL COGNITIVE NEUROSCIENCE IN ADOLESCENCE
A Developmental Psychology of Mind, Meaning, and Matter

In a weekly anger management group at a New England urban public high school, Danielle, a 16-year-old African-American student with good grades in school, speaks her mind about the best way to defend against perceived incidents of social exclusion whether at school, in her neighborhood, or in her culture: to attack verbally, certainly, but physically as well, if necessary. She locates the origins and sources of the problems she must deal with every day well outside of herself, both with respect to time and space. In time, she locates the source of her actions mainly in the (evolutionary) influences that the female sex naturally selected over time to survive ("You know girls have always …"). In space, she locates the causes of her behavior in the "out there," in what it means to be African American ("We're black") as part of the larger culture in the United States today. It is surprising, given the sophistication of her analysis, that she does not articulate any sense of her own agency, her own "executive functioning" as a factor that could account, at least partly, for her day-to-day reactions to perceived social disrespect ("What choice do we really have?") While context plays a part in her losing her temper, and it may very well be the nature of girls to focus on relational aggression, does she really think that culture and evolution fully account for her violent reactions?

It would be equally surprising, however, if Danielle located the problem deeper under her skin, say somewhere in her own lateral frontal cortex. What would we make of her saying, "You know how adolescents are. We get this pubertal passion thing coming on younger and younger, but none of the cognitive functions to control it. And then, our prefrontal cortex has not yet developed, so we make some bad decisions. And you know, violence-prone neighborhoods can impede the development of brain structures and functions, especially when you're poor. So it's not my fault. Put the blame on these neurons in my brain and the effects that this neighborhood I live in has on them and it's never gonna end."

While the empirical research may be heading in such a direction, we are not yet at the point where we can describe with certainty how the brain's developmental changes manifest as behavior, let alone try to describe the intersection of neuroscience and societal forces.

Danielle's case exemplifies the promise, and some of the challenges, of translating the recent scientific evidence on brain-behavior connections to the world of practice. For instance, at a recently scheduled case conference, the Student Support Team in Danielle's urban high school referred her, along with five other female students, to participate in a school-based counseling group to "treat her anger problems" as well as to prevent, or at least limit, her aggressive and violent behavior with classmates on school grounds. The school staff located this problem "within Danielle," or at least as potentially under her own control, and hoped that the anger management discussion group, designed to help the girls control their hair-trigger tempers, could be of some support.

Even as scientific knowledge about adolescent development through social cognitive neuroscience is burgeoning, there is not much of a way for prevention practitioners working with youth in schools to connect to either knowledge base. For instance, what can practitioners do with evidence that suggests, in general, adolescent "reward anticipation" emerges before the maturity of their "outcome processing," as noted by Casey, Getz, & Galvan (2008), or that "response inhibition" is weaker in adolescents than in adults (Stevens, Kiehl, Pearlson & Calhoun, 2007)? Or, on the other hand, how do they factor in cautions to inhibit comments such as "the prefrontal cortex develops last" (Bunge, 2007), because the structural maturation of different components of the developing brain may not automatically include functional maturation? The translation of such research into practice is contingent upon the solidification of the literature.

Nevertheless, news of advances in social cognitive neuroscience brings hope to the extent that evidence strongly suggests brain development is an important correlate, if not a causal factor, in adolescent social behavior. Policies about how adolescents are expected to navigate their physical and biological worlds, from when to drive to when to drink (Steinberg, 2003), are being informed by this body of research. Perhaps new ways to support students like Danielle to negotiate their social world may also be found in the bridge between neuroscience and social relationships. A piece of that bridge, we believe, lies in the building of better methods to assess adolescents' social cognitions, in particular their awareness of social relationships.

The Role of the Developing Social Mind in the Mediation of Brain and Behavior

In June 2007, as social developmental psychology "tourists" at the Jean Piaget Society Developmental Social Cognitive Neuroscience Conference, we were introduced to the evolving theory and accumulating evidence in this field. We already knew these discoveries had been facilitated greatly by technical innovations in the measurement of brain development and the correlates of social cognitions in related regional brain functioning. We learned that while the vocabulary in this unfamiliar territory was new to us, some of the issues were not so foreign. Brain development, like social cognitive development, seems to be framed by the juxtaposition of structural and functional analyses (terms). We also learned that the development of structural capacity (competence) may not translate into functional implementation (performance) in either discipline.

Most striking of all, however, was a notion not alien to us, but something we have long concerned ourselves with in our own research: the way the social cognitive aspects of adolescents' social behavior were being assessed. Here, we thought, there was some room for making a contribution to the integration of the fields. The experience we had at the conference became a warrant for considering how the measures we have spent a good deal of time constructing and validating in our study of adolescent social awareness might be used in partnership with the methods of social cognitive neuroscience.

Method and Inquiry into Adolescent Social Awareness

At this point, our research focuses on how adolescents make social choices, not just in the snap judgment of a particular moment, but also over time. Time, then, means both as adolescents mature, in the chronological sense of the term development, and also as they are better able to reflect more deliberately and critically, and give more time to thinking about their social actions and the actions of those around them. "Better," in our research, references both an analysis of the conceptual complexity of their social thought and the quality of their interpretations of what they think good social choices are likely to be in their context. "Context," in our research, usually means in their neighborhoods and schools, but can also mean in their culture.

The focus on context means our laboratory includes the social and civic space of the schools as well as the lab space at the university, and our research is designed to interrogate what it means to get better at dealing with the meaning of common, yet especially challenging, difficult, or uncomfortable social experiences in school, both from the perspective of those who are observers of adolescent social relationships and of the adolescents themselves. We are interested in understanding how adolescents make sense of the harassment and exclusion, drug use and abuse, and favoritism they may experience, as well as social issues such as ostracism, sexism, or racism they may perceive in society. From a disciplinary perspective, our interest in the personal meaning of social relationships requires that our primary research home base from which we explore adolescents' social choices is their minds.

Because culture and context play such a large role in the psychological processes we explore from this home base, we have yet to focus our attention on the links between the ways adolescents make meaning and develop interpretations of their social world in either brain development or architecture. Yet we think we can connect some of our methods to important issues in brain research if we focus on measurement, the ways adolescents make and justify choices when they face difficult, complex, ongoing, and evolving everyday social issues.* Just as functional magnetic resonance imaging (fMRI) techniques have demonstrated to the field of social cognitive neuroscience that measurement is both technical and possibly transformative, the same can be true of measures of the social mind. If these methods can achieve unprecedented insight into adolescents' social choices, they can point us in the direction we need to take to make theoretical advances in understanding the connection of social behavior and the brain.

Danielle's Choices: Selected Challenges for the Field

There are two methodological problems in the field of social cognitive development we regularly contend with. One is empirical, concerning the lack of confidence in the reliability and validity of self-reports adolescents give about social relationships or of their responses to hypothetical social dilemmas. Do people do what they say they will do? The other is theoretical, concerning the inherently interpretative nature of understanding social relationships. Do people say what they mean or mean what they say, and how do we understand each other when so much of social interaction is open to subjective interpretation? The study of social cognitive development relies upon an objective view of reality that may be elusive.

Theoretically and empirically speaking, what do we make, for instance, of Danielle's justification for how she deals with other students when she perceives they are talking about her behind her back? It is puzzling that, in the same breath, she heatedly claims that attack is an inevitable response and that escalation is a necessary form of retributive justice and self-protection, and yet coolly demonstrates an understanding of the complicated relationship between her own behavior and the way she (and other members of the peer group she identifies with) perceives her own (and their) lived social experience. It seems difficult to determine which perspective is more valid to Danielle, and impossible to know how to reconcile the two.

Do we think Danielle's verbal analysis of "what works" in the social context she inhabits is an accurate portrait of the way she actually deals with social relationships around her? To press further, do we think Danielle's verbal self-report and social analysis will predict her own actions in those future "next times, next moments," when she experiences the humiliation that feeling excluded or disrespected by others engenders for her? Is it enough to know that girls like Danielle say they actually prefer to choose fighting as their social problem-solving

---

* That is, rather than the immediate, one-shot, or rare ones.

strategy to understand or be able to predict their actual actions? If not, what else might we need to consider? All of these questions hinge upon the notion that self-awareness might be equated with self-prediction, an exploration that is the domain of theory and not practice.

Of course, most social scientists, along with their biologically oriented colleagues, do not believe that Danielle's espoused social choices—whether or not she would actually fight, either physically or verbally, in future situations like the one she described—are influenced solely by her ability to understand and interpret the social relationships around her. Sociologists point to structural issues in society, such as racism and poverty, as the unseen engines that drive human values and conduct. Anthropologists, much the way Danielle does, point to the importance of shared meanings, norms, and values within the "identity group" as a lens through which to understand normative social actions.

On the other hand, psychologists—at least up until recently the primary caretakers of human intentionality and agency—do not expect Danielle, and girls like her or those in similar situations, to lack personal agency totally when it comes to the social choices they make. Yet even psychologists disagree. In particular, developmental psychologists who focus on the social-cognitive foundations of social action will attempt to locate Danielle's choice as resting on some theoretically defined developmental level of social competence (Selman & Dray, 2006). Social psychologists, conversely, are deeply suspicious or skeptical of either words or reasons as predictors of actions, considering them to be only pale rationalizations of the actions humans, especially adolescents, take (Haidt, 2001). They are more likely to emphasize the power of position or the position of power (Haste, 1999). Our perspective is that there is validity to both sides of the psychological arguments.

Can Social Cognitive Neuroscience Be Helped by In-Depth Analysis of Social Awareness?

We revise our opinion [of their cognitive maturity] when we turn from the examination of the adolescent's intellectual processes themselves to consider how they fit into the general picture of his life. We are surprised to discover that this fine intellectual performance makes little or no difference to his actual behavior. His empathy into the mental process of other people does not prevent him from displaying the most outrageous lack of consideration toward those nearest to him.

—**Anna Freud (1966, p. 160)**

The Equivalence of Social Thought and Action

Anna Freud knew adolescence as a period of rapid—but not necessarily synchronous—cognitive and emotional development, and now, neurological development (Dahl, 2007; Steinberg, 2003). With admiration for the impact of fMRI technology on the field of social cognitive neuroscience, we wonder if we can develop a similar, or parallel, technological advance, some kind of "mental MRI" that can probe the minds of adolescents as they ponder the meaning of their own social relationships. To be practical, a measure of this kind would need to be cost-effective, able to quantify the qualities of social awareness, and connect to the social-cognitive neuroscience of the developing adolescent brain in a way current social cognitive measures do not.

We start with the assumption that what adolescents think about their espoused social choices, both those that they think they have and those they say they will make, are important pieces of scientific evidence, even if they do not have a one-to-one correspondence with how or why adolescents act at any given moment. Rather than invalidating what adolescents say, as Anna Freud suggests, it is scientifically important and philosophically interesting to know if and when espoused social choices do not align with actions as well as when they do (Thagard, 2000)! While many social scientists tend to discard data about what people say they would do in a difficult social situation because they do not think it will predict, or say much about, what they actually do, others feel forced to use this approach for want of access to other methods (Baumeister, Vohs, & Funder, 2007). Social philosophers, however, enjoy pointing out to those who are skeptical that how people act under difficult social circumstances is not very illuminating unless we know why, including their own reasons (Habermas, 1986–1989). It seems discouraging to consider the future of a field in which solid advances may be reliant upon a level of self-awareness that is perhaps possible only in the realm of literature.

Current State of Social Cognitive Neuroscience

To learn more about how emotions and behaviors are connected to different areas in the brain, one common neuroscientific approach involves using fMRI to look at brain functioning while study participants are presented with various types of social scenarios and are asked to respond to, interpret, or to suggest choices of dealing with them. This methodology allows researchers to see which areas of the brain "light up" by detecting changes in blood flow, thereby gaining a better understanding of the location of neurological reactions to different types of social or interpretive experiences in normative (i.e., nonclinical) participants. The images that are generated reflect that brain structures are activated during the performance of different social tasks, and provide information about which brain structures and processes are associated with affect, perception, thought, and action.

A recent study by Eisenberger, Lieberman, and Williams (2003) provides a good example of research on the interface of brain imaging with the analysis of the way social relationships are experienced that foregrounds the neuroscientific aspects, and relies on behavioral science for measures of psychological correlates. In this study, adult participants' brain activity was recorded while they were exposed to hypothetical scenarios depicting incidents of social exclusion. One of these involved the narration of a situation of intentional exclusion (individuals were actively prevented by a group of peers from participating in a social activity), while the other involved a narration of circumstantial exclusion (participants were not able to join a group in a social activity because of external circumstances beyond their control). Participants then completed surveys about their feelings while hearing about each situation. Under both scenarios, fMRI data showed activity in the same area of the brain that regulates physical pain. In other words, not only is it painful to be excluded, but that the experience of social exclusion may be neurologically similar to that when feeling physical pain. Moreover, Eisenberger et al. also found that the intensity of participants' brain activity was correlated with their self-reported levels of distress as they considered the incident.

While not specifically focused on brain functioning during the period of adolescence, this study, and those like it that are beginning to focus in on this age range (Baird, 2007; Beckman, 2004), nevertheless raises several important issues for discussion with researchers who focus on adolescence in cultural and contextual terms. First, we see it as relevant that even the virtual, or hypothetical, experience of social exclusion generates genuine feelings of pain. Even though participants in Eisenberger et al.'s study knew that the situations they were assigned to experience were artificially induced, their brains still demonstrated activity in the regions that have been found to control pain and personal distress. We believe this validates to some limited extent the use of hypothetical situations to study the way individuals react and respond to social situations.

More specifically, these findings begin to address the criticism that the use of self-report surveys in psychology are not valid measures of what people would actually do or how they feel when actually faced with them (Kagan, 2007), or the concern that self-report measures are being overly, or expediently, used for these purposes (Baumeister et al., 2007). For example, Kagan (2007) argues that findings from surveys and questionnaires about personal conduct, are largely "inaccurate" because people make personalized connections between the words used in surveys and their own actions, making them highly limited in generalizability. Since each person will bring a specific association to any social incident, the survey is less able to capture a universal phenomenon. In addition, Kagan argues, people always seek to reflect their "ego ideal," or their desire to choose the more socially acceptable action, and so will not report their "true" feelings or behaviors.

Generally, we agree that this is a concern, but only a fatal one if social conduct can itself be accurately measured and is seen as the absolute standard to which espoused social choices and reflections on social actions are applied. In other words, if social actions are viewed as being factual, rather than interpretive or socially constructed, and if the reports can actually be verified as accurate or inaccurate. For example, if you say you will vote for your friend as class president, whom you actually vote for is a case of such verifiable accuracy. How or why one actually rejects a friend, however, is not as precisely reported through either observation or self-report.

Crucial to this issue is how deeply into the "social" mind, the self's reflection and interpretation of social experiences, the social cognitive methodology allows exploration. For instance, in most social cognitive neuroscience studies, the social behavioral measures tend to skim the surface of an individual's social cognitive functions and structures, probably because validity needs to be sacrificed for reliability. The qualitative richness of highly inferential social cognitive measures is sacrificed on the altar of operationally defined behavioral measures—in other words, the scoring does not allow for much inference but the construct being measured invites interpretation. Consider the Eisenberger et al. study, in which the quantitative variation in the intensity of self-reported reactions to the exclusion theme in the vignette corresponded to the level of neuronal excitation. What we do not know, however, is what the qualities of the experience of rejection or exclusion were, as interpreted by the participants.

According to a recent news focus in *Science*, caution is also advised since the current state of fMRI technology is still primitive relative to the inferences that some researchers have put forth using brain scans alone (Miller, 2008). Although a voxel, the unit of brain tissue being imaged, is only a few millimeters, it contains millions of neurons, making it difficult for one to observe precise blood flow in the brain. Inferring brain activity based upon brain location, Miller points out, is tenuous at best because many areas have been connected to multiple functions. Activity in the amygdala, for instance, might be inferred as related to anxiety, but that same region is also activated by sexual arousal. It seems that fMRI is currently restricted to revealing correlations between cognitive processes and brain activity (Miller, 2008). New efforts to improve the validity of fMRI research tend to be focused on the noninvasive manipulation of neural activity and the observation of any behavioral changes, on comparing human brain scans to simian brain scans, and on distinguishing brain activity patterns from overall levels of activity. It would appear that fMRI would benefit from other avenues of seeing.

Split-Second Snapshots and Sequential Scenarios: Social Interactions

Social cognitive neuroscience is well suited to the study of snapshots of brain function at the particular moment in which a social choice (hypothetical or otherwise) is made or a social experience is felt. For the purposes of discussion, we call this the "moment of (measured) social choice." When hypothetical measures of social choices or reasons are dismissed, it is usually on the premise that they do not capture what people would really or actually do in that bounded moment of choice. However, questionnaires, surveys, and interviews can be thought of not only, or even primarily, as momentary self-report, but as part of a developmental process, in which only the first step is to see—as Kagan (2007) says—what personal meaning is being made in a hypothetical or reflective social context, and what people say or how they reason about it in the moment. Our work focuses on analyses that rely on the additional steps in the process.

IN-GROUP ASSESSMENT: A STEP TOWARD A CONTEXTUAL–DEVELOPMENTAL MEASURE OF THE MEANING IN MINDS

The most recent evolution of this work is of the analysis of responses adolescents give to familiar yet socially difficult situations or incidents that reach out for–if not require—critical reflection. The complete battery is called the choices in context approach. The "in-group incident" is one such case; it challenges adolescents to assume different vantage points about a specific evolving situation of peer social exclusion, as well as inquires into their thinking about the issues of interpersonal relationships and social exclusion more generally (Barr, 2005). This assessment includes a personal story written by a high school student about a situation of social exclusion she experienced in early adolescence:

> My 8th grade consisted of 28 students, most of whom knew each other from the age of 5 or 6. Although we grew up together, we still had class outcasts. From 2nd grade on, a small, elite group spent a large portion of their time harassing two or three of the others. I was one of those two or three, though I don't know why... The harassment was subtle. It came in the form of muffled giggles when I talked, and rolled eyes when I turned around. If I was out in the playground and approached a group of people, they often fell silent. Sometimes someone would not see me coming, and I would catch the tail end of a joke at my expense.
>
> There was another girl in our class who was perhaps even more rejected than I. One day during lunch ... one of the popular girls in the class came up to me to show me something she said I wouldn't want to miss. We walked to a corner of the playground where a group of three or four sat. One of them read aloud from a small book, which I was told was the girl's diary. I sat down and, laughing till my sides hurt, heard my voice finally blend with the others. Looking back, I wonder how I could have participated in mocking this girl when I knew perfectly well what it felt like to be mocked myself. I would like to say that if I were in that situation today I would react differently, but I can't honestly be sure.
>
> —**Eve Shalen, a ninth-grade student**
(***Facing History and Ourselves*, 1994, pp. 29–30**)

This narrative is followed by a set of survey questions driven by our analysis about how children and adolescents develop the capacity to coordinate various social perspectives (Selman, 2003; Selman, & Adalbjarnardottir, 2000). They move from situations involving others, like Eve, to situations they themselves find challenging. Here, we will focus on Eve. The measures involve reflections on Eve in the role of victim (her past), witness, and invitee to participation in the harassment and bullying. With respect to the witness phase, for example, we ask:

1. List at least two (2) different ways that Eve could have acted when she witnessed her classmates picking on other students.
2. Which would be the best way?
3. Why would that be the best way?

The Assessment of Adolescents' Choices: Step 1, Social Strategies

Most analyses of social situations tend to focus on what adolescents say they, or Eve, should do in this situation. We went beyond this kind of report in two further analytic steps, each revealing a deeper layer below the choices. The first step was to classify the types of strategies adolescents consider before they select one as "best." The second step was to analyze the justifications they gave.

To analyze the strategies adolescents recommend in this situation, we used a semigrounded approach (Glaser & Strauss, 1967),[*] and three strategy categories emerged. (See Table 12.1 for the codebook and sample responses.) Upstand responses suggest actions that imply helping the victim is the goal. Some responses suggest direct intervention, such as standing up to the group of students doing the

Table 12.1 Codebook for Strategy Categories

	Code Description	Anchor Response
Upstand	Requires intervention in the existing situation of exclusion	"She could of told them to stop, and that they were being mean."
	Articulates an action that assists the victim, such as standing up to the group or comforting the victim	"She should make friends with that girl."
	Aligns against the exclusion	"She should of just told a teacher what was going on."
Perpetrate	Aligns with the group doing the excluding	"She could play along and make fun of the kids, too."
	Implies that the invitation to join the excluding has been accepted	"She should go with the girls."
	Contributes to the existing situation of exclusion	
Bystand	Aligns with neither the victim nor the group doing the excluding	"Ignore them and just go on with her business."
	Avoids involvement with the existing situation of exclusion	"Make an excuse and walk away."
	Detaches from the situation actively or passively	

---

[*] We use the term "semigrounded approach" to mean that we allowed the data, rather than theory, to guide our analyses. However, because the In-Group Assessment Interview is structured around a developmental theory, we recognize that the analyses cannot be entirely grounded.

excluding (active upstanding), while others recommend aligning with the victim by making friends with her or going over to see if she's okay (passive upstanding). The upstand category also includes responses that suggest seeking help from a teacher or another adult figure.

In contrast, perpetrate responses recommend strategies that most often align with the group excluding the girl. These responses explicitly state that the best choice would be to join the crowd by laughing along with them. We also include responses in this category that suggest initiating a new conflict in response to the exclusion—attacking the girls or being mean back, for example—because these types of actions perpetuate, in fact, often escalate, the exclusion.

The bystand response includes responses that suggest strategies that avoid siding with either the victim or the group doing the excluding. One type of response in this group indicates that getting involved in the exclusion should be avoided, usually by walking away or somehow actively detaching from the situation. Another set of responses express passive bystanding actions, e.g., a recommendation to just stand by while the exclusion occurs.

Assessment of Adolescent Choices: Step 2, Justifications for Strategies
While the "strategies suggested" coding tells us about the types of choices individuals consider from each vantage point (witness, bystander, invitee), our second analytic lens, justifications, attempts to capture the factors in the social environment that individuals perceived to be most salient when choosing a strategy. We used an inductive approach (Boyatzis, 1998) to allow both the data and theory to guide the coding. Four kinds of justifications emerged. (The codebook and sample responses can be found in Table 12.2.)

The safety category captures adolescents' justifications that emphasize and prioritize the physical or emotional welfare of one of the potentially vulnerable people involved in the situation. These responses articulate an immediate need for protection—for either the victim or the self—in the face of a perceived threat as the primary motivation. Though these responses highlight the urgent need for an end to the current situation, long-term consequences are not mentioned as the driving force. Ultimately, in these responses, the chosen strategy is seen as a necessary response to provide prompt shelter from harm.

Responses in the conventional category stress the importance of rules, whether formalized through the school or simply reflecting informal social norms. Such responses typically justify a strategy by relying on moral codes that point to the "right" thing to do. Conventional

Table 12.2 Codebook for Justification Categories

	Description	Anchor Response
Safety	Indicates protection as a priority	"So they don't start picking on her, too."
	Perceives an immediate threat to one's emotional or physical well-being	"That way nothing bad happens to the victim."
	Indicates that the main goal is to stop the current situation of exclusion	"To make sure she doesn't get hurt."
	Does not reference long-term consequences or implications of recommended strategy	
Conventional	References social norms, conventions, or rules (formal or informal)	"It is the right thing to do."
	Highlights efficiency or expediency of the recommended strategy	"It would be easier."
	Does not explicate reasoning beyond simple explanations of cost-benefit analyses that one action is "better" than another	"Because I think it's the best."
		"It would keep everyone out of trouble."
Relational	Highlights the formation or maintenance of interpersonal relationship(s)	"And I could be considered the 'popular' girl."
	Articulates desire for belonging or connectedness with another person or with a group of people	"Because she would have a friend."
		"Because she'd feel like she fit in."
	Identifies a connection between people's experiences or emotions	"Because she knows what it feels like."
Transformational	Explains connections between the recommended action and possible future consequences or implications	"They might realize they're doing the wrong thing and not do it again."

Table 12.2 Codebook for Justification Categories (Continued)

Description	Anchor Response
Speculates about the possible development of or changes in other people's thinking or beliefs	"That way the problem would not exist anymore."
Articulates opportunities for group dynamics to shift as a result of the recommended action	"She would make a good influence on other people."
Implies that the recommended action could serve as a catalyst for these changes"	"That way they would understand what it's like, too."

responses also frequently refer to pragmatics by pointing to what most adolescents would do. These responses often do not include much explanation beyond the expediency or efficiency of the chosen strategy, highlighting that one choice might simply work better than another.

The relational justification category includes responses that emphasize a sense of belonging or point to connections between people. These responses place a high level of importance on interpersonal or intergroup relationships. Relational justifications also highlight a connection between people's emotions when they have had similar experiences, such as being the victim of social exclusion. These responses also articulate an awareness of the benefits (or dangers) of group affiliations.

The pro-social transformational category of justification includes responses that view the chosen action as a catalyst for change across time and into the future. Responses in this cluster of justifications demonstrate an awareness that change is possible and articulate opportunities for something to be different because of the current situation. The chosen strategy is explained as helping prevent further (or future) exclusion by either a change in the way the excluding group thinks or feels, or from a shift in larger group dynamics that would make these types of situations cease to exist.

Findings: Reliability, Validity, and Their Importance   Using this two-step analytic method, we found consistently high reliability for this coding framework—over 95% interrater reliability—for both the assignment of strategies and justifications. We also found, when reflecting on the case study in the role of a witness, that adolescents tended to recommend active resistance to the perpetrators of the exclusion. When asked to reflect on the incident in a different role (i.e., when asked to consider what they would do if pressed to join in), however, they either voiced their likely participation in the escalation of the predation (particularly if they were boys), or their refusal to stand up for the victim and withdraw to a bystander status (particularly if they were girls), even though they knew that neither of these actions are socially desirable or ethically admirable (Feigenberg et al., 2008). These findings did not surprise us. They align with findings from previous research about the strategies used by boys and girls (Crick, Bigbee, & Howes, 1996; Underwood, 2003). But we also found variations in strategies associated with school climate, something that did surprise us since initially we thought we were assessing individual adolescents' developmental capacities and not the broader context of the school. In other words, strategies tend to be very sensitive to shifting across roles or vantage points and across contexts and culture (if we allow gender to be a proxy for culture).

However, we found there to be much more stability in the kinds of justifications adolescents gave in the different positions they were asked to consider than the strategies they chose as best across positions. This suggests that even if a prosocial transformational justification may sound more socially mature than one oriented to safety, choices really do depend to a significant degree on contexts and individuals' perceptions (Feigenberg et al., 2008).

Some social psychology theorists interested in the connection between moral thought and action, such as Jonathan Haidt, might argue that participants taking a measure like this one simply provide justifications after they have already selected their action. He argues that people engage in "moral rationalization," as opposed to moral reasoning, which is to say that they work backwards to find a reason that rationalizes their final choice (Haidt, 2001). To come to this conclusion, he has devised dilemmas that deal with taboo subjects, but also include rational caveats (such as consensual protected sexual relations between a brother and sister). Haidt has found that as people struggle to explain why they think the situation is wrong, they are ultimately unable to find a logical reason. Emotion, he concludes, plays a bigger role in morality than cognitive developmental researchers may typically discuss. In this sense, the door is open for social cognitive neuroscience to help us better understand the emotional arousal critical social incidents trigger.

We think, however, this ignores an important focus—self and social reflection over time, the capacity to gain awareness of the personal meaning of choices in the context of ongoing and meaningful relationships, as much as in the emotional reaction in the moment. Such is the case of Eve, who even notes that despite looking back with disappointment

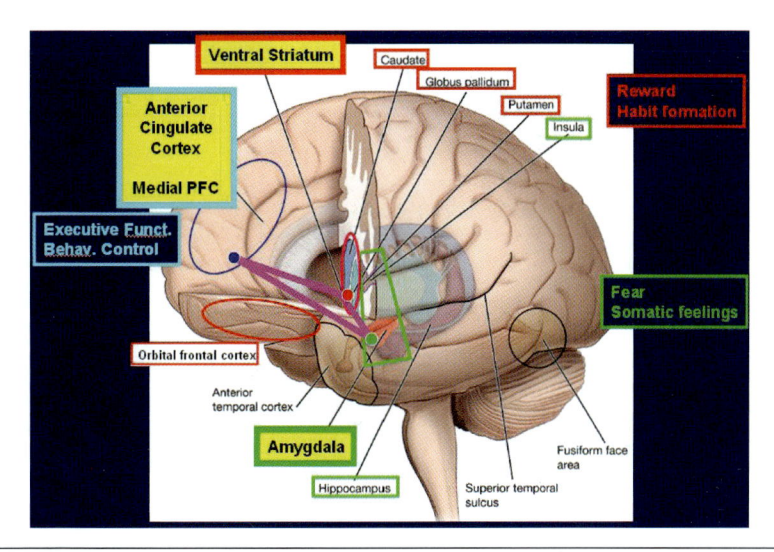

Figure 9.2 Schematic representation of a sagittal view of the brain, with the lateral slab of the prefrontal region being removed, exposing its medial section. Regions associated with the three functional systems of the Triadic Model are highlighted. The approach system includes striatum and orbitofrontal cortex. The avoidance system includes the amygdala, hippocampus, and insula. The modulatory system includes the medial prefrontal cortex, including the anterior cingulate cortex. The triangle delineates the Triadic Model. Of note, this formulation is a simplification of the function of these regions, as discussed in the text.

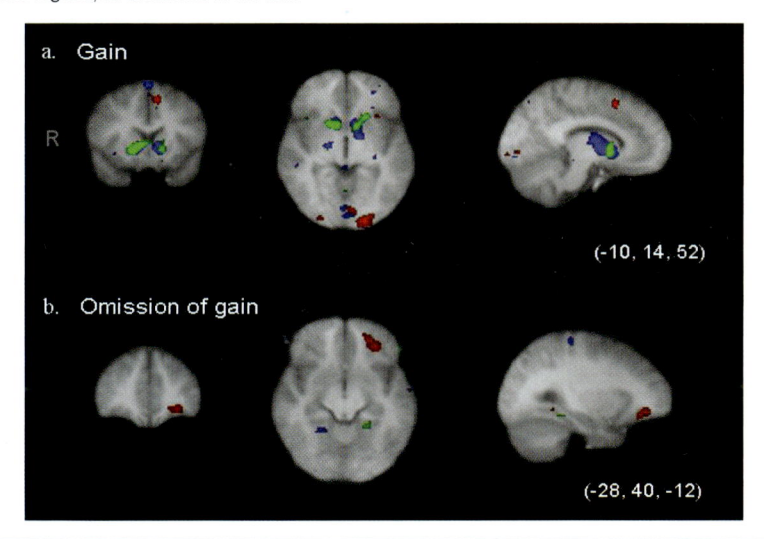

Figure 11.1 Brain activation patterns related to the processing of (a) rewards, and (b) the omission of expected rewards. Activation for 10- to 12-year-olds in green, 14- to 15-year-olds in blue, and 18- to 25-year-olds in red. Winning was associated with activation of the bilateral striatum in 10- to 12-year-olds and 14- to 15-year-olds, whereas processing of not winning was associated with left lateral OFC in 18- to 25-year-olds.

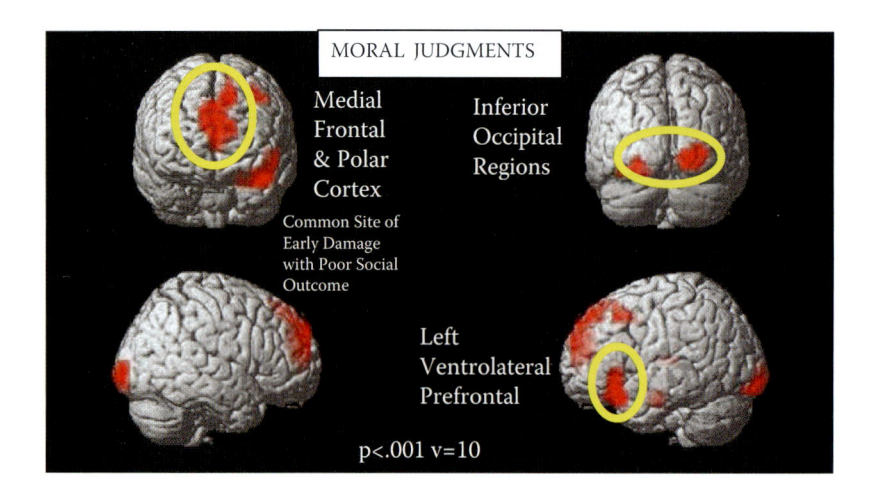

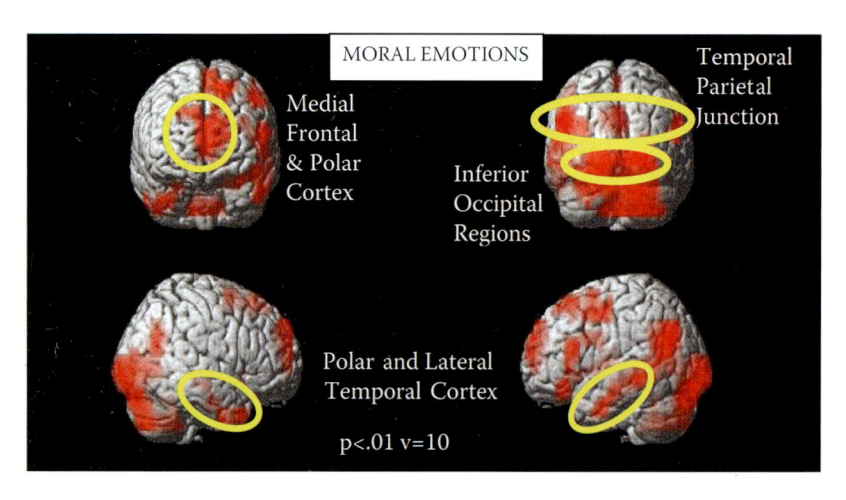

Figure 13.3 3D renderings of brain activations associated with moral judgments and moral emotions in a typical developmental sample 10–17 years of age.

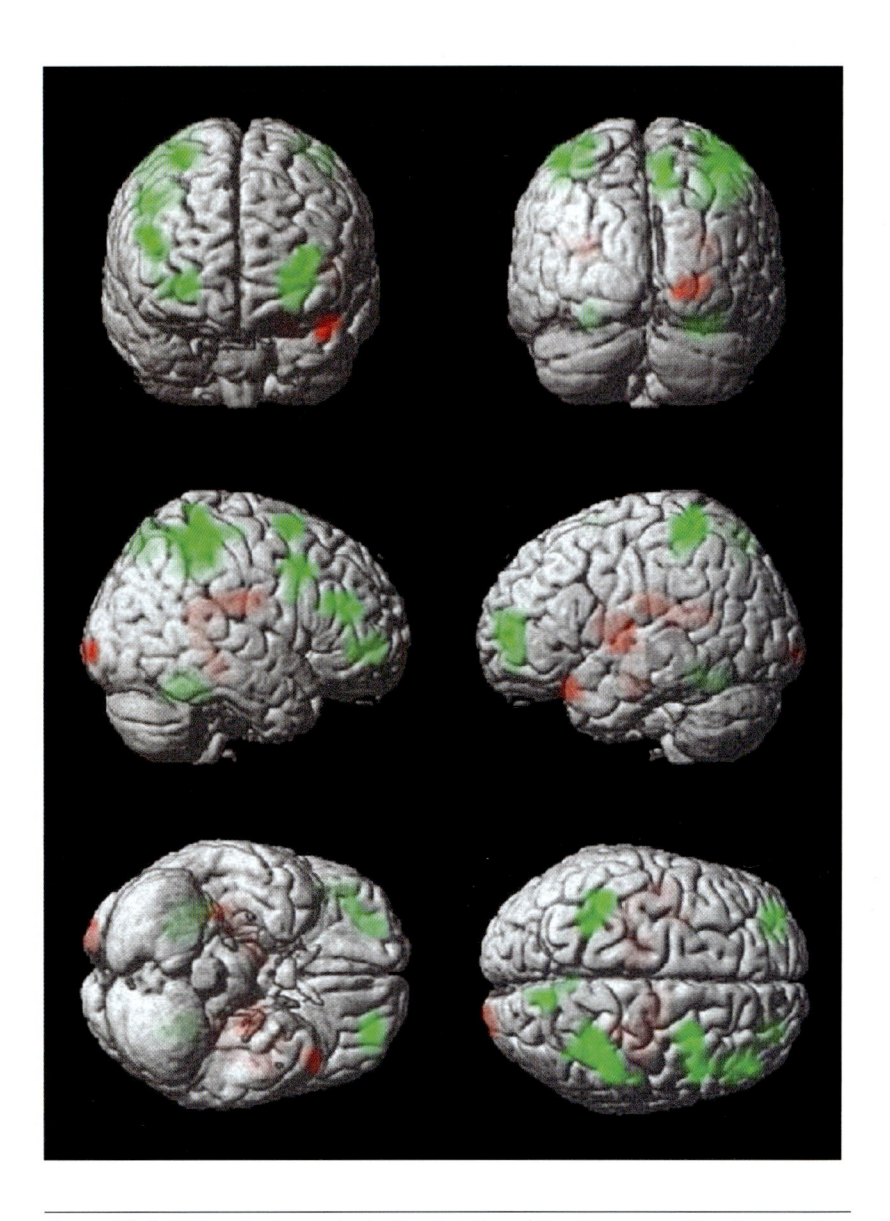

Figure 13.4 fMRI contrast analysis showing that the moral ambiguous condition (in comparison to the more rule-based moral right/wrong condition) recruited distinctive activations in the frontal polar and superior parietal regions bilaterally as well as the right dorsolateral prefrontal region (shown in green) while the more rule-based moral right/wrong condition recruited activations primarily in the lateral temporal and temporo-parietal junction regions where general social knowledge and perception have been identified (shown in red).

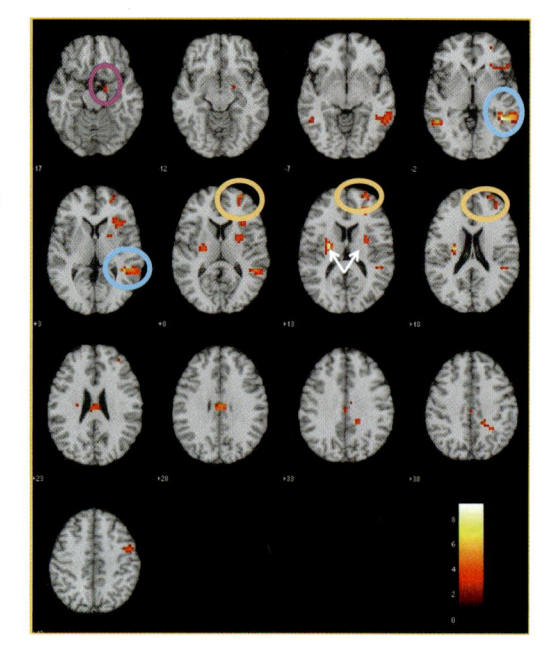

Figure 13.5 fMRI brain map showing areas of significant decreasing activity during the moral judgment tasks as a function of age in the 10–17-year-old developmental sample.

at her own actions, she might today still take the same course of action because the power to belong was so strong in her. This is not just rationalization, it is not just reasoning, it is also critical reflection. How one understands, let alone justifies, one's own actions requires an awareness of all three.

CONNECTIONS
Ah, Neuroscience!

If by early adolescence, students have available the capacity and capability for a solid understanding of all the possible social strategy categories (upstand, bystand, perpetrate) and for all the social justification categories (safety, rules, relationships, prosocial transformations), we would like to know how this social awareness in a development-contextual analysis could be clarified by neuroscience evidence. Suppose we gave a measure like the In-Group Assessment each year in a longitudinal design, beginning at age 10, and studied participants' brain activity as they rated items using the latest neuroscience methods available. Would we get a more comprehensive glimpse of the relationship between the intensity and location of their brain functioning and their responses on the measure? Would there be differences in the neuroscience picture as a function of their chronological age, especially if the age range were wide enough and began prior to adolescence, earlier in ontogenesis? Would there be any differences in neurological functioning related to their different (hypothetical) social contexts in which they participate (school, neighborhood, etc.)? One wonders whether such a strategy is our best hope of at last uncovering not just what brain development consists of, but how this information could transform practice.

Questions like these, even though they are rudimentary, could provide interesting ways to integrate our social developmental-contextual methods and models, which connect functional social awareness to the quality of the social environment, with those of neuropsychology where the natural inclination is to put culture and context aside for a while. Theories of social conduct built upon emerging evidence of how neurological systems for executive functioning of emotions, consciousness, and rationality develop are changing the face of the field of social cognition. Keep in mind, however, that most social choices are not made "in the moment," on impulse, or "without thinking." They develop as the mind develops, reflecting on them. How does the developing brain help? How is the developing brain helped by these experiences?

Coda

"I believe the best thing to do would have been to tell them that what they were doing was wrong, and that they should stop it. I chose this reason because I believe that if enough of the people are willing to do something about a problem, the problem would not exist anymore."

Ann, a ninth grader in an upper middle class suburban neighborhood

Ann and Danielle provide responses to the Eve Shalen dilemma that differ both in strategy and justification. Ann's answer suggests an upstand strategy ("Tell [the bullies] what they were doing was wrong"), whereas Danielle recommends perpetration ("You got to fight back") as the way to deal with exclusion. Ann's response demonstrates a prosocial transformational orientation ("If enough people are willing to do something about a problem, the problem would not exist anymore") and Danielle relies on a safety justification ("Protect yourself"). The difference between these two girls' responses raises several questions for us in terms of method and measurement of social awareness, as well as their connection to theories of social, developmental, and cultural psychology, and, by implication, developmental social cognitive neuroscience.

Do these responses provide evidence that Danielle and Ann differ in their ability to understand social relationships? Although we may, in our first reading, perceive Ann's use of a transformational justification as more "developed," it is important to consider that she may be embedded within a social context that promotes and supports responses about possibilities for transformation. Since her primary concern is not safety (as it is for Danielle), this priority implies that she has the privilege, probably more than simply the courage, to incorporate these concerns into the consideration of how social situations might be different. Her use of a transformational justification can just as legitimately be interpreted as an indicator of a supportive context that fosters and encourages such expression as it can of healthy individual development. Alternatively, one can also interpret these responses to suggest that while Danielle might be excessively nihilistic and pessimistic, Ann may be overly idealistic and optimistic.

In this sense the variation between Ann and Danielle is not a function of their cognitive development, or their social competence, let alone their neural development. By ninth grade, almost all adolescents, normatively speaking, have the competencies to understand each type of strategy and each type of justification. Sometimes we will want to orient to safety and other times to social change, while sometimes we will think following rules is preferable, and sometimes focusing on ongoing relationships is the most caring option. Sometimes all these concerns emerge or are in play simultaneously.

Adolescence is a period when youth need to develop the functional wisdom to appreciate and harness their capacity for critical reflection. Danielle needs to temper her street-savvy cynicism with more optimism and opportunity to see the value of prosocial transformative actions. Ann needs to temper her idealism with more realistic experience in the world so she can survive the emergence from her cloistered environment into a world where the major concern for many people she does not know is safety and survival. These two girls suggest that we need research that integrates thought and action, mind and brain, neuron and neighborhood.

REFERENCES

Baird, A. A. (2007). Moral reasoning in adolescence: The integration of emotion and cognition. In W. Sinnott-Armstrong (Ed.), *Moral psychology*. Cambridge, MA: MIT Press.

Barr, D. J. (2005). Early adolescents' reflection on social justice: Facing history and ourselves in practice and assessment. *Intercultural Education, 16*(2), 145–160.

Baumeister, R. F., Vohs, K. D., & Funder, D. C. (2007). Psychology as the science of self-reports and finger movements: Or, whatever happened to actual behavior? *Perspectives on Psychological Science, 2*(4), 396–403.

Beckman, M. (2004). Crime, culpability, and the adolescent brain. *Science, 305*, 596–599.

Bunge, S. A. (2007, June). Tutorial: Developmental changes in brain structure and function. In S. A. Bunge and J. Beer (Co-Chairs), *The development of social rule use: Implications of work in neuroscience*. Symposium conducted at the Thirty-Seventh Annual Meeting of The Jean Piaget Society, Amsterdam, The Netherlands.

Boyatzis, R. E. (1998). *Transforming qualitative information*. Thousand Oaks, CA: Sage Publications.

Casey, B. J., Getz, S., & Galvan, A. (2008). The adolescent brain. *Current Directions in Risk and Decision Making, 28*(1), 62–77.

Crick, N. R., Bigbee, M. A., & Howes, C. (1996). Gender differences in children's normative beliefs about aggression: How do I hurt thee? Let me count the ways. *Child Development, 67*(3), 1003–1014.

Dahl, R. E. (2007, June). Adolescence: A period of vulnerability and opportunity. Presented at the meeting of the Thirty-Seventh Annual Meeting of The Jean Piaget Society, Amsterdam, The Netherlands.

Eisenberger, N. I., Lieberman, M. D., & Williams, K. D. (2003). Does rejection hurt? An fMRI study of social exclusion. *Science, 302*(5643), 290–292.

Facing History and Ourselves. (1994). *Holocaust and human behavior.* Brookline, MA: Facing History and Ourselves National Foundation, Inc.

Feigenberg, L. F., King, M. S., Barr, D. J., & Selman, R. L. (2008). Belonging to and exclusion from the peer group in schools: Influences on adolescents' moral choices. *Journal of Moral Education, 37*(2), 165–184.

Freud, A. (1966). *The Writings of Anna Freud, Volume II: The ego and mechanisms of defense.* Revised edition. New York: International Universities Press, Inc.

Glaser, B. G., & Strauss, A. L. (1967). *The discovery of grounded theory: Strategies for qualitative research.* Chicago: Aldine.

Habermas, J. (1986–1989). *The theory of communicative action* (T. McCarthy, Trans.). Cambridge: Polity Press.

Haidt, J. (2001). The emotional dog and its rational tail: A social intuitionist approach to moral judgment. *Psychological Review, 108*(4), 814–834.

Haste, H. (1999) Moral understanding in socio-cultural context; lay social theory and a Vygotskian synthesis. In M Woodhead, D Faulkner and K Littleton (Eds.), *Making sense of social development.* London: Routledge.

Kagan, J. (2007). A trio of concerns. *Perspectives on Psychological Science, 2*(4), 361–376.

Miller, G. (2008). Growing pains for fMRI. *Science, 320,* 1412–1414.

Selman, R. L. (2003). *The promotion of social awareness: Powerful lessons from the partnership of developmental theory and classroom practice.* New York: Russell Sage Foundation.

Selman, R. L., & Adalbjarnardottir, S. (2000). A developmental method to analyze the personal meaning adolescents make of risk and relationship: The case of "drinking." *Applied Developmental Science, 4*(1), 47–65.

Selman, R. L., & Dray, A. J. (2006). Risk and prevention. In W. Damon & R. Lerner (Series Eds.) & K. A. Renninger, & I. E. Sigel (Vol. Eds.), *Handbook of child psychology: Vol. 4. Child psychology in practice* (6th ed., pp. 378–419). Hoboken, NJ: John Wiley & Sons.

Selman, R. L., & Schultz, L. H. (1990). *Making a friend in youth: Developmental theory and pair therapy.* Chicago, IL: University of Chicago Press.

Steinberg, L. (2003). Is decision making the right framework for research on adolescent risk taking? In D. Romer (Ed.) *Reducing adolescent risk: Toward an integrated approach.* Thousand Oaks, CA: Sage Publication.

Stevens, M. C., Kiehl, K. A., Pearlson, G. D., & Calhoun, V. D. (2007). Functional neural networks underlying response inhibition in adolescents and adults. *Behavioural Brain Research, 181*(1), 12–22.

Thagard, P. (2000). *Coherence in thought and action.* Cambridge, MA: The MIT Press.

Underwood, M. K. (2003). *Social aggression among girls.* New York: Guilford Press.

IV

The Developmental Social Cognitive Neuroscience of Moral Reasoning

13

CRUCIAL DEVELOPMENTAL ROLE OF PREFRONTAL CORTICAL SYSTEMS IN SOCIAL COGNITION AND MORAL MATURATION
Evidence From Early Prefrontal Lesions and fMRI

Paul J. Eslinger
Melissa Robinson-Long
Penn State/Hershey Medical Center

The materials summarized in this chapter were presented during symposia on social rule use and moral development that generated spirited discussions between developmental psychologists and neuroscientists of how we acquire and use social rules (from psychological and neuropsychological perspectives) and how to forge a common framework for social cognition and moral processing that might accommodate both developmental and neurobiological approaches. The symposium topics share common foundations in social cognition and social emotions that we describe in this chapter. As our theoretical framework and methodologies are drawn primarily from clinical neuropsychology and cognitive neuroscience, we briefly describe our approach to several outstanding questions and issues concerning the role of brain systems in social rule learning and social emotions. The findings are part of ongoing clinical studies of children, adolescents, and adults with acquired injuries to the brain, particularly damage to the prefrontal cortex (PFC), and how these injuries affect their social behavior, cognition, and emotions. Functional brain imaging of typically developing children and adolescents has recently been incorporated as well in order to confirm and extend findings to broader developmental questions. Hence, our approach has been to search for converging evidence with regard to what brain areas are crucial for social cognition and moral development, the individual differences that occur in such neurobiology and behavior, the plasticity of these cerebral regions throughout maturation, and whether neural reorganization can support recovery of social adaptation when one or more of these areas are damaged or dysfunctional early in life.

PREFRONTAL CORTEX AND SOCIAL BEHAVIOR

Social cognition has been of increasing interest to neuropsychologists because of the great significance of social deficits for patients, their families, and larger societal and economic concerns. Often, comprehensive examinations are needed to accurately diagnose and distinguish social cognitive deficits, and there is frequent need for continued management of underlying medical conditions and the expressed behavioral deficits. Our interests historically evolved from a series of case studies in adults who developed neurological conditions causing structural damage to the frontal lobes and in particular to the PFC. This largest lobe of the brain, both phylogenetically and ontogenetically, is multimodal and highly associative in physiology and function. It is closely interconnected with the limbic system, posterior cortical association regions, and important subcortical and effector structures of the basal ganglia, thalamus, and hypothalamus (Barbas, 1995; Eslinger, 2008; Saleem, Kondo, & Price, 2008). The PFC and its related social and executive functions are probably best conceptualized as a crucial component of a large-scale neural network that mediates diverse information processing, consolidates complex types of knowledge and experience, integrates cognitive, visceral-somatic and affective streams of processing, and significantly influences behavioral self-regulation through goal-direction, inhibition, decision-making, and self-monitoring (Eslinger & Biddle, 2008; Eslinger & Tranel, 2005).

There is strong evidence that the PFC undergoes prolonged postnatal maturation that involves not only expansion of synaptic connections among neurons but also pruning and sculpting of these connections and larger pathways to advance integrative and goal-oriented processing within PFC-related networks including cortical and subcortical regions (Giedd, 2008). Two important aspects of PFC neurobiology were largely unrealized until recent research. The first is that the PFC region, as a principal mediator of executive functions, is much more actively involved in early childhood development than previously thought (Blair, Zelazo & Greenberg, 2005). Secondly, PFC maturation extends much longer than previously thought, and reaches well into early adulthood (Grattan & Eslinger, 1991). We suspect, based on clinical, developmental, and functional brain imaging data, that PFC-related maturation and plasticity likely continues throughout much of adulthood and even healthy aging. However, when PFC damage occurs in adults, it can cause a variety of disabling symptoms, perhaps most prominently alterations in social judgment and theory of mind (Stuss et al., 2001), even though sparing general intellect, language, and memory abilities. A seminal example of this change was described by Harlow (1868) in the case of Phineas Gage, who suffered an isolated injury to the most anterior or polar regions of the PFC from passage of a tamping iron through the head. From Harlow's comprehensive description and accounting of Gage's recovery, it is clear that Gage suffered a profound change in social judgment, relationships with others, balance of personality and emotions, and how he led his life. Previously a productive and well-adjusted adult who worked as foreman on a railroad blasting crew, Gage became reckless, disinhibited, callous to the welfare of others, and unable to regulate himself to return to typical goal-directed behavior. This change occurred despite the apparent preservation of his general intellect, sensory-perception, language, memory, and motor capacities. Contemporary neuropsychological studies have replicated these findings in cases of acquired damage to the orbital and medial PFC regions. Hence, these areas appear to provide essential neural resources for social adaptation, judgment, and decision making (Eslinger & Damasio, 1985; Dimitrov, Phipps, Zahn, & Grafman, 1999; Tranel, Bechara, & Denburg, 2002). Such impairments in previously healthy individuals has variously been described as a pseudopsychopathic frontal lobe syndrome, neural sociopathy, and behavioral variant of frontotemporal dementia, based on the fact that the most profound impairments occur within social contexts, that there is apparently normal knowledge of social rules and conventions (although these no longer guide decision making and social behavior), and that the impairments are not accounted for by loss of intellectual resources or by psychosocial stressors.

While these clinical observations in adults are important to a brain-behavior framework for social cognition, they do not speak to developmental issues very clearly. It is noteworthy that the maturation of PFC regions are postnatally extended, that they can integrate cognitive and affective processing, and that they provide necessary resources for inhibitory control and experiential learning—all of which are important developmental parameters. Hence, we might expect that PFC-related regions will be active in the acquisition and maturation of social cognition, which we examine in the next section.

EARLY PREFRONTAL CORTEX DAMAGE AND SOCIAL DEVELOPMENT

A clinical case that helped bridge the chasm between adult and developmental PFC functioning and social cognition arrived in our clinic in the 1980s for evaluation of what had been continuing behavioral problems over a period of 26 years. As a 33-year-old adult, DT appeared self- centered and concrete in her thinking, but nonetheless was a conversant and interactive young lady with typical appearance and without apparent neurological deficits. However, her family described a profoundly different picture of significant social impairments that disabled her from living independently, maintaining any healthy relationships, and even caring for her infant. We learned from further study that at 7 years of age, she underwent resection of a vascular malformation in the left polar and medial region of the PFC. She did not develop any evident neurological deficits after her successful surgical treatment and returned to home and school without incident. Gradually, over the next 3 to 4 years, significant social and behavioral impairments began to emerge. DT became progressively less sensitive to others, less attentive to details, inflexible in her problem solving, and unable to utilize any feedback to improve her behavior and declining school performance. Her ability to plan, estimate probabilities of outcome, and articulate clear and consistent social judgment became increasingly problematic in comparison to maturing peers. In retrospective analysis, it became clear to us that DT suffered an arrest of social, cognitive, and emotional development that became prominent in early adolescence (and has continued since that time) and that she was no longer maturing along typical developmental trajectories, appearing more childish and self-centered, and unable to handle school as well as other functional tasks and social roles. She maintained strengths in areas of initiating behavior and remaining active in everyday tasks but has never been able to maintain a job because she does not complete much of her work and has great difficulties in getting along with others. She married the first man she dated, but was unable to care for her infant and eventually returned to the care of her parents. Her measured intellect on the Wechsler scales placed her with full-scale IQ of 83, with comparable levels of memory, language, perception, and spatial abilities—certainly sufficient cognitive functions to be independent as an adult. Furthermore, her upbringing and family were devoid of psychosocial stressors and conditions that might otherwise explain her impairments. However, her measured executive functions were quite poor with cognitive inflexibility; inability to keep details in mind for planning and decision-making; difficulties in switching tasks, handling any delay in reward, abstraction, and set shifting; impulsivity; and difficulties in theory of mind judgments and socio-moral judgment situations. Brain scan and lesion localization as an adult showed a focal area of damage in the left medial PFC region, extending to the frontal pole and encompassing parts of Brodmann's areas 10, 9, 8, and 32 (see Figure 13.1A). Her pattern of deficits from early PFC damage appeared progressive in the sense that they became more evident as she failed to mature beyond late childhood–early adolescence and to acquire any further social knowledge, metacognitive skills, and self-regulation. Her ability to learn social rules, contingencies of situational variables and potential outcomes, and anticipate the behavior of others appeared to remain at early adolescent levels. This brain-behavior pattern is quite unique and in marked contrast to studies of children with perinatal stroke (typically not affecting the PFC) who show continuing adaptation in language, spatial, and affective domains (Stiles, Reilly, Paul, & Moses, 2005).

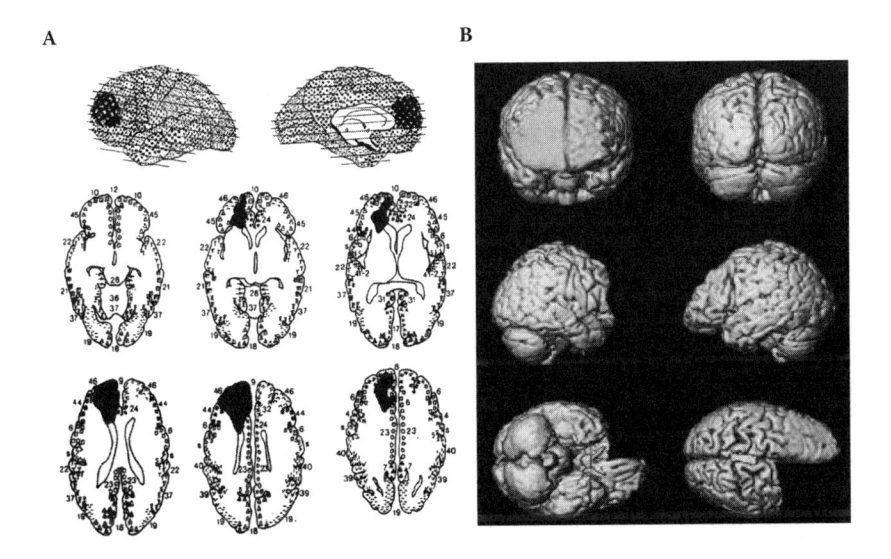

Figure 13.1 Reconstruction of early prefrontal cortex lesions in the case of DT (A), who suffered left prefrontal damage at 7 years of age, and JP (B), who suffered congenital bilateral damage of the prefrontal regions.

In support of these findings, we found an exquisite description of very early PFC damage and developmental frontal lobe syndrome from the case report of Ackerly and Benton (1948), updated by Benton in 1991. JP came to the attention of these investigators at about 13 years of age because of increasingly problematic behaviors, inability to get along with any peers, and progressive academic difficulties. There was history of neither neurological injury, psychiatric conditions, nor psychosocial stressors for such difficulties. Their analyses over several years spanned comprehensive testing, observation, and therapy as well as the results of eventual surgery to identify any treatable cause of the underlying brain pathology identified on pneumoencephalogram, which showed large atrophic areas in the PFC region. In this era prior to modern imaging, the neurosurgeon identified significant atrophy of the left PFC region and complete absence of the right PFC region, likely a congenital form of brain abnormality (see Figure 13.1B for a reconstruction of the area of atrophy based on surgical report).

The extent of JP's social impairments was extensive. His relationships were described as shallow and absent of empathy, reciprocity, theory of mind, and even agreement about game rules, particularly if he felt any disadvantage. Hence, he was profoundly disliked by his peers. His measured intellect was average, and his sensorimotor as well as language, perceptual, and spatial abilities were also within the normal range of variation. Curiously, he showed no observable anxieties, and was noted to wander long distances from home even at 3 years of age, a habit that did not respond to punishment and which continued into adulthood, sometimes entailing thousands of miles. Ackerly and Benton (1948) characterized JP's persistent neurodevelopmental difficulties as a primary social defect, which we would describe as a developmental frontal lobe syndrome today. One of the most striking features of the JP case study was his failure to benefit from experience in acquiring and elaborating on socially relevant knowledge and judgment, in relative isolation to his otherwise normal range of measured intellect and other cognitive abilities. Not only was his information processing and thinking constrained in relationship to others, but his ability to register and process emotional or value-related experiences was extremely limited. His inability to perceive and appreciate particularly negative consequences of actions appeared to be one of the determining factors in his refractory social difficulties, which never responded to any form of intervention or treatment. Another way of conceptualizing these difficulties is in terms of alterations to social emotions such as embarrassment, gratitude, humility, guilt, pity and disgust. Hence, these examples of early damage to the PFC and their echoing consequences throughout development and maturation suggest that the PFC and its extended networks provide crucial neural substrates for both acquiring and utilizing cognitive, social, and affective processing that helps guide personal actions in relationship to others in typical adolescent/adult goals and activities. Early damage to this region in the rat model has shown that while some reparative processes are possible (depending upon particular age and site of lesion), there can also be developmental morphological changes in the undamaged hemisphere in the case of a unilateral injury (see Kolb, Monfils & Sherren, 2008, for a recent summary). We suspected this occurred in the case of DT, as functional brain imaging suggested bifrontal pathophysiology even though her structural injury was unilateral (Eslinger, Grattan, & Damasio, 1992). Hence, early injury to the PFC region, whether unilateral or bilateral, may have reverberating neurobiological and functional consequences that unfold throughout development.

Outcome study of individuals with early PFC damage particularly to frontal polar and inferior-mesial regions, reveals profound deficits in learning of social rules, their elaboration into adaptive knowledge structures, and in social emotions (Eslinger et al., 2004). These appear to be refractory disabilities that severely hamper social, moral, and vocational functioning as adolescents and adults. Some rule-based learning occurs mainly around literal habits and avoidance of punishment, but levels of cooperative and reciprocal behavior are developmentally arrested, limiting their ability to abide by, negotiate, and agree with others on social rules and hence engage in adaptive relationships. The typical social emotions of gratitude, guilt, embarrassment, and pity also appear to be arrested, and most interpersonal emotions of these patients seem to revolve around getting their immediate needs met. There are many aspects of these untoward outcomes that have not been investigated but understanding such impairments and their underlying causes may inform models of brain and social-emotional maturation.

We hypothesize that at least four types of processing and knowledge impairments related to early PFC pathophysiology may be occurring: (1) primitive and literal knowledge of self and others in regard to social understanding of norms, conventions, rules, and concepts (*social knowledge dimension*); (2) role-taking, theory of mind, and perspective-taking deficiencies that alter ability to understand others' intentions, beliefs, mental states, and experiences (*social metacognitive dimension*); (3) emotional impairments in the form of inconsistent contingency-based learning based on rewards and punishments, important for registering consequences, emotional effects and outcomes; as well as shallow forms of emotional interrelatedness such as in social sensitivity, gratitude, pity, and disgust (*emotion dimension*); and (4) executive dysfunction in working memory capacities, self-awareness, and organizing interrelated experiences into stable forms of knowledge (e.g., concepts, values) that can incorporate goal-directed behavior within changing contexts, time frames, and schemas (*executive resource dimension*). In summary, case studies of individuals who suffer early PFC lesions suggest that the PFC and related cortical-subcortical networks provide key neurodevelopmental integration of cognitive, emotional, and behavioral resources for social cognition and social emotions. Hence, we might expect that typically developing children and adolescents engage PFC-related networks in these kinds of social cognitive and social emotional contexts, which can now begin to be examined with noninvasive functional brain-imaging technology.

Functional Brain Imaging of Socio-Moral Cognition and Emotions

The aforementioned clinical studies of early PFC damage suggest the hypothesis that the PFC contributes a crucial neural substrate for the acquisition and operation of social cognition and emotions. Such cognition entails the acquisition of rules and concepts that are necessary for social decision making. Social emotions are linked to the interests and welfare either of society as a whole or other persons (Haidt, 2003; Moll, Eslinger, & Oliveira-Souza, 2003). Social emotions can be prosocial (e.g., gratitude, pity, guilt) or antisocial (e.g., fear disgust, anger), and serve to foster either interpersonal concordance or establish separation and retribution toward others. Both social cognition and social emotions require mentalizing abilities that include theory of mind, perspective taking, and self-other processing as well as emotions and domain-general executive function resources such as inhibitory control, working memory, and organization (see Figure 13.2). Hence, we expect that moral judgment and moral emotion tasks will tap into these social cognition and emotion processes, and be associated with significant PFC mediation. In a larger sense, social cognition and social emotions also draw upon more basic cognitive and emotional mechanisms and resources that, if distorted in their own way, can subsequently alter social adaptation as well.

To test the PFC hypothesis for moral judgment and moral emotions, we undertook functional magnetic resonance imaging (fMRI) in a volunteer sample of 9 typically developing, healthy children and adolescents from 10 to 17 years of age (4 male, 5 female). Preliminary testing indicated that their general intellect, academic achievement, executive functions, and parent-rated inventory of social development were

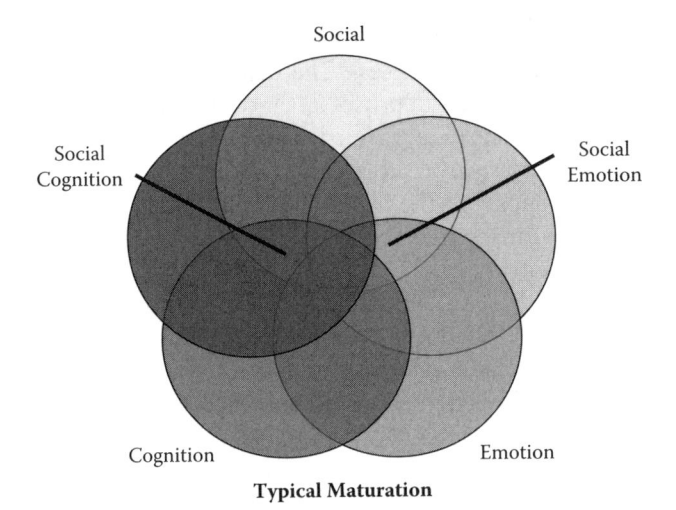

Figure 13.2 Social cognition and social emotions represent important conjunctions of cognitive and emotional resources with social-processing demands for control of actions and self-regulation.

normal. The fMRI studies were completed in a high field (3 tesla) magnet, where participants could view the 1–2 sentence moral vignettes. Participants were acclimated to the magnet and shown examples of the moral judgment and moral emotion tasks for practice ahead of time. The sentence stimuli were viewed through digital goggles in magnet and right–wrong responses in the judgment task were recorded through a handheld device. For the social emotions task, we were interested in capturing their spontaneous reactions, and hence after reading each vignette, participants pressed a button to indicate they had read the sentence. No other task instructions were given. A boxcar design was used that allowed us to present the moral vignettes in experimental blocks that could be contrasted with a baseline condition of neutral sentences that lacked a moral component, but were either factually right or wrong. In this way, brain activity related to attention, eye movements, reading, and verbal working memory could be minimized and the distinctive components of moral judgment and moral emotions could be more readily isolated.

The experimental moral judgment condition included moral actions that were judged in out-of-magnet testing to be rule-based decisions, having high agreement with regard to a moral-right or moral-wrong response. The moral-ambiguous condition presented moral actions that were less clear, and could be viewed from different perspectives, such as telling a white lie or stealing food to feed a starving family (see Table 13.1 for stimulus examples). The experimental moral emotions condition included clusters of prosocial emotions (gratitude, guilt, pity) and antisocial emotions (fear, disgust, anger). Image analysis was completed with SPM2 software and represented here in summary form as brain activation maps. For the moral judgment task, our specific interests were in determining the activations associated with making moral judgments (in contrast to the neutral baseline condition), the comparison of the rule-based judgments versus ambiguous judgments, and whether age regression effects were evident. For the moral emotions task, our interests were in determining the brain activations associated with experiencing these diverse emotions and whether age regression effects were evident.

Moral Judgment Activations

The developmental sample showed 98% agreement on judgment of the rule-based moral-right and moral-wrong vignettes as well as the

Table 13.1 Examples of the Sociomoral Stimuli Used for This Study

Moral Judgment	Example of Stimuli in fMRI Session
Moral right	At recess, I saw a classmate playing alone so I went to play with him.
Moral wrong	Everyone else was picking on Toby, so I started picking on him too.
Moral ambiguous	When my very overweight friend asked if she was fat I said "no."
Neutral right	On a clear and sunny day, the color of the sky is blue.
Neutral wrong	In the game of basketball, a soccer ball is kicked back and forth.
Moral Emotions	
Gratitude	Your Dad knew you wanted a new bike. He worked extra hours and surprised you.
Guilt	You dared your cousin to climb a tree. He fell and broke his leg.
Pity	Your friend came back from fishing in pain. You found a hook through his finger.
Fear	You were walking home from a soccer game. You noticed a stranger was following you.
Disgust	You were at a playground with friends. You saw a boy spit in a girl's face.
Anger	You stayed up late to bake cookies for a school project. You awoke to find your sister and her friends ate them.

Note: The moral judgment task required a right/wrong judgment. The moral emotion stimuli were passively viewed for spontaneous reactions.

nonmoral baseline statements, but split agreement (57% versus 43%, respectively) in judgment of the moral ambiguous actions. Mean response reaction times showed quicker responding for the moral-right and moral-wrong decisions (3821 and 3786 milliseconds, respectively), in contrast to the longer response times (5007 milliseconds) for the moral ambiguous decisions. Since the morally ambiguous actions generated more diverse judgments and engaged more decision-making time, we analyzed these blocks of trials separately in secondary analysis.

When the moral decision-making tasks were first considered together in image analysis, results showed a large cluster of activity in the superior mesial PFC region, extending from the frontal pole to the border of the anterior cingulate (encompassing portions of Brodmann's areas 10, 9, 8 & 32; see Figure 13.3 for activation map). This region of PFC is proximal to the frontal polar and orbito-medial cortical areas identified as common to the cases of early frontal lobe damage with the greatest degree of social adaptation difficulties. This was the largest cluster identified among ones that also included the left superior frontal gyrus, left ventrolateral PFC/lateroposterior orbitofrontal cortex, inferior occipital cortex bilaterally, midline thalamus and globus pallidus, and left amygdala and hippocampus. These results constitute strong support for the hypothesis that the rostral-medial PFC is an important neurobiological substrate during acquisition and maturation of social judgment abilities in typically developing children and adolescents. This is consistent with studies in adults, linking the rostral-medial PFC to processes of mentalizing, social sensitivity, moral judgments, and theory of mind (Gallagher et al., 2000; Moll et al., 2003; Saxe, Carey, & Kanwisher, 2004; Shamay-Tsoory, Tomer, Beger, Goldsher, & Aharon-Peretz, 2005).

As mentioned above, an important contrast involved comparison of the more routine and rule-based moral-right/wrong decisions versus decisions about ambiguous moral situations and actions. These analyses identified what regions were significantly more active under one condition versus the other. Results showed that the moral ambiguous condition sparked significantly greater activity in diverse regions including the frontal polar area bilaterally, the superior parietal cortex bilaterally, the right middle and superior frontal gyri, right precuneus, and fusiform/cerebellar region bilaterally. These are areas most commonly associated with aspects of higher cognition, theory of mind, perspective taking, and social cognition. In contrast, the more rule-based judgments about moral right and wrong recruited memory-related regions, including the right and left hippocampal region, together with the left insula, temporo-parietal junction, temporal

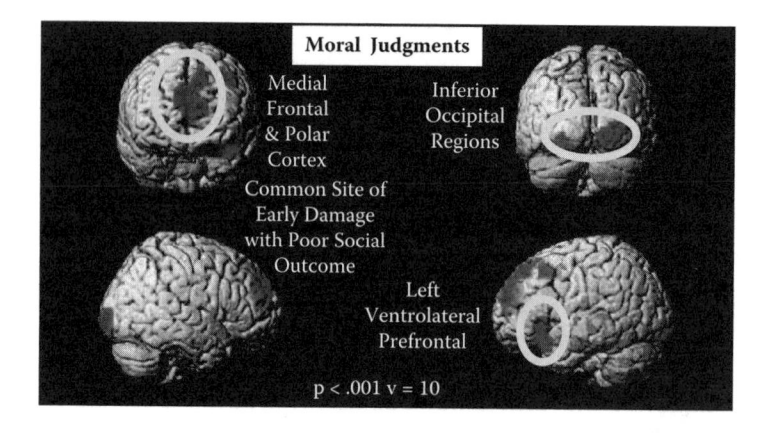

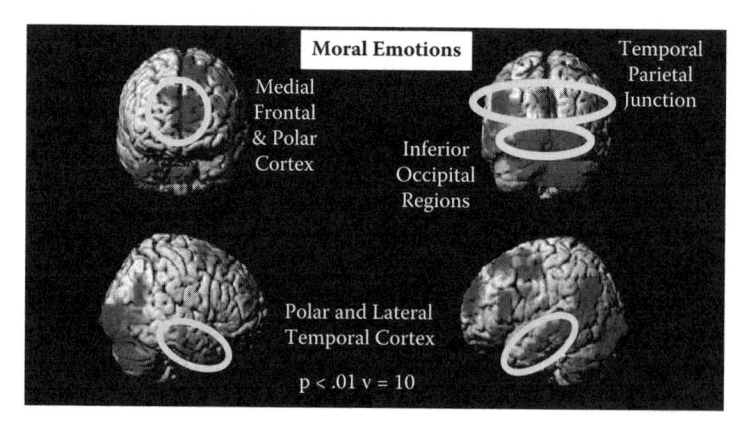

262 • Developmental Social Cognitive Neuroscience

Figure 13.3 (See color insert following p. 242) 3D renderings of brain activations associated with moral judgments and moral emotions in a typical developmental sample 10–17 years of age.

pole, and right occipital pole. These represent marked contrasts in neural activity, considering that nearly all aspects of the stimuli were similar in these conditions except that the contextual ambiguity was highlighted in one, causing recruitment of significantly greater neural resources (see Figure 13.4 for activation map).

Finally, computation of age regressions revealed that the prominent cluster of frontal polar activation did not significantly increase or decrease across the sample range of 10 to 17 years of age, confirming that this region continues to be significantly activated regardless of age, and likely continuing well into adulthood, according to available adult imaging studies. As we eased some of the statistical threshold parameters, we found that the right amygdala, right PFC/polar region,

Crucial Developmental Role of Prefrontal Cortical Systems • 263

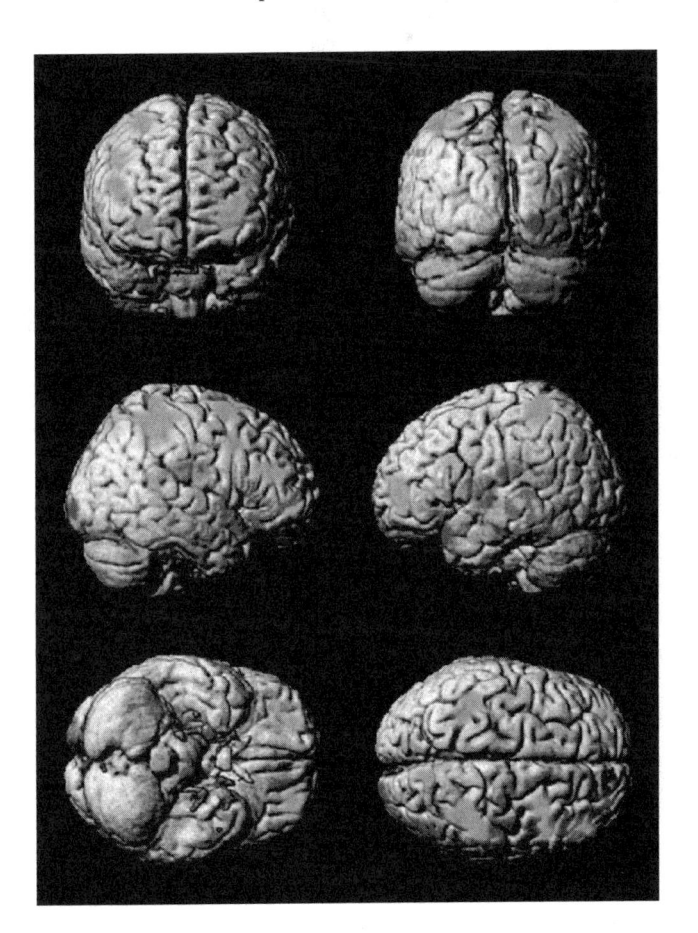

Figure 13.4 (See color insert following p. 242) fMRI contrast analysis showing that the moral ambiguous condition (in comparison to the more rule-based moral right/wrong condition) recruited distinctive activations in the frontal polar and superior parietal regions bilaterally as well as the right dorsolateral prefrontal region (shown in green) while the more rule-based moral right/wrong condition recruited activations primarily in the lateral temporal and temporo-parietal junction regions where general social knowledge and perception have been identified (shown in red).

right posterior superior temporal sulcus, and insula bilaterally became somewhat less active, suggestive possibly of increasing consolidation and automaticity of knowledge and decision-making processes regarding moral actions (see Figure 13.5).

Moral Emotions Activations

Analysis of the moral emotion trials, compiled across all types of prosocial and antisocial stimuli, revealed a significant activation in

Negative Age Regressions
Maturational Changes

Less Right Amygdala Activity

Less Right PFC/Polar Activity

Less Right Posterior STS

Less Insula Activity (Arrows)

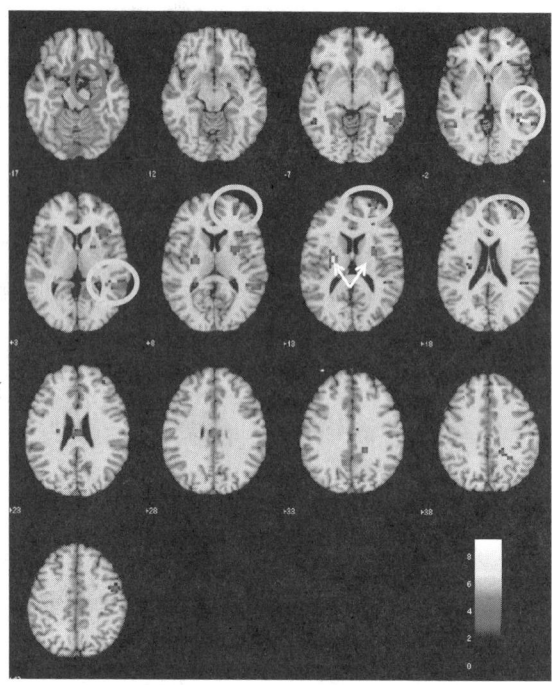

Figure 13.5 (See color insert following p. 242) fMRI brain map showing areas of significant decreasing activity during the moral judgment tasks as a function of age in the 10–17-year-old developmental sample.

the rostral-medial PFC region that overlapped to a large degree with the moral judgment results (see Figure 13.3). In addition, significant activations were detected in the lateral and temporal polar regions bilaterally, suggesting a strong limbic/paralimbic system involvement usually associated with emotion. The temporo-parietal junction, associated with some aspects of theory of mind and mentalizing, and inferior occipital cortices were also recruited. Thus, despite the passive viewing of the moral emotion vignettes, there was significant rostral-medial PFC activity in this age range, supporting an important role for this region in aspects of social emotions, quite possibly the mentalizing components that tap into theory of mind, intentionality, perspective taking, and self-other judgment processes. The additional involvement of the temporal polar areas suggests that more processing resources encompassing emotion were likely recruited. Preliminary analysis of prosocial versus antisocial emotion trials showed that both enlisted rostral-medial PFC activity, but the prosocial emotions also recruited a broader network of frontal-temporal activity in comparison to antisocial emotions. Possibly the prosocial emotions engender more cognitive appraisal, whereas the antisocial emotions engender more basic emotional reactions, but further analysis of an enlarged sample is needed. In addition, it appeared that the different moral emotions were associated with somewhat different configurations of frontal-temporal-limbic system activation patterns which we are now analyzing with an enlarged sample.

Age regression analysis revealed significantly decreasing activity in the right amygdala and superior parietal regions with age, but no significant change in PFC activity through this 10–17-year-old age range. These maturational shifts in brain activity (particularly in the right amygdala, an important emotion processing structure) along with the persistent PFC activation are similar to what we found in the moral judgment condition. It may be the case that even with moral emotional circumstances, emotional resources during socio-moral processing are invoked somewhat less strongly with experience and maturation, but nonetheless PFC regulation of decision making and reactions remains continually active.

SUMMARY

Studies of children and adults with localized damage to the brain and with functional brain imaging converge on the PFC as a crucial neural substrate for socio-moral processing and perhaps as one of the most important neural substrates mediating social adjustment in the world. This is not to conclude that all aspects of social behavior and adjustment are dependent on this region, for it is clear that several interconnected cortical and subcortical regions are necessary for social cognition and social adaptation, and are beginning to be formulated into an identifiable large scale network. Our research has found that the PFC appears to be an extremely important coordinator and contributor to social cognitive and social emotional processing and vital to adaptive self-regulation within social contexts. These roles fit well with the known anatomy and physiology of the PFC and related networks and their long postnatal maturation as well as the trajectories of social development and self-regulation. Hence, research into further linkages between the developmental psychology and developmental neurobiology of the PFC region holds great promise for understanding important aspects of human development and many of the afflictions that arise throughout maturation. Although the core function(s) of the PFC remains imprecisely specified, we suggest that several of the following are possibilities:

- Social (cognitive) knowledge including rules, conventions, and norms
- Interpersonal processing including theory of mind, perspective-taking, and self-awareness
- Executive cognitive resources that include working memory capacities, formulation of goals, planning, and associative linking of actions with consequences
- Emotional resources that are critical for motivation, contingency-based learning based on primary and secondary reinforcers, and social emotional processes such as emotional empathy

It seems likely that several of these possibilities are linked within PFC-related mechanisms, and most particularly in the rostral-medial regions where we see the most significant neurodevelopmental deficits after lesions and the most activity during socio-moral processing in typical development.

REFERENCES

Ackerly, S. S., & Benton A. L. (1948). Report of case of bilateral frontal lobe defect. *Res Publ Assoc Res Nerv Ment Dis 27*: 479–504.

Barbas, H. (1995). Anatomic basis of cognitive–emotional interactions in the primate prefrontal cortex. *Neurosci Biobehav Rev 19*: 499–510.

Benton, A. L. (1991). Prefrontal injury and behavior in children. *Dev Neuropsychol 7*: 275–281.

Blair, C., Zelazo, P. D., & Greenberg, M. T. (2005). The maturation of executive functions in early childhood. *Dev Neuropsychol 28*: 561–571.

Dimitrov, M., Phipps, M., Zahn, T. P., & Grafman, J. (1999). A thoroughly modern Gage. *Neurocase 5*: 345–354.

Eslinger, P. J. (2008). The frontal lobe: executive, emotional, and neurological functions. In P. Marien & J. Abutalebi (Eds), *Neuropsychological research: A review* (pp. 379–408). Psychology Press: Hove and New York.

Eslinger, P. J., & Biddle, K. R. (2008). Prefrontal cortex and the maturation of executive functions, cognitive expertise, and social adaptation. In V. Anderson, R. Jacobs & P. J. Anderson (Eds.), *Executive functions and the frontal lobes: A lifespan perspective* (pp. 299–316). Taylor & Francis: New York.

Eslinger, P. J., & Damasio, A. R. (1985). Severe disturbance of higher cognition after bilateral frontal ablation: Patient EVR. *Neurology 35*: 1731–1741.

Eslinger, P. J., Flaherty-Craig, C., & Benton, A. L. (2004). Developmental outcomes after early prefrontal cortex damage. *Brain Cog 55*: 84–103.

Eslinger, P. J., Grattan, L. M., & Damasio, A. R. (1992). Developmental consequences of childhood frontal lobe damage. *Arch Neurol 49*: 764–769.

Eslinger, P. J., & Tranel, D. (2005). Integrative study of cognitive, social, and emotional processes in clinical neuroscience. *Cog Behav Neuro 18*: 1–4.

Gallagher, H. L., Happe, F., Brunswick, N., Fletcher, P. C., Frith, U., & Frith, C. D. (2000). Reading the mind in cartoons and stories: An fMRI study of "theory of mind" in verbal and nonverbal tasks. *Neuropsychologia 38*: 11–21.

Giedd, J. N. (2008). The teen brain: Insights from neuroimaging. *J Adoles Health 42*: 335–343.

Haidt, J. (2003). The moral emotions. In R. J. Davidson, K. R. Scherer, & H. H. Goldsmith (Eds.), *Handbook of affective sciences*. Oxford University Press: Oxford, UK.

Kolb, B., Monfils, M., & Sherren, N. (2008). Recovery from frontal cortical injury during development. In Anderson V, Jacobs R, & Anderson PJ (Eds.), *Executive functions and the frontal lobes: A lifespan perspective* (pp. 81–104). Taylor & Francis: New York.

Moll, J., Eslinger, P. J., & Oliveira-Souza, R de (2003). Morals and the human brain: A working model. *NeuroReport 14*: 299–305.

Saxe, R., Carey, S., & Kanwisher, M. (2004). Understanding other minds: Linking developmental psychology and functional neuroimaging. *Annual Rev Psy 55*: 87–124.

Saleem, K. S., Kondo, H., & Price, J. L. (2008). Complementary circuits connecting the orbital and medial prefrontal networks with the temporal, insular, and opercular cortex in the macaque monkey. *J Comp Neurol 506*: 659–93.

Shamay-Tsoory, S. G., Tomer, R., Beger, B. D., Goldsher, D., & Aharon-Peretz, J. (2005). Impaired "affective theory of mind" is associated with right ventromedial prefrontal damage. *Cog Behav Neurol 18*: 55–67.

Stiles, J., Reilly, J., Paul, B., & Moses, P. (2005). Cognitive development following early brain injury: Evidence for neural adaptation. *Trends Cog Sci 9*: 136–143.

Stuss, D. T., Gallup, G. G., Jr., & Alexander, M. P. (2001). The frontal lobes are necessary for 'theory of mind.' *Brain 124*: 279–286.

Tranel, D., Bechara, A., & Denburg, N. (2002). Asymmetric functional roles of the right and left ventromedial prefrontal cortices in social conduct, decision-making, and emotional processing. *Cortex 38*: 589–612.

14

CONTRIBUTIONS OF NEUROSCIENCE TO THE UNDERSTANDING OF MORAL REASONING AND ITS DEVELOPMENT

R. James Blair
National Institutes of Health

INTRODUCTION

This paper is based on a talk within a symposium delivered to the Jean Piaget Society in Amsterdam 2007. Before this talk, the presenters were asked to address four questions by the chair, Ulrich Müller. These were:

1. How does neuroscientific research advance our understanding of morality/moral development?
2. Do the findings from neuroscientific research necessitate a reconceptualization of moral development (and if so, to what extent)?
3. What are limitations of neuroscientific research in advancing our understanding of moral development? Can these limitations be overcome?
4. How would a productive collaboration of the neuroscience of morality and traditional psychological research on moral development look like?

My hope in this paper is to address these questions with reference to recent work stressing the critical role of emotion and brain regions such as the amygdala and ventromedial prefrontal cortex (vmPFC) in moral reasoning. But before beginning, it is necessary to briefly sketch what I, at least, consider neuroscientific data. There are some who consider only data obtained by brain-imaging methodologies such as functional magnetic resonance imaging (fMRI) as neuroscientific. Such a restrictive definition will not be followed here. Within this chapter the definition of neuroscientific data is considerably broader and also includes both neuropsychological (on patients with neurological lesions) and neuropsychiatric studies (on patients with psychiatric conditions). While it appears obvious that neuropsychological data should be considered neuroscientific, considerable basic neuroscience work on animals involves lesion methodology; hence, some might question the brain basis of neuropsychiatric work. However, fMRI work with many psychiatric populations, including those referenced in this paper, has allowed considerable progress in specifying their neural pathophysiology (Finger et al., 2008; Frith & Frith, 2006; Marsh et al., 2008).

Importantly, each of these three types of study has important strengths and weaknesses. FMRI studies provide us with information on blood flow changes in the brain with the assumption that regions requiring greater blood are showing greater neural activity. As such, fMRI provides us with excellent information regarding the neural systems engaged by a functional process. However, fMRI does not provide us with information as to whether activity within a neural region is necessary for that functional process to occur; it could reflect spreading activation from a region that is necessary. Neuropsychological studies involve investigating the behavioral performance of patients following neurological lesions. Such studies determine whether a neural region is necessary for a particular functional process; if it is, patients will show significant impairment in their ability to engage that process. However, the (fortunate) scarcity of child neurological patients prevents much developmental neuropsychological work. Neuropsychiatric studies involve investigating the behavioral performance or blood flow through the fMRI of psychiatric populations. They are less able to provide data regarding the criticality of a particular region, given that several regions are likely implicated in the disorder. However, they can importantly allow the determination of whether a particular functional process is necessary for the development of another functional process. In short, data from both these types of study as well as functional imaging work will be cited here.

HOW DOES NEUROSCIENTIFIC RESEARCH ADVANCE OUR UNDERSTANDING OF MORALITY/MORAL DEVELOPMENT?

One of the ways in which neuroscientific research has advanced our understanding of morality has been by providing additional data to distinguish between positions on moral reasoning and its development. Two examples will be considered here: First, the role of neuroscientific data in clarifying whether moral reasoning should be considered a unitary form of processing. Second, the role of neuroscientific data in clarifying the role of rational reasoning in moral reasoning.

The early psychological positions on moral development—those of Piaget and Kohlberg (Kohlberg & Kramer, 1969; Piaget, 1932)—suggested a unitary view of morality that covered all social rules. For example, Piaget believed that it was possible to study the child's moral development by examining the development of the child's understandings of the game of marbles. Children were questioned regarding their theories about how the rules for marbles came about and Piaget argued that this provided information regarding how they believed moral/social rules to emerge. Such a view is echoed in some of the recent neuroscientific writings where all forms of rules are considered to be processed by a system which, while it may consist of multiple components, is considered to process all rules similarly (cf. Moll, Zahn, de Oliveira-Souza, Krueger, & Grafman, 2005). In other words, in the same way that Piaget believed that all social rules were learned similarly whether they concern the game of marbles or interpersonal harm, Moll and colleagues' position (Moll, Zahn et al., 2005) suggested that the same neural system processes all social rules.

In contrast, the domain theorists suggested that there were different systems (i.e., concept-based domains) that were involved in processing different forms of social rule (Nucci, 1982; Smetana, 1993; Turiel, Killen, & Helwig, 1987). In particular, a distinction was made between "moral" rules (involving harm to others and issues of justice; e.g., one person hitting another) and conventional rules (involving issues of social disruption; e.g., one person talking to another during a school lesson).

The neuroscientific data has been far more consistent with the domain theorists. Thus, there are neuropsychiatric populations who appear to show selective impairment for moral transgressions (cf. psychopathy; Blair, 1995), while others may show selective impairment for the processing of conventional items (cf. childhood bipolar disorder; McClure et al., 2005). These neuropsychiatric data indicate that different forms of social rule (moral and conventional) can be selectively impaired, which could be predicted from the domain approach but would be incompatible with accounts suggesting that all rules are learned, or processed, in the same way.

The fMRI literature has concentrated on care-based moral items (transgressions which lead to harm to others; cf. Blair, 2007) and particularly implicated vmPFC, posterior cingulate cortex, and the amygdala (Greene, Sommerville, Nystrom, Darley, & Cohen, 2001; Heekeren et al., 2005; Luo et al., 2006; Moll, de Oliveira-Souza, Eslinger et al., 2002). These regions, for example, show greater activity when the individual processes care-based moral items rather than grammatical rule violations. However, some recent fMRI work has indicated that "purity"-based transgressions (transgressions associated with disgust) are associated with insula activity (Moll, de Oliveira-Souza et al., 2005). The insula is critical for taste aversion learning (learning that particular foods are disgusting), as well as being responsive to the disgusted expressions of others (Phillips et al., 1998; Phillips et al., 1997; Sprengelmeyer, Rausch, Eysel, & Przuntek, 1998). As such, the fMRI data has effectively extended the domain theorists' position. The data suggests that rather than there being a unitary "moral" domain, there are separable "domains," or at least information-processing systems, for the processing of different types of moral rules—those that are care-based as opposed to those that are disgust-based.

In addition to providing data to distinguish between positions on moral reasoning/development, neuroscientific research has allowed additional tests of the fundamental assumptions of dominant positions. For example, although the domain theorists differed from the earlier theorists, Piaget and Kohlberg, in proposing separable systems mediating different forms of social rule, they made suggestions with respect to how moral development was achieved that were similar to those of the earlier theorists. In particular, they also stressed the importance of abstract reasoning and conceptual transformations. There was limited reference, or importance given, to emotional processing. For example, Turiel and colleagues (1987) wrote that although they did not "exclude the emotional features of individuals' moral and social orientations" they suggested that "children's judgments and conceptual transformations ... are a central aspect of moral and social convention" (p. 170; Turiel et al., 1987).

Such positions suggested three predictions that could be examined through neuroscientific research:

Prediction 1

If rational thought is critical for moral reasoning, it seems plausible that moral reasoning should be associated with activity in those regions of the brain associated with rational thought.

This prediction is not easy to test precisely. Defining rational thought and the process of conceptual transformation at the neural level is difficult. However, one potential way to consider these processes is to associate them with executive functioning. Executive functioning allows the control and regulation of other cognitive systems (cf. Shallice, 1988). As such, it can be suggested that for rational thought to occur, at least some executive functions must be implicated. In essence it is assumed that rational thought is executive functioning or that rational thought corresponds to a subset of executive functions. This assumption is important because it allows us to suggest brain regions that are likely to be implicated in rational thought because they are known to be implicated in nonemotion-based executive functioning. These regions include dorsomedial and lateral frontal, and parietal cortices (Botvinick, Cohen, & Carter, 2004; Kastner & Ungerleider, 2000; Miller & Cohen, 2001; Roberts, Robbins, & Weiskrantz, 1998; Shallice & Burgess, 1996). Thus, it might be predicted that neuro-imaging studies of moral reasoning should implicate these regions if moral reasoning requires rational thought.

For the most part, fMRI studies of moral reasoning disconfirm this prediction. Neuroimaging studies of moral reasoning consistently implicate ventromedial prefrontal cortex, posterior cingulate cortex and, to a lesser extent, the amygdala (Greene et al., 2001; Heekeren et al., 2005; Luo et al., 2006; Moll, de Oliveira-Souza, Eslinger et al., 2002); see Figure 14.1 for depiction of these regions. For example, Greene and colleagues reported increased vmPFC activity to personal as opposed to impersonal moral choices, where the difference between these two situations relates effectively to the salience of the victim (Greene et al., 2001). Similarly, in more recent work, Luo et al. demonstrated increased amygdala and vmPFC activity to more severe, relative to less severe, moral transgressions (Luo et al., 2006). These regions are not implicated in nonemotion-based executive functioning. Instead, they are consistently implicated in emotional processing (Blair, 2007; LeDoux, 2007; Schoenbaum & Roesch, 2005; Valentin, Dickinson, & O'Doherty, 2007).

It is worth considering here the extremely interesting work on "trolley problems" (Greene et al., 2001; Koenigs et al., 2007; Mikhail, 2007). In these problems and their variants, a circumstance (e.g., a runaway trolley) is described as about to cause substantial loss of life (five people

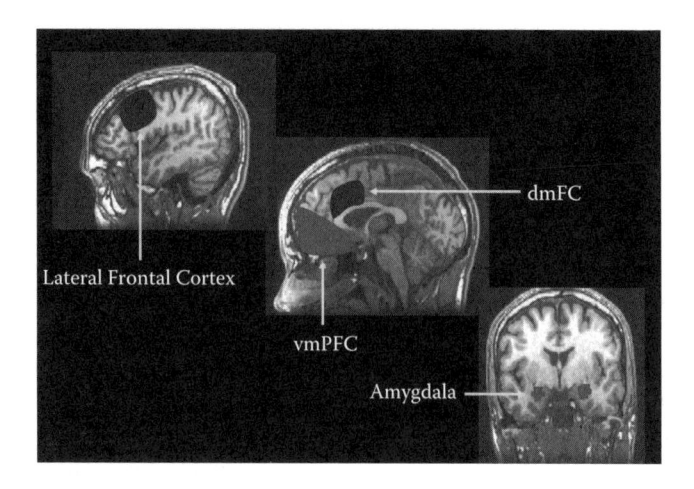

274 • Developmental Social Cognitive Neuroscience

Figure 14.1 Neuroimaging studies of moral reasoning implicate VmPFC, posterior cingulate cortex (dmFC), and the amygdala.

will be killed if it proceeds on its present course). Individual variants on the problem provide the participant with different options for saving the five. These options typically have implications for at least one other individual. For example, one option is to hit a switch that will turn the trolley onto an alternate set of tracks where it will kill one person instead of five (impersonal dilemma). Another option is to push a stranger off the bridge onto the tracks below (personal dilemma). This person will die but his body will stop the trolley from reaching the five. Most people choose to save the life of the five, if the option to do so involves hitting the switch. Most people choose not to save the five, if the option involves pushing another individual on to the tracks (Hauser, 2006).

The big difference between whether one saves the five or the one may relate to the salience of the suffering of the one. In the personal dilemmas, the single victim's distress appears more salient than in the impersonal dilemmas. FMRI work has shown that regions implicated in emotional responding (e.g., ventromedial prefrontal cortex) are significantly more activated to personal relative to impersonal dilemmas (Greene, Nystrom, Engell, Darley, & Cohen, 2004; Greene et al., 2001). Interestingly, lesions of vmPFC increase the probability that the individual will save the five rather than the one (Koenigs et al., 2007). VmPFC is implicated in representing reinforcement outcome expectancies; i.e., representing emotional information relevant for decision making (Blair, 2007; Schoenbaum & Roesch, 2005; Valentin et al., 2007). It can be argued that by representing the expectation of the one salient victim, the individual's decision making is biased to avoid hurting the one salient victim at the expense of the nonsalient five individuals (Blair, 2007).

Of course, the above data on trolley problems further emphasizes the importance of emotion in moral reasoning. However, it is worth noting the neural systems engaged when individuals perform trolley problems where the salience of the one victim is not so stressed; the impersonal items in the terminology of Greene (Greene et al., 2001). When the individual decides to save, by pressing the button, the five rather than the one, the regions implicated are, in fact, those regions of frontal and parietal cortices that are implicated in executive functioning. In other words, it is possible that some forms of moral reasoning (i.e., reasoning when considering impersonal dilemmas) might implicate an abstract reasoning process akin to that suggested by the early theorists. However, it is difficult to make more concrete conclusions about this, given the absence of detail regarding the nature of the potential reasoning process. Moreover, it is important to remember that this, at least currently, appears to reflect a fraction of the generated examples of moral reasoning.

Prediction 2

Individuals who show impaired moral reasoning should show some form of impairment in rational thought.

There are two populations, one neurological and one neuropsychiatric, who show dramatically impaired moral reasoning. The first population is neurological patients with acquired lesions of vmPFC. As noted above, such patients show anomalous performance on trolley problems, choosing to save the five rather than one, even in cases where the salience of the one's victim status is high (Koenigs et al., 2007). The second population is individuals with psychopathy.

Psychopathy is a developmental disorder (Lynam, Caspi, Moffitt, Loeber, & Stouthamer-Loeber, 2007) that involves emotional dysfunction, characterized by reduced guilt, empathy, and attachment to significant others, and antisocial behavior including impulsivity and poor behavioral control (Hare, 1991). Individuals with psychopathic tendencies show significantly less of a moral or conventional distinction than healthy individuals (Blair, 1995).

According to the domain theorists, moral transgressions (e.g., one person hitting another) are defined by their consequences for the rights and welfare of others (Nucci, 1982; Smetana, 1993; Turiel et al.,

1987). Social conventional transgressions are defined as violations of the behavioral uniformities that structure social interactions within social systems (e.g., dressing in opposite-gender clothing). Healthy individuals distinguish conventional and moral transgressions in their judgments from the age of 39 months (Smetana, 1981) and across cultures (Song, Smetana, & Kim, 1987). In particular, moral transgressions are judged less rule contingent than conventional transgressions; individuals are less likely to state that moral, rather than conventional, transgressions are permissible in the absence of prohibiting rules (Turiel et al., 1987).

When individuals with psychopathy are presented with the moral/conventional distinction test, they are more likely than healthy individuals to allow moral transgressions in the absence of rules (Blair, 1997). In addition, they make less distinction between moral and conventional transgressions in the absence of prohibiting rules (Blair, 1995, 1997). Similar impairment has been reported in other aggressive populations who use antisocial behavior to achieve their goals, such as behaviorally disruptive adolescents (Arsenio & Fleiss, 1996; Nucci & Herman, 1982).

According to the second prediction, these two populations who show such dramatically impaired moral reasoning should show some form of impairment in rational thought/conceptual transformation. However, patients with vmPFC lesions are remarkable in that they fail to show impairment on nonaffect-based executive function tasks (unless the lesion extends beyond vmPFC) (Damasio, 1994; Shallice & Burgess, 1991). Similarly, individuals with psychopathy, who show amygdala and vmPFC dysfunction (Blair, 2007), show no indications of impairment in nonaffect-based executive function (K. S. Blair et al., 2006; LaPierre, Braun, & Hodgins, 1995; Mitchell, Colledge, Leonard, & Blair, 2002). Moreover, there are no indications of significant impairment in semantic reasoning (Blair, Mitchell, & Blair, 2005). In short, Prediction 2 appears disconfirmed.

Prediction 3
Individuals with some form of impairment in rational thought must show impairment in moral reasoning.

There is a neuropsychiatric condition that is not only associated with significant executive dysfunction (and thus presumably impairment in rational thought; Hughes, Russell, & Robbins, 1994; Pennington & Ozonoff, 1996) but also impairment in conceptual transformation, at least by some accounts (Gopnik & Meltzoff, 1997). This condition is autism. Autism is a severe developmental disorder described by the American Psychiatric Association's diagnostic and statistical manual (DSM-IV) as "the presence of markedly abnormal or impaired development in social interaction and communication and a markedly restricted repertoire of activities and interests" (American Psychiatric Association, 1994, p. 66). Individuals with autism show significant impairment in executive functioning (Hughes et al., 1994; Pennington & Ozonoff, 1996) and Theory of Mind (Frith & Happe, 2005).

According to Prediction 3, it might be expected that individuals with autism should show impairment in moral reasoning. However, individuals with autism do not show impairment in those aspects of moral reasoning that do not require the representation of another's intentions (i.e., those aspects that do not require theory of mind). Thus, individuals with autism make the moral/conventional distinction (Blair, 1996; Leslie, Mallon, & DiCorcia, submitted; Steele, Joseph, & Tager-Flusberg, 2003). They even, like typically developing children, fail to consider it wrong to commit an action that resulted in another's distress if the other individual's distress was unreasonable (e.g., they were crying because, although they had already eaten a cookie, they were not being allowed to eat the protagonist's cookie) (Leslie et al., submitted). In short, Prediction 3 appears disconfirmed.

Summary

Perhaps, the main contribution of neuroscientific research with respect to advancing our understanding of morality/moral development is that it allows greater scope regarding hypothesis testing. It is very difficult to test whether rational thought/conceptual transformation is necessary for moral development when working only with typically developing children. Indeed, the claim was not really tested, probably because of this difficulty. However, the three predictions that can be generated by the claim can, and have been, tested using fMRI as well as with neuropsychological and neuropsychiatric populations. All three predictions, at least as well as they can be tested, have been disconfirmed. Moreover, the process of generating these predictions fuels demands with respect to greater precision in researchers' theories. For example, it could be argued that the above tests of the claims are misdirected, that the original authors' did not conceptualize rational thought in this way. However, at the very least, the above data demands greater detail from rational thought positions as to what precise computational process is being considered.

DO THE FINDINGS FROM NEUROSCIENTIFIC RESEARCH (AND IF SO, TO WHAT EXTENT) NECESSITATE A RECONCEPTUALIZATION OF MORAL DEVELOPMENT?

There are three main areas where recent neuroscientific findings appear to have necessitated a reconceptualization of ideas regarding moral development and moral reasoning. These are:

1. The importance of rational thought. This first has been principally covered in the section above. In contrast to earlier positions, the neuroscientific data has not supported claims regarding the role of rational thought in moral reasoning and moral development. This research has suggested that rational thought is not central to all aspects of moral reasoning.
2. The importance of conceptual knowledge. According to the domain theorists, the child generates "understandings of the social world by forming intuitive theories regarding experienced social events" (Turiel et al., 1987, p. 170). The differing social consequences of moral and conventional transgressions result in the child's development of separate theoretical structures (domains) for moral and conventional transgressions. In short, the domain theorists regard an individual's verbal reasoning as critical data in understanding their moral reasoning as it allows access to the theoretical structures upon which moral reasoning is thought to rely.

In contrast, the majority of the recent moral neuroscience positions have denied the importance of verbalizable theories arguing that judgments of permissibility are reliant on emotional responses. This claim was first made in Blair's account of the development of the moral/conventional distinction, the violence inhibition mechanism model (Blair, 1995), and then particularly well articulated by Haidt (2001). Blair's model suggested the existence of a basic emotional learning system that, when activated by distress cues—cues indicating the fear and sadness of others—associated the aversiveness of the activation elicited by these cues with representation of the actions that had given rise to those cues. Thus, the individual would come to find actions that harmed others aversive and regard them as bad. This position thus suggested no role for an individual's verbalizable theories about actions in the decision-making process. In fact, the model suggested that the basis of the individual's verbalizable theories about care-based transgressions were a function of the operation of this basic emotion learning system. When activated, it increased attention (the representational strength) of the stimuli that activated it; i.e., distress cues. Thus, if the individual engaged in a causal analysis, they would be significantly more likely to regard the reason for their sense of wrong as the victim's harm. In the future, when asked why the action was wrong to do, they would thus be more likely to say that it was wrong because it harmed others.

A caveat should be mentioned here, however. In an important criticism of Blair's (1995) model, a criticism applicable to most other emotion-based positions on morality, Nichols (2004) showed that while emotional systems might generate permissibility judgments, they did not allow an individual to make immorality judgments. The same aversive affect should be present whether a victim hurt themselves or was hurt by another; i.e., this affect can generate judgments of "badness" but not "immorality" (Nichols, 2004). It is possible that judgments of immorality, rather than simple permissibility, may require access to verbalizable theories, knowledge that a particular action is considered within the society to be wrong.

In short, it can be argued that the moral neuroscience literature has led to a reconceptualization regarding the importance of the individual's theories. These may be necessary for judgments of immorality (cf. Nichols, 2004). But such judgments also require appropriate emotional responses (see also Blair, 2007).

3. The importance of emotion. Before the mid-90s, for the most part, emotion was not considered central with respect to the moral development literature. There were some who considered emotion to be important. For example, Kagan argued that differences between moral and conventional transgressions were due to moral transgressions being associated with emotional reactions while conventional ones were not (Kagan & Lamb, 1987). There was also consideration regarding the importance of empathic reactions for curtailing antisocial behavior (e.g., P. A. Miller & Eisenberg, 1988; Perry & Perry, 1974). But the consensus view on moral reasoning was certainly grounded in ideas of rational thought and conceptual transformation.

However, if aspects of emotional responding were necessary for the development of morality, then there was a clear prediction: a neuropsychiatric population impaired in those aspects of emotional processing should show impaired moral reasoning. Specifically, Blair argued that if individuals with psychopathy were impaired in emotional responding to the distress of others, then they should show problems with care-based moral reasoning and might perform atypically on the moral conventional distinction task (Blair, 1995, 1997). This prediction was confirmed, as were later predictions concerning the nature of the emotional impairment in psychopathy. This impairment reflected their reduced ability to respond to the distress of others (Aniskiewicz, 1979; Blair, Colledge, Murray, & Mitchell, 2001; Blair, Jones, Clark, & Smith, 1997; Dadds et al., 2006) and impairment in stimulus-reinforcement learning (Blair, Leonard, Morton, & Blair, 2006; Flor, Birbaumer, Hermann, Ziegler, & Patrick, 2002), capacities thought to be critical for moral socialization (Blair, 2007; Blair, 1995). Importantly, parenting studies demonstrated that the children showing the emotional dysfunction seen in psychopathy were less easy to socialize through standard socialization practices (Oxford, Cavell, & Hughes, 2003; Wootton, Frick, Shelton, & Silverthorn, 1997).

Following on from this work, a series of neuroimaging studies examining moral reasoning tasks involving responses to trolley problems (Greene et al., 2004; Greene et al., 2001), passive responding to scenes of moral transgressions (Harenski & Hamann, 2006; Moll, de Oliveira-Souza, Eslinger et al., 2002), judging descriptions of behaviors as moral or immoral (Borg, Hynes, Van Horn, Grafton, & Sinnott-Armstrong, 2006; Heekeren et al., 2005; Heekeren, Wartenburger, Schmidt, Schwintowski, & Villringer, 2003; Moll, De Oliveira-Souza, Bramati, & Grafman, 2002), or performance of a morality implicit association task (Luo et al., 2006). Many of these studies report amygdala activity (Greene et al., 2004; Harenski & Hamann, 2006; Heekeren et al., 2005; Luo et al., 2006; Moll, de Oliveira-Souza, Eslinger et al., 2002); all of the picture-based studies (Harenski & Hamann, 2006; Luo et al., 2006; Moll, de Oliveira-Souza, Eslinger et al., 2002) though only 2/6 text-based studies (Greene et al., 2004; Heekeren et al., 2005). All report vmPFC activity. In an interesting convergence of the neuropsychiatric and neuroimaging literatures, both the amygdala and vmPFC are considered to be dysfunctional in psychopathy (for a review of this literature, see Blair, 2007).

Summary

The neuroscientific research has necessitated a major reconceptualization of our understanding of moral development. Specifically, it is now clear that emotion plays a central role in moral development and reasoning. Indeed, it now appears likely that different emotional learning systems play roles in the development of different forms of social prohibition; i.e., victim-based "moral" transgressions, hierarchy-based social conventions, disgust-based prohibitions regarding eating and sexual habits, and justice-based rules (for discussion of these systems are their roles, see Blair, 2007).

WHAT ARE LIMITATIONS OF NEUROSCIENTIFIC RESEARCH IN ADVANCING OUR UNDERSTANDING OF MORAL DEVELOPMENT? CAN THESE LIMITATIONS BE OVERCOME?

All methodologies have limitations. For example, it was extremely difficult for the developmental theorists to test their assumptions regarding the central role of rational thought in moral development when only studying typically developing children. With respect to fMRI, it is extremely difficult to be sure that a region activated during moral reasoning is necessarily involved in moral reasoning. It could simply reflect the receipt of input from another region that is necessarily involved. In other words, because superior temporal cortex (STC) and vmPFC are frequently activated in moral reasoning paradigms, it could be that only STC is necessary for moral reasoning and that the vmPFC activity results only as a consequence of input from STC.

Neuropsychological studies can identify whether a region is necessarily involved; if the function is disrupted after the lesion, then it can be assumed that the region is necessary for the function. Indeed, we can discount the possibility raised immediately above because the impact of damage to vmPFC on moral decision making has been documented (Koenigs et al., 2007). However, neuropsychological studies have not been so informative with respect to development. Neuropsychiatric studies have been considerably informative developmentally. Thus, work with individuals with autism has allowed us considerable insight into the development of theory of mind (Frith & Frith, 2006). Moreover, work with both individuals with autism and individuals with psychopathy has informed us about the development of morality (Blair, 2007). We know from work with individuals with autism, for example, that theory of mind is not necessary for the development of the fundamental distinction between care-based moral and conventional transgressions (Blair, 1996). Similarly, we can infer from work with individuals with psychopathy that the basic emotional response to the distress of others is necessary for the development of moral and conventional distinctions (Blair, 2007). However, work with neuropsychiatric populations, when not accompanied by neuroimaging, says little about critical brain regions.

In short, all research methodologies have limitations but all of these limitations can be at least partially overcome by using counterpart methodologies. As such, with the increased variety of experimental techniques now available, progress in understanding moral reasoning and moral development is rapid.

WHAT WOULD A PRODUCTIVE COLLABORATION OF THE NEUROSCIENCE OF MORALITY AND TRADITIONAL PSYCHOLOGICAL RESEARCH ON MORAL DEVELOPMENT LOOK LIKE?

In some respects, it may be possible to answer this question by examining the field at the moment. For example, Joshua Greene began his scientific career as a philosopher. His initial work examined the trolley problem, a form of problem that has exercised moral philosophers for some time (Greene et al., 2001). However, more recently, his work has made direct contact with more traditional psychological research (Greene et al., 2004). Mark Hauser has been recently combining the cutting edge neuroscience techniques of fMRI and transmagnetic stimulation (TMS) with developmental psychological theory concerning theory of mind in his recent investigations of morality (Young, Hauser, & Saxe, 2007). My own work began by applying traditional developmental psychology measures of moral development, the moral/conventional distinction test, to neuropsychiatric populations (Blair, 1995, 1997). More recently, the focus has shifted to determining the functional contributions of the regions implicated in prior fMRI work on moral reasoning using novel paradigms (Luo et al., 2006). However, I am also currently engaged in collaborative work with well-known domain theorists, Larry Nucci and Judith Smetana, specifically examining regions implicated in mediating the moral/conventional distinction test. Incidentally, this collaboration was a direct result of the symposium of the Jean Piaget Society that generated the questions discussed in the current paper.

CONCLUSIONS

The goal of this paper was to consider the usefulness of the neuroscientific approaches for the understanding of moral development and moral reasoning. I believe these approaches have been highly informative. They have allowed a series of predictions of early moral developmental positions concerning rational thought and theory transformations to be, in principle, tested. Perhaps, almost as important, they have forced greater precision on the field: What is rational thought? How should we consider theoretical transformations?

By allowing the testing of these predictions, they have enabled our understanding of moral development and moral reasoning to significantly advance. This advance has been coupled with some degree of reconceptualization of the field. On the basis of the neuroscientific data, the earlier emphasis on rational thought and theoretical transformation has been found wanting. Instead, it has become clear that to understand both moral development and moral reasoning it is necessary to consider the role of emotion. Here again, the advantage of the addition of other methodologies and approaches with respect to great precision of theoretical positions can be seen. There are grounds for thinking that rational thought, as least when conceptualized as nonaffect-based forms of executive functioning, is important for impersonal moral reasoning, to use the terminology of Greene. Similarly, judgments of immorality may require accessing a semantic memory concerning the individual's knowledge of what actions are socially prohibited.

Of course, there are limitations to any of the neuroscientific methodologies. There are considerable limitations to only relying on fMRI data. But these limitations can be overcome by combining these data with those obtained through neuropsychological and neuropsychiatric studies—as well as classic developmental psychology work. Moreover, it is important to remember that any of these methodologies, including developmental psychology work, have limitations if they are considered in isolation.

The last question asked by our symposium chair Ulrich Müller—"What would a productive collaboration of the neuroscience of morality and traditional psychological research on moral development look like?"—I would like to think would be answered that it resembled the current field. I believe this collaboration is occurring. Without doubt, it could become more extensive; the practice of ignoring data from other methodologies is still followed but, I hope, this practice is becoming the exception rather than the rule.

ACKNOWLEDGMENTS

This research was supported by the Intramural Research Program of the National Institutes of Health: National Institute of Mental Health.

REFERENCES

Aniskiewicz, A. S. (1979). Autonomic components of vicarious conditioning and psychopathy. *Journal of Clinical Psychology, 35*, 60–67.

Arsenio, W. F., & Fleiss, K. (1996). Typical and behaviourally disruptive children's understanding of the emotion consequences of socio-moral events. *British Journal of Developmental Psychology, 14*, 173–186.

Blair, K. S., Leonard, A., Morton, J., & Blair, R. (2006). Impaired decision making on the basis of both reward and punishment information in individuals with psychopathy. *Personality and Individual Differences, 41*, 155–165.

Blair, K. S., Newman, C., Mitchell, D. G., Richell, R. A., Leonard, A., Morton, J. et al. (2006). Differentiating among prefrontal substrates in psychopathy: Neuropsychological test findings. *Neuropsychology, 20*(2), 153–165.

Blair, R. (2007). The amygdala and ventromedial prefrontal cortex in morality and psychopathy. *Trends in Cognitive Science, 11*(9), 387–392.

Blair, R. (1995). A cognitive developmental approach to morality: Investigating the psychopath. *Cognition, 57*, 1–29.

Blair, R. (1996). Brief report: Morality in the autistic child. *Journal of Autism and Developmental Disorders, 26*, 571–579.

Blair, R. (1997). Moral reasoning in the child with psychopathic tendencies. *Personality and Individual Differences, 22*, 731–739.

Blair, R., Colledge, E., Murray, L., & Mitchell, D. G. (2001). A selective impairment in the processing of sad and fearful expressions in children with psychopathic tendencies. *Journal of Abnormal Child Psychology, 29*(6), 491–498.

Blair, R., Jones, L., Clark, F., & Smith, M. (1997). The psychopathic individual: A lack of responsiveness to distress cues? *Psychophysiology, 34*, 192–198.

Blair, R., Mitchell, D. G. V., & Blair, K. S. (2005). *The psychopath: Emotion and the brain.* Oxford: Blackwell.

Borg, J. S., Hynes, C., Van Horn, J., Grafton, S., & Sinnott-Armstrong, W. (2006). Consequences, action, and intention as factors in moral judgments: An FMRI investigation. *Journal of Cognitive Neuroscience, 18*(5), 803–817.

Botvinick, M. M., Cohen, J. D., & Carter, C. S. (2004). Conflict monitoring and anterior cingulate cortex: An update. *Trends in Cognitive Science, 8*(12), 539–546.

Dadds, M. R., Perry, Y., Hawes, D. J., Merz, S., Riddell, A. C., Haines, D. J. et al. (2006). Attention to the eyes and fear-recognition deficits in child psychopathy. *British Journal of Psychiatry, 189*, 280–281.

Damasio, A. R. (1994). *Descartes' error: Emotion, rationality and the human brain.* New York: Putnam (Grosset Books).

Finger, E., Marsh, A. A., Mitchell, D. G. V., Reid, M. E., Sims, C., Budhani, S. et al. (2008). Abnormal ventromedial prefrontal cortex function in children with psychopathic traits during reversal learning. *Archives of General Psychiatry, 65*(5), 586–594.

Flor, H., Birbaumer, N., Hermann, C., Ziegler, S., & Patrick, C. J. (2002). Aversive Pavlovian conditioning in psychopaths: Peripheral and central correlates. *Psychophysiology, 39*, 505–518.

Frith, C. D., & Frith, U. (2006). The neural basis of mentalizing. *Neuron, 50*(4), 531–534.

Frith, U., & Happe, F. (2005). Autism spectrum disorder. *Current Biology, 15*(19), R786–790.

Gopnik, A., & Meltzoff, A. N. (1997). *Words, thoughts, and theories.* Cambridge, MA: The MIT Press.

Greene, J. D., Nystrom, L. E., Engell, A. D., Darley, J. M., & Cohen, J. D. (2004). The neural bases of cognitive conflict and control in moral judgment. *Neuron, 44,* 389–400.

Greene, J. D., Sommerville, R. B., Nystrom, L. E., Darley, J. M., & Cohen, J. D. (2001). An fMRI investigation of emotional engagement in moral judgment. *Science, 293,* 1971–1972.

Haidt, J. (2001). The emotional dog and its rational tail: A social intuitionist approach to moral judgment. *Psychological Review, 108*(4), 814–834.

Hare, R. D. (1991). *The hare psychopathy checklist—revised.* Toronto, Ontario: Multi-Health Systems.

Harenski, C. L., & Hamann, S. (2006). Neural correlates of regulating negative emotions related to moral violations. *Neuroimage, 30*(1), 313–324.

Hauser, M. (2006). *Moral minds: How nature designed our universal sense of right and wrong.* New York: Harper Collins.

Heekeren, H. R., Wartenburger, I., Schmidt, H., Prehn, K., Schwintowski, H. P., & Villringer, A. (2005). Influence of bodily harm on neural correlates of semantic and moral decision-making. *NeuroImage, 24,* 887–897.

Heekeren, H. R., Wartenburger, I., Schmidt, H., Schwintowski, H. P., & Villringer, A. (2003). An fMRI study of simple ethical decision-making. *Neuroreport, 14,* 1215–1219.

Hughes, C., Russell, J., & Robbins, T. W. (1994). Evidence for executive dysfunction in autism. *Neuropsychologia, 32,* 477–492.

Kagan, J., & Lamb, S. (1987). *The emergence of morality in young children.* Chicago: University of Chicago Press.

Kastner, S., & Ungerleider, L. G. (2000). Mechanisms of visual attention in the human cortex. *Annual Review of Neuroscience, 23,* 315–341.

Koenigs, M., Young, L., Adolphs, R., Tranel, D., Cushman, F., Hauser, M. et al. (2007). Damage to the prefrontal cortex increases utilitarian moral judgements. *Nature, 446,* 908–911.

Kohlberg, L., & Kramer, R. (1969). Continuities and disconuities in childhood and adult moral development. *Human Development, 12,* 93–120.

LaPierre, D., Braun, C. M. J., & Hodgins, S. (1995). Ventral frontal deficits in psychopathy: Neuropsychological test findings. *Neuropsychologia, 33,* 139–151.

LeDoux, J. (2007). The amygdala. *Current Biology, 17*(20), R868–R874.

Leslie, A. M., Mallon, R., & DiCorcia, J. A. (submitted). Transgressors, victims and cry babies: Is basic moral judgment spared in autism?

Luo, Q., Nakic, M., Wheatley, T., Richell, R., Martin, A., & Blair, R. (2006). The neural basis of implicit moral attitude—an IAT study using event-related fMRI. *NeuroImage, 30*(4), 1449–1457.

Lynam, D. R., Caspi, A., Moffitt, T. E., Loeber, R., & Stouthamer-Loeber, M. (2007). Longitudinal evidence that psychopathy scores in early adolescence predict adult psychopathy. *Journal of Abnormal Psychology, 116*(1), 155–165.

Marsh, A. A., Finger, E. C., Mitchell, D. G. V., Reid, M. E., Sims, C., Kosson, D. S. et al. (2008). Reduced amygdala response to fearful expressions in children and adolescents with callous–unemotional traits and disruptive behavior disorders. *American Journal of Psychiatry, 165*(6), 712–720.

McClure, E. B., Treland, J. E., Snow, J., Schmajuk, M., Dickstein, D. P., Towbin, K. E. et al. (2005). Deficits in social cognition and response flexibility in pediatric bipolar disorder. *American Journal of Psychiatry, 162*(9), 1644–1651.

Mikhail, J. (2007). Universal moral grammar: Theory, evidence and the future. *Trends in Cognitive Science, 11*, 144–152.

Miller, E. K., & Cohen, J. D. (2001). An integrative theory of prefrontal cortex function. *Annual Review of Neuroscience, 24*, 167–202.

Miller, P. A., & Eisenberg, N. (1988). The relation of empathy to aggressive and externalizing/antisocial behavior. *Psychological bulletin, 103*, 324–344.

Mitchell, D. G. V., Colledge, E., Leonard, A., & Blair, R. (2002). Risky decisions and response reversal: Is there evidence of orbitofrontal cortex dysfunction in psychopathic individuals? *Neuropsychologia, 40*, 2013–2022.

Moll, J., De Oliveira-Souza, R., Bramati, I. E., & Grafman, J. (2002). Functional networks in emotional moral and nonmoral social judgments. *Neuroimage, 16*, 696–703.

Moll, J., de Oliveira-Souza, R., Eslinger, P. J., Bramati, I. E., Mourao-Miranda, J., Andreiuolo, P. A. et al. (2002). The neural correlates of moral sensitivity: A functional magnetic resonance imaging investigation of basic and moral emotions. *Journal of Neuroscience, 22*(7), 2730–2736.

Moll, J., de Oliveira-Souza, R., Moll, F. T., Ignacio, F. A., Bramati, I. E., Caparelli-Daquer, E. M. et al. (2005). The moral affiliations of disgust: A functional MRI study. *Cognitive & Behavioral Neurology, 18*(1), 68–78.

Moll, J., Zahn, R., de Oliveira-Souza, R., Krueger, F., & Grafman, J. (2005). Opinion: The neural basis of human moral cognition. *Nature Reviews Neuroscience, 6*(10), 799–809.

Nichols, S. (2004). *Sentimental rules: On the natural foundations of moral judgment*. New York: Oxford University Press.

Nucci, L. P. (1982). Conceptual development in the moral and conventional domains: Implications for values education. *Review of Educational Research, 52*, 93–122.

Nucci, L. P., & Herman, S. (1982). Behavioral disordered children's conceptions of moral, conventional, and personal issues. *Journal of Abnormal Child Psychology, 10*, 411–425.

Oxford, M., Cavell, T. A., & Hughes, J. N. (2003). Callous–unemotional traits moderate the relation between ineffective parenting and child externalizing problems: A partial replication and extension. *Journal of Clinical Child and Adolescent Psychology, 32*, 577–585.

Pennington, B. F., & Ozonoff, S. (1996). Executive functions and developmental psychopathology. *Journal of Child Psychology and Psychiatry, 37*, 51–87.

Perry, D. G., & Perry, L. C. (1974). Denial of suffering in the victim as a stimulus to violence in aggressive boys. *Child Development, 45*, 55–62.

Phillips, M. L., Young, A. W., Scott, S. K., Calder, A. J., Andrew, C., Giampietro, V. et al. (1998). Neural responses to facial and vocal expressions of fear and disgust. *Proceedings of the Royal Society London B Biological Sciences, 265*(1408), 1809–1817.

Phillips, M. L., Young, A. W., Senior, C., Brammer, M., Andrews, C., Calder, A. J. et al. (1997). A specified neural substrate for perceiving facial expressions of disgust. *Nature, 389*, 495–498.

Piaget, J. (1932). *The moral development of the child*. London: Routledge and Kegan Paul.

Roberts, A. C., Robbins, T. W., & Weiskrantz, L. (1998). *The prefrontal cortex: Executive and cognitive functions*. Oxford, UK: Oxford University Press.

Schoenbaum, G., & Roesch, M. (2005). Orbitofrontal cortex, associative learning, and expectancies. *Neuron, 47*(5), 633–636.

Shallice, T. (1988). *From neuropsychology to mental structure*. Cambridge: Cambridge University Press.

Shallice, T., & Burgess, P. W. (1991). Deficits in strategy application following frontal lobe damage in man. *Brain, 114*, 727–741.

Shallice, T., & Burgess, P. W. (1996). The domain of supervisory processes and temporal organization of behaviour. *Philosophical Transactions of the Royal Society of London B, 351*, 1405–1412.

Smetana, J. G. (1981). Preschool children's conceptions of moral and social rules. *Child Development, 52*, 1333–1336.

Smetana, J. G. (1993). Understanding of social rules. In M. Bennett (Ed.), *The child as psychologist: An introduction to the development of social cognition* (pp. 111–141). New York: Harvester Wheatsheaf.

Song, M., Smetana, J. G., & Kim, S. Y. (1987). Korean children's conceptions of moral and conventional transgressions. *Developmental Psychology, 23*, 577–582.

Sprengelmeyer, R., Rausch, M., Eysel, U. T., & Przuntek, H. (1998). Neural structures associated with the recognition of facial basic emotions. *Proceedings of the Royal Society of London B, 265*, 1927–1931.

Steele, S., Joseph, R. M., & Tager-Flusberg, H. (2003). Brief report: Developmental change in theory of mind abilities in children with autism. *Journal of Autism and Developmental Disorders, 33*, 461–467.

Turiel, E., Killen, M., & Helwig, C. C. (1987). Morality: Its structure, functions, and vagaries. In J. Kagan & S. Lamb (Eds.), *The emergence of morality in young children* (pp. 155–245). Chicago: University of Chicago Press.

Valentin, V. V., Dickinson, A., & O'Doherty, J. P. (2007). Determining the neural substrates of goal-directed learning in the human brain. *Journal of Neuroscience, 27*(15), 4019–4026.

Wootton, J. M., Frick, P. J., Shelton, K. K., & Silverthorn, P. (1997). Ineffective parenting and childhood conduct problems: The moderating role of callous-unemotional traits. *Journal of Consulting and Clinical Psychology, 65*, 292–300.

Young, L., Hauser, M., & Saxe, R. (2007). The neural basis of the interaction between theory of mind and moral judgment. *Proceedings of the National Academy of Sciences, 104*, 8235–8240.

15

IS A NEUROSCIENCE OF MORALITY POSSIBLE?

Jeremy Carpendale
Simon Fraser University

Bryan W. Sokol
Saint Louis University

Ulrich Müller
University of Victoria

In one fairly trivial sense, the answer to the question posed in the title to our chapter is obviously "yes." By virtue of being biological creatures, thinking about morality, just as thinking of any kind, must involve brain activity. We take it as a matter of course that morality is causally dependent on neural activity, and so it naturally follows that some neurological level of analysis of morality is possible. Taking a page from the pragmatist philosopher John Dewey, we would agree that "moral science is not something with a separate province. It is physical, biological, and historic knowledge placed in a human context where it will illuminate and guide … [human] activities" (Dewey, 1922, p. 296). Research on morality and neuroscience has certainly brought to the fore the biological aspect of Dewey's claim, and recent trends in neuroscience have the potential for opening up questions and topics that interest moral psychologists, particularly developmentalists (Killen & Smetana, 2008).

But, the question still remains, should we be quick to jump from the uncontroversial conclusion that a brain is necessary when dealing with moral matters to the altogether different expectation that it is useful to search for the location(s) of morality in the brain? The issue then becomes not whether neural activity is *necessary* for moral reasoning and behavior, but whether such activity is *identical* with dedicated moral structures in the brain. On this prospect, the open question is really, Can morality be naturalized by reducing it to purely biological, or neurophysiological, categories? In this chapter, we suggest a cautious approach to dealing with this more reductionistic question. Ultimately we argue that although neurophysiological activity is necessary (a necessity, *sine qua non*) for morality to emerge, it alone is insufficient to understand what morality is and how it operates in peoples' lives. Instead of reducing morality to neurophysiological activity, we suggest that a more productive approach begins by understanding morality as being rooted in day-to-day social interactions.*

The thrust of our argument against the neurophysiological reductionism of morality is based on what we take to be a key feature of morality—its normative status. Normative accounts of morality argue for the centrality of value-laden reasons, not mechanical causes, in the explanation of human conduct. Although, generally speaking, norms serve various functions in regulating social groups, from the personal level of an individual, norms are used to justify one's actions to others—that is, to provide a meaningful backdrop against which actions may be evaluated. Moral norms, in particular, turn on the notion that individuals' actions are their own, freely chosen responses to social situations, as opposed to being behaviors that are mechanically caused by factors outside of one's control. This is one of the reasons, for instance, why coerced moral behavior is seen as suspect; it is unclear whether an individual chose to act morally or was compelled by other forces.

The normative status of morality has been a major conceptual obstacle to naturalizing ethics—at least in philosophy—and has gained prominence through various rehearsals of the "naturalistic fallacy" (Moore, 1903). Technically speaking, this fallacy describes the alleged illogical leap from descriptive "is" statements, common in the empirical sciences, to prescriptive "ought" statements, most frequently seen in the study of ethics (MacIntyre, 1998). Beyond this technical meaning, the naturalistic fallacy has also been used to characterize more

---

* Social interaction, of course, is also a natural process. To be clear, however, we do not mean to suggest here that *biological* reductionism should be replaced with a form of *social* reductionism—an issue that also arises in the neuroscience literature.

loosely—but usually with derision—any attempt to bridge objective "facts" with subjective "values." Although moral psychologists, perhaps most famously Lawrence Kohlberg (1981), have worked to legitimize the bridge between facts and values, recent neuroscience research can also be seen as engaging the so-called "is-to-ought" gap (Casebeer, 2003, pp. 842–843; Greene, 2003). The tack taken in most of this research, however, is to narrow the gap by discounting or denying the place of value in human affairs and focusing only on the causal, neurophysiological level of explanation. Unless or until, this work extends to the level of reasons and value-laden nature of human conduct, it cannot, in principle, provide a satisfactory account of morality.

Our chapter is organized as follows. First, and without laying claim to being exhaustive, we briefly review three different lines of empirical evidence on attempts to localize morality in the brain, and critically discuss the merits of this enterprise. In the second section, we sample the neuroscience literature for definitions of morality. We show that these definitions are wanting because they fail to distinguish between two very different kinds of norms, or reasons, for acting: moral norms and social conventions. As it happens, reasons of any kind prove to be problematic within reductionistic accounts, but, as we elaborate in the third section, the normative status of moral rules, in particular, cannot be reduced to the causality of neurophysiological functioning. Finally, in the fourth section, we continue to take issue with the largely nativist and mechanist approach to morality that has become commonplace within the neuroscience literature, and contrast it with a developmental systems approach that emphasizes the importance of multiple, interacting explanatory levels, including the level of meaningful and value-laden social interactions.

NEUROSCIENCE AND MORALITY: "NATURALIZED" METHODOLOGIES

In the field of moral neuroscience, the attempts to localize moral functioning are pursued along three general lines. One approach is to study people with damage to particular parts of their brain (i.e., lesion data). Here, the inevitable mention of Phineas Gage's gruesome accident with a railroad tamping iron is often cited as evidence that certain intact brain structures are necessary lest, like Mr. Gage, we are all to fall into moral decline. Such 'acquired sociopathy,' as some researchers (e.g., Damasio, 1994) say, is the natural result of damage to, or impaired functioning within, the prefrontal cortex.* Lending some support to this view, developmental researchers have found that lesions of the prefrontal cortex early in life are linked to later impairments in moral reasoning and moral behavior (Anderson, Bechara, Damasio, Tranel, & Damasio, 1999; Eslinger, Flaherty-Craig, & Benton, 2004; Price, Daffner, Stone, & Mesulam, 1990; see Eslinger & Robinson-Long, chapter 13, this volume).

A second general approach involves neuroimaging studies of people engaged in making moral judgments and decisions (e.g., Heekeren, Wartenburger, Schmidt, Schwintowski, & Villringer, 2003). This essentially involves examining what parts of the brain are active when participants are asked to think about various morally relevant topics or conundrums, perhaps the most famous ones also involving railroad calamities—a trolley on a collision course with five people on the tracks (e.g., Greene, Nystrom, Engell, Darley, & Cohen, 2004; Greene, Sommerville, Nystrom, Darley, & Cohen, 2001). Researchers working in this area are particularly interested in what parts of the brain show increased activity when people are presented with hypothetical scenarios involving the making of life-and-death decisions.†

Finally, the third approach to localizing moral processing uses functional and structural imaging techniques in individuals diagnosed with antisocial personality disorder or psychopathy (for reviews, see Blair, Marsh, Finger, Blair, & Luo, 2006; Raine & Yang, 2006). This line of research has generally highlighted deficits in brain areas (e.g., amygdala, ventromedial portion of prefrontal cortex) that engage emotions that are considered to be central for moral decision making and behavior (for a more complex model, see Blair et al., 2006; Chapter 14, this volume). In reviewing these three approaches, the research on early brain damage, in particular, suggests that intact anterior (prefrontal) portions of the brain are necessary for the development of morality—a necessity *sine qua non*.

Although these various research avenues have helped identify important areas of neuronal activation, the growing consensus among neuroscientists is actually that many different brain regions are active in moral

---

* However, it should be noted that there are considerable differences between sociopaths and patients with lesions to the prefrontal cortex (see Damasio, 2000).
† An issue that cannot be appropriately discussed here concerns the tasks or dilemmas used in neuroscientific research (for a critical discussion see Killen & Smetana, 2008; Turiel, 2006; Chapter 16, this volume). In addition, the dilemmas used in neuroscientific research are often complex and the aspects of the problem which are linked to the activation of particular areas of the brain may be misidentified (Waldmann & Dieterich, 2007). Clearly, a more thorough analysis of the dilemmas used in this research is necessary.

thinking. As Casebeer (2003) points out, "Moral cognition is distributed ... it involves, more or less, the entire brain" (p. 846).* Making a similar point, Greene and Haidt (2001) state that "Many brain areas make important contributions to moral judgment" (p. 517). Moreover, as studies in this area work to achieve greater ecological validity, it appears that more and more of the brain becomes involved (Casebeer & Churchland, 2003). Altogether, such conclusions make the case for localization a difficult prospect to accept, particularly given that other plausible explanations exist for linking the brain to morality. Specifically, damage to particular brain areas could impair social interactions, which, in turn, could compromise moral reasoning and behavior.

Compounding such difficulty, a number of issues have also been raised concerning the interpretation of neuroimaging data. Kagan (2006), for example, has cautioned against the assumption that the activation of a particular brain region while a person performs a particular task indicates that this region has a single function (e.g., see Miller [2008] regarding Greene et al.'s [2001] conclusions). A further caution to consider when interpreting functional magnetic resonance imaging (fMRI) data concerns the indirect means by which neuronal activation is assessed. That is, although fMRI has a number of advantages, including being noninvasive and allowing relatively high spatiotemporal resolution, it also has limitations because the imaging relies on changes in blood-volume or oxygen concentration as a surrogate signal to reflect mass activation of neurons. Imaging results, then, can be ambiguous and do not always indicate increased firing of neurons (Logothetis, 2008).

Beyond these empirical concerns, however, lies a more fundamental issue referred to as the "mereological fallacy" (Bennett & Hacker, 2003, pp. 68–107). This is the practice of ascribing psychological attributes to some body part, such as a part of the brain, rather than to the whole person. "Psychological predicates are predicates that apply essentially to the whole living animal, not to its parts. It is not the eye (let alone the brain) that sees, but we see with our eyes (and we do not see with our brains, although without a brain functioning normally in respect of the visual system, we would not see)" (Bennett & Hacker, 2003, pp. 72–73). Thus, we should be cautious about statements such as "human brains are obviously products of natural selection, adapted to solve problems that faced our hominid ancestors for millions of years" (Haidt & Joseph, 2004, p. 56), or "the brain reads facial expressions extremely

---

* This conclusion has led Casebeer and Churchland (2003, p. 172) to argue that "it is highly unlikely, given the data, that we shall find anything like a well-defined 'moral reasoning module' in the mind/brain."

rapidly" (Blakemore, Winston, & Frith, 2004, p. 219). It is not the brain that solves problems and reads facial expressions, but rather persons. Morality is not easily analyzed at the purely neurophysiological level because social norms, particularly those related to morality, operate principally on a personal level. Interpersonal or social phenomena, such as morality, require, as the earlier quotation from Dewey (1922) suggested, analytic methods that extend well beyond the biological or neurophysiological domain. Even though it is a person's hand that holds the pen, swings the sword, or guides the plowshare, we do not search for a moral muscle in the hand. We do not claim that all neuroscientists fall into this problematic way of thinking, only that it is essential to be aware of these problems if neurophysiological data are going to be used effectively to broaden the study of morality.

DEFINING MORALITY OUT OF EXISTENCE?

Regarding our second concern, we need to ask how morality is defined in most neuroscience research. Although clear-cut definitions of morality are actually difficult to find, when explicit definitions can be located in the neuroscience literature they tend to offer fairly broad, conventionalized views of morality, sometimes even indicating that morality is simply rule-following or compliance. In other words, the reductionistic tendency that is present in much of this work now manifests as a form of social reductionism, instead of biological reductionism. Moll, Zahn, de Oliveira-Souza, Krueger and Grafman (2005; see also Moll, Oliveira-Souza, & Zahn, 2008), for example, state that "morality is considered as the set of customs and values that are embraced by a cultural group to guide social conduct, a view that does not assume the existence of absolute moral values" (p. 799). Casebeer and Churchland (2003) have similarly argued that "moral reasoning deals with *cognitive acts and judgments associated with norms* ... [including among various social animals] the groups' local conventions, pecking order, division of labor, and even who has what kind of knowledge" (p. 171). In other places, Casebeer (2003) has been, as he admits, even more "deflationary" by claiming that "moral cognition comprises any cognitive act that is related to helping us ascertain and act on what we should do" (p. 842).

Finally, still other researchers, such as Raine and Yang (2006, p. 203), arguing that "the overlap between morality and antisocial disorders is substantial," tend to reduce immoral thinking and feeling to essentially any "'antisocial tendency' [or] (rule-breaking behavior)" (p. 203), that is, "behavior that breaks the moral guidelines set down by society"

(p. 206). These researchers typically add that morality supplies "the brakes on rule-breaking behavior" (Raine & Yang, 2006, p. 209), in the form of moral emotions, such as shame, guilt, or remorse, that seem to be absent in most clinically diagnosed antisocial individuals (e.g., psychopaths). The more reductionistic members of this camp (e.g., Greene & Haidt, 2002; Haidt, 2008) also suggest that morality simply is the emotional, or "gut," reaction associated with the violation of norms. On this view, moral thought is merely the discourse that individuals use to justify to others the decisions that have already been made by feelings or intuitions (Greene & Haidt, 2002).

Although there is certainly precedent for these kinds of broad definitions in other research literatures in psychology (e.g., Baumeister & Exline, 1999; Kochanska & Aksan, 1995), the notion that morality is essentially rooted in compliance to societal convention has hardly gone unchallenged. Prominent developmental theorists, Piaget (1932/1965), Kolhberg (1981), and Turiel (2008) among them, have voiced strong opposition to "socialization" and "conventionalist" views of morality. They have argued, in particular, that viewing morality as conformity or compliance to culturally-specific rules not only promotes moral relativism, but also fails to capture the generative, and often subversive, characteristics of moral thinking. That is, many exemplars of moral action (e.g., members of the U.S. civil rights movement in the 1960s) have actively resisted conforming to local conventions and rules, particularly when they are understood to violate fundamental moral principles involving justice and fairness. Morality, then, involves not just passively following the social norms imposed by a culture, but rather the construction (or reconstruction) of normative principles that allows individuals to infuse their personal experiences with value and meaning (Forst, 2005). The very notion of a moral person implies such autonomous meaning-making and stands in stark contrast to the reduction of human conduct to mechanical, neurophysiological causes (or, conversely, broad social currents). As Forst (2005, p. 67) has argued:

> For if we were not individual members of a "space of reasons" in which we must provide each other with justifications capable of withstanding intersubjective normative examination, then we would be like machines that operate within certain tolerances laid down by "norms" but could not hold each other accountable for violations of these norms. Such machines do not have "reasons" for operating in conformity with, or in violation of, the relevant norms. Accordingly, to be autonomous means to be situated in a space of norms and to be capable of acting in accordance with reasons.

Although the autonomy of moral conduct, as well as other similar definitional matters (e.g., the universality of moral prescriptions), may seem more like a concern for philosophers than psychologists, in fact, Turiel and his colleagues have structured an entire developmental research program around children's growing understanding of the criteria that distinguish moral issues from other kinds of social norms (Nucci, 2001; Smetana, 2006; Turiel, 2002). The definitional criteria of morality employed in neuroscience research are rarely so carefully conceived (for an exception, see Blair, 1995). Some (e.g. Greene, 2007) have even militated against using any conceptual models of morality to guide their research (for discussion, see Killen & Smetana, 2007; Smetana & Killen, 2008).

NEUROSCIENCE AND NORMATIVITY

Central to the neurophysiological explanation of morality is the issue of whether or not normativity can be explained causally, by brain activity. Even though it may be difficult to provide an inclusive definition of morality, it is hardly disputable that moral judgments are prescriptive claims of the kind "I ought to do this," and that they express this normativity, or appeal to correctness (Wright, 1982). Morality, then, can be understood as "a system of categorically binding norms, and corresponding rights and duties, which hold reciprocally and universally among human beings, qua human beings, in their character as moral persons" (Forst, 2005, p. 66, emphasis in original). Moral normativity, according to Forst (2005, p. 67), is what "grounds the validity and binding power of moral norms and makes them worthy of recognition, so that one 'has a reason' to act in accordance with these norms from the right motives." Reasons are normative and motivating in the sense that they make claims on the agents to act in accordance with these reasons (Forst, 2005; Korsgaard, 2005). Wright (1982, p. 211) illustrates the distinctiveness of such motivating claims in the following:

> [I]t does not follow from the assertion "my parents tell me I ought to do this" that I ought to do it unless a second premise is added, namely, "I ought to do what my parents tell me to do." ... The assertion "I ought to do what my parents tell me" is one which logically I alone can make; it must originate from me. To put the point more generally, the universe of moral discourse is not one that I can be logically compelled to enter, but must, so to speak, step into voluntarily. Then the concept of ought itself, which is central to prescriptive utterance, is not reducible to, or translatable into, concepts of fearing, desiring, or wanting and the like.

It exists *sui generis* and the kind of work it does is distinctively different from other motivational concepts.

To elaborate on the distinction between norms, values, and meaning, on the one hand, and causality, on the other, it is helpful to draw on Piaget's discussion of the relation between action, or conduct, and neurophysiological mechanisms. According to Piaget, forms of conduct are the subject matter of psychology (Piaget, 1950, pp. 137–138). Conduct consists of an ongoing exchange between an organism and its environment over the course of which both undergo modifications (Piaget, 1950, p. 138). Piaget (1950, p. 139) acknowledged that human conduct has a neurophysiological aspect, and that neurophysiological mechanisms are a causal condition for such conduct. Nevertheless, he also argued that simply acknowledging neurophysiological mechanisms did not imply that conduct could be reducible to the level of neurophysiological explanation.

There are important limits to neurophysiological explanation, and these are set by the consciousness of logical-mathematical necessity and the awareness of moral obligation ("It has to be," see Piaget, 1954, 1968). For example, logical–mathematical necessity is manifest in arithmetic judgments:

> ... [the] truth of $2 + 2 = 4$ is not the "cause" of the truth of $4 - 2 = 2$ in the same way that a cannon causes the movement of two billiard balls, [rather] the truth ... of $2 + 2 = 4$ "implies" that of $4 - 2 = 2$, which is quite a different matter. In the same way the value attributed to an aim or moral obligation is not the "cause" of the value of the means or of an action connected with the obligation; one of the values implies the other in a way similar to logical implication, and one can call this implication between values [Piaget, 1968, pp. 187–188; see also, e.g., Piaget, 1950, 1970, 1971].

Piaget concludes that material or physical causality and logical-mathematical implications are irreducible to each other. Piaget (1950, p. 141) also argued that it is not just later developing cognitive and affective structures that are irreducible to causal, neurophysiological explanation. All states of consciousness are related by implication (i.e., by relations between meanings).

Piaget's claims are very similar to von Wright's (1971) distinction between causes and reasons. According to von Wright, actions are brought into the world by an agent: Had the agent not chosen to do, or undertake, what we ascribe to her or him as an action, it would not have occurred. Furthermore, our actions are guided by practical orientations, which in normal and simple cases are called reasons (von Wright, 1971). Human action is to be understood as intentional, free action, oriented by reason, which differs from causal explanation, which von Wright characterizes as the general norm of scientific research. According to von Wright, both causal and intentional explanations are compatible: "The explanandum of a teleological explanation is an action, that of a causal explanation an intentionalistically noninterpreted item of behavior, i.e., some bodily movement or state. Since the explananda are different, the question of compatibility does not arise at this level" (von Wright, 1971, p. 124).

Leaving aside the question of whether compatibilist theories (Habermas, 2007) are coherent, it might be argued that the idea of free will and reasons is illusory (Greene & Cohen, 2004). From this perspective, human beings operate on the basis of a badly mistaken folk psychology. Much like B. F. Skinner's infamous claims (e.g., Skinner, 1972) regarding human freedom, contemporary neuroscientists have argued that the "New neuroscience will undermine people's common sense, libertarian conception of free will.... The net effect of this influx of scientific information will be a rejection of free will as it is ordinarily conceived" (Greene & Cohen, 2004, p. 1776). The take-home message conveyed by neuroscience is that "we are all puppets. The combined effects of genes and the environment determine all of our actions" (Greene & Cohen, 2004, p. 1780).

Although folk psychological distinctions are not sacrosanct, the question remains whether such an approach is not self-contradictory (i.e., "sawing off the branch on which one sits," Bennett & Hacker, 2003, pp. 376–377). Here, it is important to remember that science itself is a normative enterprise. Scientists' laws are claims to truth, and scientists observe ethical guidelines in conducting their work. If everything is determined, this not only makes folk-psychological distinctions between intentional actions and nonintentional reflexes obsolete, it also annihilates scientific norms. Reductionistic-minded scientists must somehow extricate themselves from determinism (see Turiel, 2006, for a critical discussion of this point) or they commit a performative self-contradiction (i.e., their practice of doing science contradicts the propositional content of their theories [Habermas, 2007]). That is, if scientists are puppets whose decisions are determined, then there would seem little point in selecting scientific positions based on the weight of reasons and evidence, nor is it clear why Greene and Cohen (2004) appear to be appealing to their readers' reasoning. Moreover, it is difficult to see how it can be argued that science is progressive, given that progress is also a normative category.

In the end, because normativity is not reducible to causally law-governed phenomena (McDowell, 1994, 1998; Tanney, 1999; Wingert, 2006), attempts to reduce morality to the functioning of the brain ultimately fail to capture its most critical features: autonomy, normativity, and reasons.* As the field currently stands, then, it is not clear what neuroscience contributes to understanding the specific character of morality. One way that neuroscientists might begin to change this assessment is to explicate some of their assumptions underlying the goal of mapping brain areas that are associated with morality. There are, in fact, various possible motivations for attempting to localize morality; however, most of these can be organized around two competing views regarding the nature of evolution and development.

EXPLICATING ASSUMPTIONS: MODULARITY VERSUS DEVELOPMENTAL SYSTEMS

There are various possible relations between neuroscience and morality, but the assumptions underlying neuroscience approaches are often left unspecified. Titles such as, "The neural basis of human moral cognition" (Moll et al., 2005), are ambiguous because at least two possible sets of assumptions could be operating. First, "modularity approaches" (e.g., Mikhail, 2007) begin with the assumption that human beings have evolved very specific, innate ways of thinking about moral issues. We will consider two difficulties with this assumption: its problematic view of meaning and its approach to innateness. As an alternative, we review a second, "developmental systems approach" (e.g., Gottlieb, 2007; Lickliter, 2008; Oyama, 2000; Griffiths & Stotz, 2000), according to which general evolutionary adaptations provide a basis for developing specific abilities over an individual's lifetime (as opposed to providing those specific abilities at the onset), such as thinking about moral matters. This latter, developmental account does not discount genetic influences; rather, it situates these among the many other levels of interacting factors that influence individuals' development, including their social worlds.

Instead of suggesting how forms of thinking emerge developmentally, some modularity approaches assume that it is the "forms of thought" themselves that are innate. For example, consider the following quotation:

---

* Neuropsychological approaches continue the tradition of behaviorist (e.g., Skinner), sociological (e.g., Durkheim), and psychoanalytic approaches (e.g., Freud) that also fail to realize the *sui generis* character of the "ought" motive and reduce it to natural or social contingencies (Wright, 1982). The novelty of the neuropsychological approach is simply that these contingencies are conceived of in terms of patterns of brain activity.

It seems that the *form* of moral thought is highly dependent on the large-scale structure of the human mind. Cognitive neuroscience has made it increasingly clear that the mind/brain is composed of a set of interconnected modules. ... The form of human moral thought is to a very great extent shaped by how the human mind happens to have evolved. In other words, our moral thinking is not the product of ... experience.... As for the content of human morality, there are good reasons to think that genes play an important role here as well. (Greene, 2005, p. 351; see also Hauser, 2006a, 2006b; Hauser, Cushman, Young, Jin & Mikhail, 2007; Mikhail, 2007).

Some authors explicitly endorse a modularity approach and suggest that "humans possess an innate moral faculty that is analogous ... to the language faculty that has been postulated by Chomsky and other linguists" (Mikhail, 2007, p. 143). Here the assumptions are that the mind processes information like a computer and that specialized neurological circuits have evolved through natural selection. This means that "we inhabit mental worlds populated by the computational outputs of battalions of evolved, specialized neural automata" (Tooby & Cosmides, 1995, p. xii). Even the ability to understand "the meaning of a glance" is an evolved specialized neural mechanism (Tooby & Cosmides, 1995, p. xi); and some neurological circuits or algorithms (often referred to as computation) are specialized for morality.

Proponents of modularity approaches, such as Mikhail (2007, see also Hauser, 2006a, 2006b), tend to overlook the extensive criticism of Chomsky's view of language (e.g., Baker & Hacker, 1984; Tomasello, 2003). The same problems with a modular approach to language are present with an approach to morality. Here, however, we will focus on the problematic view of meaning on which this approach depends.*
Specifically, this approach treats meaning as being fixed; that is, the rules embodied in the neural mechanism compute a particular outcome given particular input. Against this view, many philosophers (e.g., Goldberg, 1991; Wittgenstein, 1968) have argued that meaning is nonmechanistic or not fixed. That is, meaning cannot be simply attached to a word or gesture, and cannot be computed mechanically. For example,

---

* To the extent that modularity approaches endorse a reductionistic neurophysiological approach to meaning, they face the same problems as the latter (see section on Neuroscience and Normativity). This becomes particularly evident in Mikhail's (2007) modular theory. According to Mikhail, normativity ("moral" and "deontic" structure) is instantiated in the mechanical operations of a modular computational device. Mikhail does not seem to realize that derivation of normativity from the mechanical workings of such a "machine" is somewhat problematic, to say the least.

there are many possible meanings that can be conveyed with a glance (Tooby & Cosmides' example); its meaning cannot be fixed. Similarly, a pointing gesture can be used to convey many different meanings, and therefore, a module devoted to computing the meaning of pointing gestures could not have evolved.

Meaning depends on the context, the sequence of interactions in which particular acts are embedded. In some contexts, for instance, politeness can be rude and in other contexts insults can be endearing. The same event or utterance can be funny in some circumstances but not in others. In some contexts lying or stealing might even be moral. Imagine, for instance, a young man running and forcefully pushing an old woman to the ground. At first this might seem like a cruel act, but then imagine a broader scenario in which the woman, unaware of an oncoming truck, was about to step into its path. Now the act can be seen in its proper light as being moral, even heroic. Cases like these make claims about evolved moral behaviors difficult to accept because morality cannot be simply reduced to specific algorithms in the brain—to neural circuits. If we are looking for cases of evolved systems involving fixed meaning, a better example might be the dances of honeybees because the meaning of the dance (i.e., the way the other bees response to it) appears to be relatively fixed and is not flexible. However, a human sense of meaning cannot be computed in a fixed manner.

Mikhail's (2007) initial evidence for his claim that morality is modular, which he acknowledges is inconclusive (p. 148), includes the fact that 3- to 4-year-old children already know something about morality and that every natural language seems to have words for moral concepts. We agree with his assessment regarding the inconclusiveness of this evidence. Children of this age have already learned about many aspects of their social and cultural worlds that we do not resort to explaining in terms of innate modules, and the fact that languages have words for important aspects of human experience says nothing about whether that area of experience is modular. Bates (1984) illustrated problems with this way of thinking by pointing out that just because humans eat with their hands and cook their food does not mean that we need to explain these facts in terms of modules.

Psychologists endorsing modular approaches have a tendency to use terms like *innate* and *hardwired* without clearly specifying what they mean. For example, Haidt and Joseph (2004, p. 56, emphasis in original) "propose that human beings come equipped with an *intuitive ethics*, an innate preparedness to feel flashes of approval or disapproval toward certain patterns of events involving other human beings." They further claim that "since infant brains hardly vary across cultures and races, it is reasonable to suppose that many psychological facts (e.g., emotions, motivations, and ways of processing social information) are part of the factory-installed equipment that evolution built into us to solve those recurrent problems." This jump from infants' brains to emotions, motivation, and thinking exemplifies a common view of innateness.

The commonsense concept of innateness is employed in cognitive sciences, but its use can lead to overlooking the role of many levels of interaction in development (Mameli & Batson, 2006). Without explication, a default assumption seems to be that genes contain information in the form of a "genetic program" that results in the creation of neural circuits or modules. This tends to overlook the importance of resources in addition to genes in the process of development (Lickliter, 2008). Genes code for proteins, and there is still a long way from molecules to minds. There are many elements in the developmental system other than genes: "The phenotype of any individual is the consequence of a unique web of interactions between the genes it carries, the complex, multidetermined molecular interactions within and across individuals cells, and the nature and sequence of the physical, biological, and social environments through which it passes during development" (Lickliter, 2007, pp. 361–362.). Furthermore, "Genes are interpreted differently in different cells and at different times, as are all other factors which make up the developmental system" (Griffiths & Stotz, 2000, p. 35).

Regularities in development do not have to be regarded as preprogrammed; they can arise through the interaction of various levels. Instead of thinking of genes as carrying information, consider the metaphor of ecological transitions (Griffiths & Stotz, 2000, p. 33). For example, after a forest fire in the Northwest of North America, one of the first plant species to appear is fireweed, followed by bushes, and later deciduous trees, then Douglas fir, and finally Western hemlock and red cedar. The development of a mature hemlock and cedar forest after a major fire moves through a regular series of phases. Genes are, of course, an essential component in this progression but they are part of a developmental system involving many other essential components, and the outcome should not be thought of as stored information within a genetic program. Instead this is an expectable outcome due to interactions between characteristics of the various plant species and the available resources such as sunlight. This is a relational approach because explanations cannot be found solely in the individual or in a collective but rather within relations.

Clearly, there must be an important neurophysiological and evolutionary side of the story in accounting for moral development, and there are many levels of interacting factors. Attempting to naturalize ethics by rooting human moral development within natural processes is a laudable goal, at least from the perspective of most developmental scientists, as well as various pragmatist philosophers. Nevertheless, "naturalizing" in this context does not automatically necessitate defecting to the "nature," or biological, side of the nature–nurture wars in psychology or committing oneself to a form of moral realism. Jean Piaget's (1932/1965) seminal work, *The Moral Judgment of the Child*, shows this quite clearly in its explication of the kinds of social interactions that shape children's moral reasoning. More contemporary work, such as Jesse Prinz's "Against Moral Nativism" (2009), also illustrates one of the naturalizing currents in philosophy that explicitly rejects any form of biological reductionism, showing instead a more sociocultural pathway for the emergence of morality. In the field of moral neuroscience, however, most of this critical work has been bypassed in attempts to localize moral cognition in the brain.

Rather than reducing morality to the causal processes of the functioning of the brain, we suggest a nonreductive way of naturalizing morality that is more fertile, and consists in accounting for the development of morality in the context of social interaction (see Cash, 2009; MacArthur, 2004; McDowell, 1994, 1998; Piaget, 1932/1965). From our perspective, humans have evolved adaptations that establish the forms of interaction, or relational patterns, out of which human modes of thought and conduct, or "forms of life," emerge. Human infants are biologically prepared to begin life and develop within social interaction. Infants' brains are shaped by social interaction, and it is in the course of these social interactions that morality gradually emerges. Such an account comes in two parts. First, it needs to describe and explain how in each individual person, born with certain capacities, the ought-motive emerges in the context of social practices. Second, a phylogenetic–evolutionary account of the evolution of biological adaptations that make possible the development of social practices is required (see Cash, 2008, 2009).

From our perspective, morality develops within relations with others because it is due to the coordination of social interaction—that is, "morality is the logic of action" (Piaget, 1932/1965, p. 398). Moral norms emerge through this social process. It is within particular forms of social interaction, however, that agreement is facilitated. Relations involving an imbalance of power or status do not help in the process of all parties reaching mutual understanding and agreement because one perspective can simply be imposed on the others. In contrast, relations among equals involving mutual respect are well suited to achieving mutual understanding, and it is within these relationships that children become competent at interacting with each other in a way that is structured by implicit moral principles. Children, then, gradually become able to verbally reflect on the implicit morality of their practical activity (Carpendale, in press; Piaget, 1932/1965).

Rooting the development of morality in the process of social interaction directs our attention to those biological adaptations that make human forms of social interaction possible. For example, the extended period of helplessness in human infancy necessitates social interaction and means that development occurs within interaction with caregivers, and evolved emotional reactivity could lead to the forms of relations in which morality can develop. Biological adaptations that result in emotional engagement and interest in others set up the infant for the kinds of social interaction in which human development occurs (Carpendale, Lewis, Müller, & Racine, 2005). Human infants and their caregivers are adapted to enter into emotionally saturated social engagement with each other, and infants are able to profit from this experience by acquiring expectations about routine social activities. Within the context of this social interaction, neurological structures are shaped (Hrabok & Kerns, in press). Such biological roots are necessary but not sufficient for the development of morality, which turns on the coordination of action with others.

CONCLUSION

As we have mentioned, at a trivial level it is a truism that there must be a neural basis to any form of human activity, including morality. As pointed out by Stekeler-Weithofer (2007), it is important to distinguish the neurophysiological conditions that are necessary for the typical working of the brain, without which morality might be impossible, from the additional and sufficient conditions that may be required for morality.

> To say, as Joseph LeDoux does, that synapses underlie personality, is, therefore, as true as to say that oxygen or water underlies personality, since oxygen or water also underlies everything the brain does. To say that charitable cooperation and mutual respect underlie personality would, in a sense, be much closer to the facts [Stekeler-Weithofer, 2007, p. 89].

In our opinion, a reductionistic neuroscientific approach to morality that is based on Cartesian substance dualism makes human life unintelligible, and eventually is self-contradictory. Instead of reducing morality to the physical structures inside an organism, we believe that morality is irreducible to the functioning of the brain, and that it is a more fertile approach to look for the origins of morality in human action and practices (Stekeler-Weithofer, 2007, p. 90). Indeed, science, including neuroscience, is a human practice following certain reasons, aims, and norms. Furthermore, science itself is grounded in and "standing on the shoulders of an intentionally understood human world" (Kambartel, 2005, p. 121), and the "intentional and practical concepts of our life world are *semantically* prior to scientific constructs and situations in whatever way parallel or connected with them" (Kambartel, 2005, p. 122). Not only are the events of interest to moral neuroscience saturated with normative practices (Cash, 2009), the practice of doing neuroscience itself always presupposes our intentional understanding of ourselves, as rationally acting individuals who follow reasons in pursuit of the goals. Indeed, without this understanding, science itself would become meaningless. Therefore, arguing that normativity can be reduced to causal processes while doing science is a performative contradiction in that the activity of doing science is itself normative (Habermas, 2007).

Furthermore, we have argued that the search for evolved, specialized neural mechanisms of morality is misplaced. This does not mean that we disagree with the idea that morality is the result of adaptations, but rather, with the nature of those adaptations. We do not inherit morality; rather we inherit a broader capacity to develop morality. Although this might appear to be a truism, we encourage attention to the necessary processes in development. These capacities set up the forms of human interaction in which morality can develop. From our perspective, there are several ways in which neuroscience is linked to morality. First, the evolution of human forms of emotional reactivity, for example, would be required as one aspect involved in setting up the human forms of interaction in which development occurs. This interaction forms the context in which infants' brains develop. When we turn to the adult brain, which is the focus of most neuroscience research on morality, then the pathways laid down in development related to thinking about moral matters could also be studied. But the complex developmental process involved must not be forgotten; it is a long way from genes and molecules to minds. Although there will be a neuroscience side to the story of moral development, this development will be rooted in interpersonal engagement. We have objected, in particular, to a modularity approach to morality, because, as Tomasello (1999) argued, this approach skips from the first page involving genetics to the last page, and in the process overlooks a complex process of biological and social development. Instead, we have argued for a relational approach, both at the level of biological development and social development.

Let us conclude with an extended quotation from Forst (2005, pp. 84-85), which, we think, nicely sums up our argument:

> The normativity of morality is a normativity *sui generis*. .... It owes its existence to an autonomous insight into an original responsibility toward others in accordance with the principle of justification, which is acknowledged at the moment this insight dawns on us. There is nothing mysterious about this insight, for this being responsible simply is our basic way of being in the world as finite beings who use reasons, that is, who can give reasons and who demand them of others. Morality is merely the form of 'justifying' existence in a particular practical context. To become part of such contexts means to learn to recognize what justifications are, when one owes them, and to whom. Such processes of *formation* do not 'ram' an 'absolute must' into us in an inexplicable manner. ... Rather, they constitute the way in which we are as fellow human beings and through which we become individual persons. At the same time, being a moral person is only one component of our identity among others, albeit a special one on which we rely when we have moral confidence in ourselves and in others. In this sense, the normativity of morality can be explained naturalistically, in terms of our 'second nature' as reasonable social beings. The recognition of an 'original ought' is part of our nature as *animalia rationalia*—as justifying beings.

ACKNOWLEDGMENT

We thank the editors for inviting us to contribute a chapter to this volume, and Michael Chandler for many excellent comments on an earlier draft of this chapter.

REFERENCES

Anderson, S. W., Bechara, A., Damasio, H., Tranel, D., & Damasio, A. R. (1999). Impairment of social and moral behavior related to early damage in human prefrontal cortex. *Nature Neuroscience, 2*, 1032–1037.

Baker, G. P., & Hacker, P. M. S. (1984). *Language, sense and nonsense*. Oxford: Blackwell.

Bates, E. (1984). Biograms and the innateness hypothesis. *Behavioral and Brain Sciences, 7*, 188–190.

Baumeister, R. F., & Exline, J. J. (1999). Virtue, personality, and social relations: Self-control as the moral muscle. *Journal of Personality, 67,* 1165–1194.

Blair, J., Marsh, A. A., Finger, E., Blair, K. S., & Luo, J. (2006). Neurocognitive systems involved in morality. *Philosophical Explorations, 9,* 13–27.

Blair, R. J. R. (1995). A cognitive developmental approach to morality: Investigating the psychopath. *Cognition, 57,* 1–29.

Cacioppo, J. T., Berntson, G. G., Sheridan, J. F., & McClintock, M. K. (2000). Multilevel integrative analyses of human behavior: Social neuroscience and the complementing nature and biological approaches. *Psychological Bulletin, 126*(6), 829–843.

Carpendale, J. I. M., Lewis, C., Müller, U., & Racine, T. P. (2005). Constructing perspectives in the social making of minds. *Interaction Studies, 6,* 341–358.

Carpendale, J. I. M. (in press). Piaget's theory of moral development. In U. Müller, J. I. M. Carpendale, & L. Smith (Eds.), *Cambridge companion for Piaget.* New York: Cambridge University Press.

Casebeer, W. D. (2003). Moral cognition and its neural constituents. *Nature Reviews Neuroscience, 4* (10), 840–846.

Casebeer, W. D., & Churchland, P. S. (2003). The neural mechanisms of moral cognition: A multiple-aspect approach to moral judgment and decision-making. *Biology & Philosophy, 18* (1), 169–194.

Cash, M. (2008). Thoughts and oughts. *Philosophical Explorations, 11,* 93–119.

Cash, M. (2009). Normativity is the mother of intention: Wittgenstein's normative account of intentionality, and is relevance for talk of neurological representation. *New Ideas in Psychology, 27,* 133–147.

Damasio, A. (1994). *Descartes' error: Emotion, reason, and the human brain.* New York: Putnam.

Damasio, A. R. (2000). A neural basis for sociopathy. *Archives of General Psychiatry, 57,* 128–129.

Dewey, J. (1922). *Human nature and conduct.* New York: The Modern Library.

Dupoux, E., & Jacob, P. (2007). Universal moral grammar: A critical appraisal. *Trends in Cognitive Sciences, 11,* 373–378.

Eslinger, P. J., Flaherty-Craig, C. V., & Benton, A. L. (2004). Developmental outcomes after early prefrontal cortex damage. *Brain and Cognition, 55,* 84–103.

Flanagan, O. (1996). Ethics naturalized: Ethics as human ecology. In L. May, M. Friedman, and A. Clark (Eds.), *Mind and morals: Essays in cognitive science and ethics* (pp. 19–43). Cambridge, MA: MIT Press.

Forst, R. (2005). Moral autonomy and the autonomy of morality: Toward a theory of normativity after Kant. *Graduate Faculty Philosophy Journal, 26,* 65–88.

Goldberg, B. (1991). Mechanism and meaning. In J. Hyman (Ed.), *Investigating psychology: Sciences of the mind after Wittgenstein* (pp. 48-66). London: Routledge.

Gottlieb, G. (2007). Probabilistic epigenesis. *Developmental Science, 10,* 1–11.

Greene, J. (2003). From neural "is" to moral "ought": What are the moral implications of neuroscientific moral psychology? *Nature Reviews Neuroscience, 4*, 847–850.
Greene, J. (2007). The biology of morality: Neuroscientists respond to Killen and Smetana. *Human Development: Letters to the Editor*.
Greene, J. (2005). Cognitive neuroscience and the structure of the moral mind. In P. Carruthers, S. Laurence & S. Stich (Eds.), *The innate mind: Structure and contents* (pp. 338–352). New York: Oxford University Press.
Greene, J., & Cohen, J. (2004). For the law, neuroscience changes nothing and everything. *Philosophical Transactions of the Royal Society London B, 359*, 1775–1785.
Greene, J., & Haidt, J. (2002). How (and where) does moral judgment work? *Trends in Cognitive Sciences, 6*(12), 517–523.
Greene, J. D., Nystrom, L. E., Engell, A. D., Darley, J. M., & Cohen, J. D. (2004). The neural bases of cognitive conflict and control in moral judgment. *Neuron, 44*, 389–400.
Greene, J. D., Sommerville, R. B., Nystrom, L. E., Darley, J. M., & Cohen, J. D. (2001). An fMRI investigation of emotional engagement in moral judgment. *Science, 293* (5537), 2105–2108.
Griffiths, P. E., & Stotz, K. (2000). How the mind grows: A developmental perspective on the biology of cognition. *Synthese, 122*, 29–51.
Habermas, J. (2007). The language game of responsible agency and the problem of free will: How can epistemic dualism be reconciled with ontological monism? *Philosophical Explorations, 10*, 13–50.
Haidt, J. (2008). Morality. *Perspectives on Psychological Science, 3*, 65–72.
Haidt, J., & Joseph, C. (2004). Intuitive ethics: How innately prepared intuitions generate culturally variable virtues. *Daedalus*, 55–66.
Hauser, M. D. (2006a). *Moral minds: How nature designed our universal sense of right and wrong*. New York: HarperCollins Publishers.
Hauser, M. D. (2006b). The liver and the moral organ. *Social Cognitive and Affective Neuroscience, 1*, 214–220.
Hauser, M., Cushman, F., Young, L., Jin, R. K.-X., & Mikhail, J. (2007). A dissociation between moral judgments and justifications. *Mind and Language, 22*, 1–21.
Heekeren, H. R., Warertenburger, I., Schmidt, H., Schwintowski, H.-P., & Villringer, A. (2003). An fMRI study of simple ethical decision-making. *NeuroReport, 14*, 1215–1219.
Hrabok, M., & Kerns, K. A. (in press). The development of self-regulation: A neuropsychological perspective. In B. Sokol, U. Müller, J. I. M. Carpendale, A. Young, & G. Iarocci (Eds.), *Self- and social-regulation: Exploring the relations between social interaction, social cognition, and the development of executive functions*. New York: Oxford University Press.
Kagan, J. (2006). Biology's useful contribution: A comment. *Human Development, 49*, 310–314.

Kambartel, F. (2005). On the incompatibility of intentional and causal explanation. Thoughts after re-reading von Wright. *Acta Philosophica Fennica, 77*, 115–125.

Killen, M., & Smetana, J. (2007). The biology of morality: Human development and moral neuroscience. *Human Development, 50*, 241–243.

Killen, M., & Smetana, J. (2008). Moral judgment and moral neuroscience: Intersections, definitions, and issues. *Child Development Perspectives, 2*, 1–6.

Kochanska, G., & Aksan, N. (1995). Mother–child mutually positive affect, the quality of child compliance to requests and prohibitions, and maternal control as correlates of early internalization. *Child Development, 66*, 236–254.

Kohlberg, L. (1981). From is to ought: How to commit the naturalistic fallacy and get away with it in the study of moral development. In L. Kohlberg, *Essays in moral development: The philosophy of moral development.* San Francisco: Harper & Row Publishers.

Korsgaard, C. M. (2005). Acting for a reason. *Danish Yearbook of Philosophy, 40*, 11–36.

Lickliter, R. (2008). The growth of developmental thought: Implications for a new evolutionary psychology. *New Ideas in Psychology, 26*, 353–369.

Logothetis, N. (2008). What we can and what we cannot do with fMRI. *Nature, 453*, 869–878.

MacArthur, D. (2004). Naturalizing the human or humanizing nature: Science, nature, and the supernatural. *Erkenntnis, 61*, 29–51.

Mameli, M., & Batson, P. (2006). Innateness and the sciences. *Biology & Philosophy, 21*, 155–188.

MacIntyre, A. (1998). *A short history of ethics (2nd ed.).* London: Routledge.

McDowell, J. (1994). *Mind & world.* Cambridge, MA: Harvard University Press.

McDowell, J. (1998). *Mind, value, and reality.* Cambridge, MA: Harvard University Press.

Mikhail, J. (2007). Universal moral grammar: Theory, evidence, and future. *Trends in Cognitive Sciences, 11*, 143–152.

Miller, G. (2008). Growing pains for fMRI. *Science, 320*, 1412–1414.

Moll, J., de Oliveira-Souza, R., & Zahn, R. (2008). The neural basis of moral cognition: Sentiments, concepts, and values. *Annals of the New York Academy of Sciences, 1124*, 161–180.

Moll, J., Zahn, R., de Oliveira-Souza, R., Krueger, F., & Grafman, J. (2005). The neural basis of human moral cognition. *Neuroscience, 6*, 799–809.

Moore, G. E. (1903). *Principia ethica.* New York: Cambridge University Press.

Nucci, L. P. (2001). *Education in the moral domain.* Cambridge: Cambridge University Press

Oyama, S. (2000). *Evolution's eye: A systems view of the biology–culture divide.* Durham: Duke University Press.

Piaget, J. (1965). *The moral judgment of the child.* New York: Free Press. (Original work published in 1932.)

Piaget, J. (1950). *Introduction à l'èpistemologie gènètique, Vol. 3: La pensèe biologique, la pensèe psychologique et la pensèe sociologique* [Introduction in genetic epistemology: Biological, psychological, and sociological thinking]. Paris: Press Universitaires de France.
Piaget, J. (1954). The problem of consciousness in child psychology: Developmental changes of awareness. In H. A. Abramson (Ed.), *Conference on problems of consciousness* (Vol. 4, pp. 136–177). New York: Joisah Macy Foundation.
Piaget, J. (1968). Explanation in psychology and psychophysiological parallelism. In P. Fraisse & J. Piaget (Eds.), *Experimental psychology: Its scope and method* (Vol. 1: History and method, pp. 153–191). London: Routledge & Kegan Paul. (Original work published in 1963.)
Piaget, J. (1970). *The place of the sciences of man in the system of sciences*. New York: Harper & Row.
Piaget, J. (1971). *Biology and knowledge*. Chicago: The University of Chicago Press. (Original work published in 1967.)
Piaget, J. (1978). *Behavior and evolution*. New York: Pantheon Books.
Price, B. H., Daffner, K. R., Stowe, R. M., & Mesulam, M. M. (1990). The comportmental learning disabilities of early frontal lobe damage. *Brain, 113*, 1383–1393.
Prinz, J. J. (2009). Against moral nativism. In D. Murphy and M. Bishop (Eds.), *Stich and his critics* (pp. 167–189). Malden, MA: Wiley-Blackwell.
Raine, A., & Yang, Y. (2006). Neural foundations to moral reasoning and antisocial behavior. *Social Cognitive and Affective Neuroscience, 1*, 203–213.
Skinner, B. F. (1972). *Beyond freedom and dignity*. New York: Bantam Vintage.
Smetana, J. (2006). Social–cognitive domain theory: Consistencies and variations in children's moral and social judgments. In M. Killen & J. Smetana (Eds.), *Handbook of moral development* (pp. 119–153). Mahwah, NJ: Lawrence Erlbaum Associates.
Smetana, J., & Killen, M. (2008). Moral cognition, emotions, and neuroscience: An integrative developmental view. *European Journal of Developmental Science, 2*, 324–339.
Stekeler-Weithofer, P. (2007). From individual mind to human forms of practice. In N. Parros & K. Schulte-Ostermann (Eds.), *Facets of sociality* (pp. 85–115). Frankfurt: Ontos Verlag.
Tanney, J. (1999). Normativity and judgment. *Proceedings of the Aristotelian Society Supplement, 73*, 44–61.
Tomasello, M. (1999). *The cultural origins of human cognition*. Cambridge, MA: Harvard University Press.
Tomasello, M. (2003). *Constructing a language: A usage-based theory of language acquisition*. Cambridge, MA: Harvard University Press.
Tooby, J., & Cosmides, L. (1995). Forward to *Mindblindness* by S. Baron-Cohen. Cambridge, MA: The MIT Press.
Turiel, E. (2002). *The culture of morality: Social development, context, and conflict*. New York: Cambridge University Press.

Turiel, E. (2006). Thought, emotions, and social interactional processes in moral development. In M. Killen & J. G. Smetana (Eds.), *Handbook of moral development* (pp. 7-35). Mahwah, NJ: Erlbaum.

Turiel, E. (2008). Social decisions, social interactions, and the coordination of diverse judgments. In U. Müller, J. I. M. Carpendale, N. Budwig, & B. Sokol (Eds.), *Social life and social knowledge: Toward a process account of development* (pp. 255-276). New York: Taylor Francis.

Von Wright, G. H. (1971). *Explanation and understanding*. Ithaca, NY: Cornell University Press.

Von Wright, G. H. (1994). On mind and matter. *Journal of Theoretical Biology, 171*, 101-110.

Von Wright, G. H. (2000). Valuations—or how to say the unsayable. *Ratio Juris, 13*, 347-357.

Waldmann, M. R., & Dieterich, J. H. (2007). Throwing a bomb on a person versus throwing a person on a bomb: Intervention myopia in moral intuitions. *Psychological Science, 18*, 247-253.

Wingert, L. (2006). Grenzen der naturalistischen Selbstojektivierung (Limits of naturalistic self-objectification). In D. Sturma (Ed.), *Philosophie und Neurowissenschaften* (pp. 240-260). Frankfurt: Suhrkamp.

Wittgenstein, L. (1968). *Philosophical investigations*. Oxford: Blackwell.

Wright, D. (1982). Piaget's theory of moral development. In S. Modgil & C. Modgil (Eds.), *Piaget: Consensus and controversy* (pp. 207-217). London: Holt, Rinehart and Winston.

16

THE RELEVANCE OF MORAL EPISTEMOLOGY AND PSYCHOLOGY FOR NEUROSCIENCE

Elliot Turiel

University of California, Berkeley

Research into brain processes and moral decisions has proliferated in recent years. However, the most frequent explanations of moral decisions given by neuroscience researchers have failed to take into account epistemological considerations with regard to the moral realm and have not been built on relevant findings from developmental psychology. A developmental moral neuroscience would benefit from inclusion of philosophical considerations and from findings on moral development. With the focus on brain processes, the research on morality has emphasized unconscious and emotionally driven decisions to the exclusion of the important role of reasoning in human functioning.

In what follows I begin with the premise, supported by developmental research, that reasoning integrated with emotions as evaluative appraisals is central to human functioning. I then consider how the depreciation of thought in the recent thinking of some neuroscientists represents a reversion to a road traveled by much of the field of psychology prior to the cognitive revolution of the second half of the twentieth century. I then critically examine current work on morality, emotions, and brain processes with regard to epistemological considerations, the need to integrate biology, thought, and emotions in developmental explanations that would also account for the role of social interactions. Finally, I present an alternative interpretation of results from methods typically used in neuroscience research on moral decisions. The alternative interpretation is based on the propositions that moral decisions often involve complex judgments and emotions regarding multiple components of social situations.

DEVELOPMENT AND REASONING

Martha Nussbaum, a philosopher and legal scholar, has stated that a long tradition of liberal moral philosophy "holds that human beings are above all reasoning beings and that the dignity of reason is the primary source of human equality" (Nussbaum, 1999, p. 71). Amartya Sen (1999, 2006), a Nobel Prize-winning economist and philosopher who shares Nussbaum's perspective on morality, has also asserted the view that human beings are fundamentally reasoning beings. As one means of illustrating the importance of reasoning, Sen (2006, p. 161) pointed to the paradox of persons using reasoning to deny the process of reasoning: "Reason had to be supreme, since even in disputing reason, we would have to give reasons." In the moral philosophic tradition adhered to by Nussbaum and Sen and connected to moral theorists like Mill (1863/1964), Kant (1785/2001), and in more recent times Rawls (1971), Gewirth (1982), and Dworkin (1977), reasoning does not by any means exclude emotions, which entail evaluative appraisals, so that: "the entire distinction between reason and emotions begins to be called into question, and one can no longer assume that a thinker who focuses on reasoning is excluding emotions" (Nussbaum, 1999, p. 72; for an extensive discussion see Nussbaum, 2000). Substantively, reasoning in conjunction with emotional appraisals produces the moral standpoint "that all, just by being human, are of equal dignity and worth, no matter where they are situated in society, and that the primary source of this worth is a power of moral choice within them, a power that consists in the ability to plan a life in accordance with one's own evaluation of ends" (Nussbaum, 1999, p. 54).

There is a good deal of research that, in a variety of ways, provides evidence for the propositions that reasoning, in conjunction with emotional appraisals, is centrally involved in social and moral development. One source of evidence comes from numerous studies showing that by a relatively young age (starting at 2 to 4 years and clearly by 5 or 6 years), children make judgments about different domains that are distinctly different from each other (see Smetana, 2006; Turiel, 1998, 2002, 2006 for summaries). The domain of morality is conceptualized differently from the domains of social conventions and personal jurisdiction. Moral judgments are based on concepts of welfare, justice, and rights, whereas understandings of social conventions are based on concepts about the coordination of social interactions. Moral judgments are seen to be obligatory, not contingent on existing rules, authority dictates, or common practices. Additional discriminations are made in that there are activities judged to be out of the moral and conventional realms and subject to personal choice. These domains of judgment, which constitute distinct developmental pathways, entail complex configurations and discriminations among activities and aspects of social organization. Corresponding with the development of these domains of judgment is the early emergence of associated emotions like sympathy, empathy, and respect (Eisenberg & Fabes, 1991; Hoffman, 2000; Piaget, 1932).

The findings of the early emergence of moral and social judgments should not be taken to mean that they are innately determined. In the early part of the twentieth century when many social scientists (e.g., McDougall, 1926) emphasized "instincts" as the source of much of personality, morality, and social behavior (a viewpoint that came into direct conflict with the environmentalism of the behaviorist movement), some proposed on the basis of research with young children that moral actions were innately determined. As an example, Antipoff (1928), as cited in Piaget (1932), proposed that a sense of justice was "innate and instinctive" and that its emergence did not require social experiences. Since Antipoff's research was with children from 3 to 9 years of age, Piaget (1932) succinctly noted that by the age of 3 years children have experienced a good deal of social interaction that could influence their development. Findings with young children do not necessarily speak to the question of innateness. A number of studies have shown that a variety of young children's social experiences are distinctly associated with the development of judgments in the moral, social, and personal domains (Nucci & Turiel, 1978; Nucci, Turiel, & Encarcion-Gawrych, 1983; Nucci & Weber, 1995).

The development of distinct domains of moral, conventional, and personal judgments means that individuals maintain a diversity of ways of thinking and goals. In turn, the existence of a diversity of ways of thinking means that in order to understand how individuals come to their decisions it is necessary to account for those ways of thinking, if and how they might be applied to a given situation, and how they are coordinated (weighed and balanced) in coming to a decision. Research indicates that people's thinking is flexible, such that they recognize and coordinate different types of considerations and goals in coming to decisions. Studies of concepts of rights and honesty have shown that individuals set priorities when making social decisions in particular social contexts. Although children, adolescents, and adults endorse the concept of rights (e.g., to freedom of speech or religion) in the abstract and in many situations, they also sometimes subordinate rights to matters like avoiding harm or promoting community interests (Helwig, 1995; McClosky & Brill, 1983). Similarly, concepts of honesty and trust are morally important to individuals, but they will subordinate honesty to concerns like preventing harm to persons or promoting fairness and justice (Perkins & Turiel, 2007; Turiel & Perkins, 2004). For instance, deception has been judged acceptable when it serves to prevent someone from sustaining physical harm or in order to get around restrictions unjustly imposed by persons in positions of greater power and control in the social system.

All of these studies have shown that individuals maintain understandings of the value of rights and of acting honestly, but will give priority to other considerations in given situations. These types of coordinations in judgments indicate that judgments or actions often deemed by researchers as nonrational or irrational may not be irrational at all. All too often rationality is defined as acting to maximize material or economic self-interest, deciding on the basis of benefits and losses, or acting in accord with one's stated moral goals (e.g., to uphold rights or honesty). These formulations fail to account for the multiple types of thought that people maintain, including the judgments and goals that are not solely based on maximizing benefits and minimizing costs, and for the ways people draw priorities among the variety of considerations and goals held when confronting multifaceted social situations.

The heterogeneity of systems of reasoning, along with a concomitant flexibility of mind in interpreting situations and coordinating judgments, are evident in the ways individuals commonly oppose and resist perceived injustices in their social relationships, aspects of societal organization, and certain cultural practices (Turiel, 2003). Research has shown that in Western and non-Western cultures people in subordinate positions in the social hierarchy often evaluate societal arrangements and cultural practices from a moral perspective. They judge as unfair inequalities in social relationships and privileges held by one group over another (Neff, 2001; Wainryb & Turiel, 1994). Moreover, it has been found that individuals act to oppose restrictions on their personal activities (Abu-Lughod, 1993; Wikan, 1996). These types of judgments and actions indicate that people do not respond in inflexible ways, but instead adjust to the needs of the situation.

ARE WE FOOLING OURSELVES?

The central features of the positions in the tradition of liberal moral philosophy as characterized by Nussbaum and Sen, as well as the findings of the development of domains of social judgments, coordination on social decisions, and opposition and resistance, are in contradiction with assumptions made in a line of psychological theorizing dating back to the early part of the 20th century that has once again been emphasized in the 21st century, including much contemporary work on morality. Various psychological theorists have presumed that people do not reason much (or any) of the time, that rationality (insofar as it exists) is overwhelmed by emotions, and that moral choice or any other kind of choice in planning lives are merely illusions primarily serving as a means for people to feel good about themselves. Psychoanalytic and behaviorist theories, which were prominent in the first half of the 20th century, took the approach that conscious or reflective decisions masked what was actually occurring psychologically. Psychoanalytic theory portrayed human beings as actors whose conscious thoughts, feelings, and actions for the most part disguised true but unknown, deep unconscious motivations. Equally radical behaviorist theories characterized humans as mechanistic responders whose actions were due to ways they had been shaped by environmental forces (reinforcement, conditioning). Behaviorism began with forceful denials of the existence of consciousness, which was treated as a secular version of the idea of a soul (Watson, 1924). B. F. Skinner (1971) famously proclaimed that just about all our mentalist terms, such as thinking, personality, autonomy, freedom, dignity, and choice, are illusions, with no psychological reality, that people maintain to elevate themselves. Mentalist terms are said to be illusory since the psychic state is largely empty, but with inborn drives or needs and propensities to avoid pain and repeat pleasurable events. A third position (e.g., McDougall, 1926) prominent in the early part of the 20th century emphasized the role of instincts in social interactions and social behaviors (e.g., instincts of affiliation, gregariousness, dependence or independence, aggressiveness, morality).

Psychoanalytic, behaviorist, and instinctual claims that people's decisions are not what they appear to be to laypersons or that many of their understandings are illusory have also been a mainstay of some subsequent research. Two presuppositions have been prominent in several quarters of psychological theorizing. One is that the mind continually plays tricks on us. The other is that we are driven by our emotions, not our reasoning—we are often fooled into believing that it is our reasoning and choices that produce decisions and actions. The field of social psychology, dating back at least to theories of cognitive dissonance (Festinger, 1957), is replete with experiments aimed at demonstrating the ways attitudes, thinking, and irrational processes most often determine behaviors and override reasoning. In a very thoughtful early treatise on *Social Psychology*, Solomon Asch captured these approaches in stating: "When today people speak of the new psychology they have little more in mind than the "irrationality" of motives and the mechanical character of the learning process" (Asch, 1952, p. 20). Asch criticized researchers in social psychology for their lack of concern with people as social beings and complained that "No assumption has spread more widely in modern psychology than that men are ruled by their emotions and that these are irrational. Although there is much to support this view, it has nevertheless been responsible for a systematic depreciation of intelligence and thinking in human affairs" (Asch, 1952, p. 21).

The depreciation of thinking, according to Asch, entailed "The view that the reasons men give for their actions and their convictions are usually not true causes but rationalizations … " (p. 21). The idea of rationalization had taken the place of analyses of thinking. Asch called for analyses of when individuals engage in thought processes in efforts to attain understandings of the world—including the world of social relationships. After Asch's volume appeared in 1952 the field of psychology did undergo the so-called cognitive revolution, with its emphasis on thought and explanations of development that included constructive processes stemming from individual–environment interactions and children's efforts to understand their environments. Many researchers also examined the development of moral thought, emotions, and action (the work of Kohlberg, 1969, 1971 was the most prominent at the time).

In recent years, however, we have seen a reversion to the depreciation of thinking in human affairs to which Asch referred. In the realm of morality, for instance, it has been proposed that moral evaluations and decisions are based on emotions and intuitions that occur automatically, without thought, reflection, or scrutiny (Haidt, 2001). Moreover, it is claimed that moral judgments or reasoning, in the form of understandings of matters like welfare, justice, and rights are merely after-the-fact justifications and not the fundamental bases of morality that people are deluded into thinking they are (see Turiel, 2006, 2008a for critiques of this position). The proposition that reasoning is after-the-fact rationalization poses the paradox highlighted by Sen (2006) that emerges when reasons are used to dispute reason. It is a process of reasoning and reflection that has led to the assertion that reasoning and reflection are largely irrelevant to human activities. This is not dissimilar to behaviorists' extensive use of reasoning and theory construction to dispute that human reasoning is not the process by which humans function. This is not to say that all analyses of brain processes and neuron activities pose a paradoxical situation. Rather, it is to say that the claim that human activities are not characterized by reasoning, deliberative thought, or reflection is in contradiction with the very mental activities used in making such assertions.

The proposition that moral decisions are intuitive, unexamined, and independent of conscious reasoning is part of a contemporary reversion to explanations of decisions in many realms as overdetermined by subconscious brain functions. Conscious, reasoned, or reflective decisions, as well as autonomy and choice, are seen to be as illusory as Skinner thought all mentalist explanations are illusory. One line of research has attempted to show that consciousness or awareness of decisions come after the decisions were already made. As an example, it has been found that in experimental situations, actions like pressing a button or flicking a finger occurred slightly (a half second) after the occurrence of brain signals associated with the actions (Wegner, 2002). These types of findings are interpreted to mean that conscious awareness of decisions comes after the decision has already been taken in a subconscious way. Therefore, our beliefs that we deliberatively or consciously make decisions and choose from alternatives are merely illusions. They are after-the-fact explanations for what we have no alternative to doing. (For extensive philosophical discussions of determinism, free will, and neuroscience, see Habermas, 2007 and a set of commentaries in a special issue of *Philosophical Explanations*.)

Often there is little in the way of explanation of how it is that we act in these automatic ways—except that somehow, invoking material cause, the brain does it. Perhaps the most common explanation as to why the brain does it is that of genetic inheritance. The process of evolution has resulted in some hardwired, genetic makeup that, in turn, determines thoughts and actions. Such explanations are then applied to a broad range of activities, including sexual activities, choices of sexual partners, occupational choices, emotional reactions, political affiliations, attitudes toward political issues like school prayers, property taxes, and the military draft, and morality.

Some of these propositions regarding the intuitive nature of moral evaluations and decisions fail to have even the explanatory power of psychoanalytic and behaviorist theories. Brain functioning, intuitions, and emotions that simply exist apart from awareness are said to result in actions. By contrast, at the heart of psychoanalytic theory was an attempt to explain how and why conscious perceptions and awareness become deeply unconscious and placed out of awareness. The structure and functions of unconscious processes were described, which included explication of forces accounting for conscious unavailability. For strict behaviorists, the explanation of actions as independent of conscious awareness or reasoning was based on a duality between actions and mind, with actions following laws having nothing to do with cognitive activities.

MORALITY, EMOTIONS, AND THE BRAIN

The proposition that morality is primarily intuitive and that reasoning is an after-the-fact rationalization to justify decisions to oneself or to convince others (Haidt, 2001) has been embraced by neuroscience researchers (Cushman, Young, & Hauser, 2006; Greene, Sommerville, Nystrom, Darley, & Cohen, 2001; Koenigs, Young et al., 2007). In presenting their point of view, neuroscientists typically draw a contrast with "traditional rationalist approaches to moral cognition that emphasize the role of conscious reasoning from explicit principles" (Koenings et al., 2007, p. 908). Kohlberg's theory of moral development is usually cited as the best-known example of such a traditional rationalist approach. As is well known, Kohlberg derived a model of six stages in the development of moral judgments through the use of semistructured clinical interviews in which participants were presented with a number of complex dilemmas or conflicts among competing moral and social claims (e.g., among life, property rights, and law in a situation in which a man has to decide whether to steal an unaffordable drug that could save the life of his wife who is dying of cancer). Kohlberg did indeed view reasoning and reflection (but not to the exclusion of emotions) as central in moral evaluations and decisions. The question of whether he viewed reasoning as "conscious" is more complicated than it has been presented in the neuroscience discussions. Following Piaget (1970), Kohlberg proposed that his six stages of moral development reflected "structural" processes at levels of understanding deeper than surface content (but which could be held in consciousness and reflected upon).

However, one most important feature of Kohlberg's approach relevant to an analysis of intuitive and neuroscience approaches lies in the fact that he attempted to provide definitional-philosophical groundings for psychological analyses of morality (Kohlberg, 1971; Turiel, 2008a). In research on cognition (Piaget, 1970; Werner, 1957), language (Chomsky, 1979), morality (Kohlberg, 1971; Turiel, 1983), and other realms, it has been recognized that psychological studies need to be informed by epistemological analyses and definitional criteria for the realm under study.

Otherwise, both theory and research are simply guided by intuitions about the realm of investigation and ad hoc choices of empirical tasks.

For many years researchers of moral development have taken seriously the importance of grounding their studies on epistemological features of the moral realm. Those now emphasizing intuitions and subconscious biologically based processes have paid scant attention to epistemological issues. In some case, morality is taken almost entirely out of the realm of philosophy and "psychologized" (that is, treated as only a phenomenon to be analyzed from the viewpoint of psychological processes). Neuroscience research also pays little attention to epistemological issues in favor of explanations of the biological sources of morality. This is consistent with the call in the 1970s by E. O. Wilson, one of the leaders of the movement in sociobiology, to devalue philosophical analyses: "Scientists and humanists should consider together the possibility that the time has come for ethics to be removed temporarily from the hands of the philosophers and biologicized" (Wilson, 1975, p. 562). At the time, Wilson's proclamation of independence from philosophy received a fair amount of criticism, not for the call for inclusion of biological analyses of morality, but for the call for exclusion of philosophy. Wilson's call went largely unheeded until recently when evolutionary psychologists and neuroscientists turned their attention to the biological side of morality, using difficult and complex moral dilemmas to study people's responses and brain activity, without systematic or rigorous grounding in definitions of the moral realm. Empiricist approaches that fail to account for substantive analyses of the realms under investigation run the risk of examining phenomena that are too broad or not entirely relevant to the supposed realm under investigation. In the study of moral development it has often been the case that broad assumptions about morality, such as that it is the learning of cultural values, have resulted in the inclusion of moral and nonmoral issues (Kohlberg, 1971). In turn, empiricist approaches that start with identification of experimental tasks for the study of moral behavior and acquisition have produced inadequate data because the tasks often were not perceived to be in the moral domain by participants in the studies (Turiel, 1983). As Chomsky (1979, p. 43) has put it: "No discipline can concern itself in a productive way with the acquisition or utilization of a form of knowledge, without being concerned with the nature of that system of knowledge."

Two prominent lines of research illustrate, in different ways, the lack of sufficient concern with definitional-epistemological issues. One line stems from researchers who study people with brain impairments and who generally do not provide systematic or well-formulated definitions of the realm of judgments or behaviors they are studying (e.g., Damasio, 1994; Koenigs et al., 2007). Although these researchers have appropriately stated an interest in studying connections among biology, rationality, emotions, and feelings (Damasio, 1994), the analyses of loosely defined morality leave much to be desired regarding the types of emotions involved. As one example, Damasio lists a hodgepodge of emotions presumably associated with morality: sympathy, embarrassment, shame, guilt, pride, jealousy, envy, gratitude, admiration, indignation, contempt, and disgust. No effort is made to distinguish among these very different types of emotions, some of which are positive and some aversive, or to specify which play what role in moral acquisition or moral decisions. Similarly, undifferentiated definitions are provided, which fail to account for the types of domain distinctions evident in the judgments of children and adults (Turiel, 2006; see also Killen & Smetana, 2007). Damasio reverts to vague notions, lumping together terms like social conventions, ethical rules, religious beliefs, law, and justice. Analyses are not provided of these categories or of how they are formed. Other references by Damasio (2003) to the moral realm include the idea that social conventions and ethical rules are manifestations of homeostatic and cooperative relationships regulated by culture. However, morality involves goals other than equilibrium and cooperation since it importantly involves struggle, opposition, and resistance of perceived unfair practices of inequality and hierarchy that are regulated by culture. By contrast, Damasio sees conventions and social rules as mechanisms for achieving homeostasis within social groups in the context of social hierarchies and inequalities: "It is not difficult to imagine the emergence of justice and honor out of practices of cooperation. Yet another layer of social emotions, expressed in the form of dominant or submissive behaviors within the group, would have played an important role in the give and take that define cooperation" (Damasio, 2003, p. 163). The types of critical scrutiny of relationships of dominance and subordination documented in research mentioned above has little place in these formulations.

A second and related line of research has focused on neuroimaging studies of people making moral decisions. These researchers usually take an empiricist approach. As put by Greene (2007), who has conducted neuroimaging studies: "As an empiricist, I believe that we can study things like life without defining them … This strategy, I believe, works just as well for the aspect of life that we call 'morality.' For empiricists, rigorously defining morality is a distant goal, not a prerequisite. If anything, I believe that defining morality at this point is more of a hindrance than a help, as it may artificially narrow the scope of inquiry .... Rather than seeking out morality by the light of a philosopher's definition (Kantian or otherwise), I and like-minded scientists choose to study decisions that ordinary people regard as involving moral judgment."

The strategy that has emerged from this empiricist stance involves the identification of tasks or problems that would generally be regarded as part of the moral realm; that is, problems pertaining to life and death. Actually borrowing from one philosophical tradition, researchers have used tasks that entail utilitarian calculations, such as whether it is preferable to sacrifice one life to save a greater number of lives. Ironically, many of the neuroscientist researchers who, as noted above, have set themselves in opposition to Kohlberg, use complex and difficult hypothetical dilemmas to study moral responses that have commonalities with the types of dilemmas used by Kohlberg. Dilemmas that place issues like theft in conflict with saving a life are paralleled in even more difficult and supposedly utilitarian dilemmas posing a conflict between sacrificing one life to save five other lives. Although "ordinary" people are likely to regard these issues of life as involving moral judgments, contemplating the idea of sacrificing one life to save five others is not likely to be what they regard as straightforward or everyday moral decisions.

The tasks used in the neuroscience research are anything but straightforward regarding the issues of the value of life and decisions on acting to save lives since participants are essentially posed with the problem of whether it is permissible for them to act as executioners. In addition, the tasks are set up so as to maximize the possibility that people would make what appear to be contradictory decisions (the gotcha part of proclivities in some psychological research). The participants are asked to "act as executioners" by presenting them with what are referred to as trolley car bystander and trolley car footbridge scenarios. The bystander scenario is as follows: "A runaway trolley is about to run over and kill five people, but a bystander can throw a switch that will turn the trolley onto a side track, where it will kill only one person." Participants are posed with the following type of question: "Is it permissible to throw the switch?" The scenario involves a utilitarian calculation as to whether it is better to sacrifice one life to save more lives. The "trick" part of this research comes in the footbridge version: "A runaway trolley is about to run over and kill five people, but a bystander who is standing on a footbridge can shove a man in front of the train, saving the five people but killing the man." Still acting as utilitarian executioners, participants need to decide if it is permissible to save five lives by sacrificing one, but in this case by actually pushing a man to his death.

In some of these studies using the trolley car scenarios (e.g., Greene et al., 2001) it was hypothesized that generally some moral dilemmas engage emotions more than others and that the differences affect judgments. More specifically, it was proposed "the crucial difference between the trolley dilemma and the footbridge dilemma lies in the latter's tendency to engage people's emotions in a way that the former does not" (Greene et al., 2001, p. 2106). They stated that the thought of pushing someone to his death is more emotionally salient than the thought of throwing a switch and that "it is this emotional response that accounts for people's tendency to treat these cases differently" (p. 2106).

Various studies have shown that people do treat these cases differently. Most state that it is permissible to throw the switch in order to save five people, but most state that it is not permissible to push a man even though that act would save the same number of people. Accordingly, it has been presumed that the findings confirm the hypothesis that the difference between the two scenarios is the emotions evoked by the footbridge problem. Moreover, it is asserted that many moral decisions are nonrational, intuitive, and determined by unconscious processes. The argument is supposedly buttressed by findings that different parts of the brain associated with rationality and emotions are activated in the two scenarios.

There are several serious problems with this type of research and the interpretations of the findings. One problem lies in the idea that because the footbridge scenario evokes emotions greater than or different from the bystander scenario the differences in decisions are accounted by emotions. No doubt, the footbridge problem does evoke different and/or more intense emotions. This is an obvious and perhaps trivial observation since the idea of physically pushing someone to his death is going to be more emotional than throwing a switch. What makes it less trivial in these analyses is the unexamined, and in my view inaccurate, presumption that the two situations are otherwise essentially the same; that is, they involve the utilitarian calculation that saving five lives is better than taking one life. Hence, the emotional component in the footbridge scenario is regarded as the sole difference between the two; the emotional component overwhelms the rational choice in ways unknown to the responders.

For an adequate analysis of these situations it is necessary to take into account the emotions involved in each, why one might evoke more or different emotions, that rationality does not entail only utilitarian calculations, and the generalizability of these types of decisions to other moral decisions. In the first place, it is misleading to say that the footbridge scenario is more emotionally salient (it probably is) and leave it at that because it ignores the emotions likely to be involved in the bystander scenario as well and the unique features of these types of scenarios. Although the researchers treat the trolley car tasks as representative examples of moral problems involving utilitarian calculations, in actuality a decision as to whether to take a life in these ways is highly unusual for most people, poses complex considerations with regard to the nightmarish problem of whether it is acceptable to act as an executioner, and is particularly charged with emotion because of the very issues that make it moral—the perceived value and sacredness of life and prohibitions against taking a life. Surely, the idea of taking an action like throwing a switch that will cause the death of an individual even to save more lives evokes strong emotions. The researchers largely ignore the emotions likely to be invoked in the bystander scenario. Both types of scenarios are complex and emotionally laden because a strongly held value—the value and perceived sacredness of life—must be violated in order to preserve that very value. These situations are unusual, complex, and involve dilemmas for people because they are forced to repudiate morality with morality.

Assuming that the footbridge scenario evokes more intense emotions than the bystander scenario, a crucial question left unaddressed by neuroscience researchers is why it does so? What is it about the situation that makes it more emotional? If we do not simply disembody emotions from judgments, then the likely answer is that people do make judgments about the act of physically pushing someone to his death. The footbridge scenario constitutes a different context of evaluation from the bystander scenario (what Asch, 1952, referred to as "objects of judgment"). In addition to the dilemma of saving lives by repudiating the prohibition on taking lives embedded in the bystander scenario, the footbridge scenario entails judgments and emotions about actively and physically causing another's death (it also raises the possibility in participants' minds that the actor could jump onto the track instead of pushing another). Unlike the bystander scenario, the footbridge scenario includes the component of what might be interpreted as inflicting physical assault on another person and thereby directly executing him. This is to say that rationality does not solely entail utilitarian calculations in these situations, that judgments are made about the fundamental conflict in values in the situations, that judgments are made about the means used to achieve ends, and that people do take into account the different features of social situations and attempt to coordinate different types of judgments relevant to those features.

To a greater extent than the bystander scenario, the footbridge scenario presents a compounded problem involving the saving of lives, taking a life, the natural course of events, the responsibility of individuals altering natural courses, and causing someone's death very directly. The emotions and coordination of judgments involved are more complex in one than the other. The diversity of features embedded in social situations can be even more complex, as is evident in another scenario that has been used in this line of research. It is a scenario in which it is stated that a doctor can save five patients who are dying from organ failure by cutting up and killing a sixth healthy patient to use his organs for the others. Not very surprisingly, very few judged it permissible for the doctor to use a healthy patient's organs to save five others. Although this scenario also includes the five versus one calculation, it is seldom the case that physicians act this way; it is rarely condoned, and never seems to be contemplated as legitimate medical or social policy. It is not solely emotions that determine responses to this situation. In an even more complex way than the footbridge scenario, the "transplant" scenario also raises issues about a doctor's duties and responsibilities, the power granted to individuals to make life-and-death decisions, the legal system, and societal roles and arrangements.

The differences among the scenarios used in neuroscience studies seem to be designed to confound people by presenting significantly different features in the guise of similar features (i.e., the utilitarian calculation). However, all these scenarios constitute what philosophers call "hard cases." Another example of a hard case is the question of whether it is permissible to torture someone (say a known terrorist) in order to obtain information that would save the lives of many people (say from a terrorist attack). The difficulties of using hard or extreme cases to generalize to other issues were articulated by Walzer (2007, p. 302):

> Back in the early 1970s, I published an article ... that dealt with the responsibility of political leaders in extreme situations, where the safety of their people seemed to require immoral acts. One of my examples was the "ticking bomb" case, where a captured terrorist knows, but refuses to reveal, the location of a bomb that is timed to go off soon in a school building. I argued that a political leader in such a case might be bound to order the torture of the prisoner, but that we should regard this as a moral paradox, where the right thing to do was also wrong. ... This was widely criticized at the time as an incoherent position ... But I am inclined to think that the moral world is much less tidy than most philosophers are prepared to admit. Now Dershowitz has cited my argument in his

defense of torture in extreme cases (though he insists on a judicial warrant before anything at all can be done to the prisoner). But extreme cases make bad law. Yes, I would do whatever was necessary to extract information in the ticking bomb case .... But I don't want to rewrite the rule against torture to incorporate this exception. Rules are rules, and exceptions are exceptions.

Extreme cases can also make for inadequate science. Walzer's assertions are relevant to neuroscience research on morality because hard, extreme, and complex cases are used to generalize to most moral decisions. The trolley car (and transplant) scenarios pose particular types of emotionally laden problems with multiple considerations that are difficult to reconcile without violating serious moral precepts in order to achieve serious moral goals. The trolley car situation may indeed be useful for philosophical analyses (Waldmann & Dieterich, 2007), but they are lacking as starting points for the psychological or even biological studies analyses of people's moral decision making. These types of situations might be useful for the study of moral decisions in complex and extreme situations with a solid background of data and theory on people's judgments and decisions regarding more straightforward situations. In this regard, the scenarios used in the neuroscience studies have affinities with the types of moral situations used by Kohlberg (1969) in his research. As already noted, Kohlberg used complex dilemmas without separating variables or examining how individuals think and feel about the different components. Research into judgments about straightforward situations that do not involve complex coordination has shown that the six stages of moral development proposed by Kohlberg failed to account for the judgments of children, adolescents, and adults about the moral domain, how they distinguish morality from social conventional and personal domains, and how people coordinate different domains or different facets of the moral domain in their decision making (Turiel, 2008b). In the neuroscience research, too, there has been a lack of attention to how individuals think and feel about the various components in complex situations, and their coordination. Gross generalizations are made from unusual situations to moral functioning in general. The seeming inconsistencies in responses to the supposedly same (i.e., the utilitarian calculations) situations have been taken to mean that morality is due to evolutionarily determined, emotionally based intuitions and that reasoning is merely rationalization of and justification for subconscious decisions. These approaches to moral acquisition and functioning entail one-dimensional, mechanistic, causal explanations rather than ones that attempt to integrate biology, individual–environment interactions, thought, and emotions.

CONCLUSION

I have maintained that theoretical, epistemological, and methodological considerations render the conclusions about the subconscious, emotionally based, and intuitive nature of moral decisions and actions inaccurate. There are also conceptual and methodological difficulties in the most unique, and in some ways most central, part of the neuroscience research: the use of functional magnetic resonance imaging (fMRI). The proposition that responses to scenarios like the footbridge and transplant cases are mainly determined by emotion is supported by fMRI findings that activated brain regions are associated with emotional processes. On the methodological side, participants in fMRI studies are in a severely constrained, uncomfortable, and passive position. The situation is likely to interfere with how people react to tasks and, therefore, could very well produce responses that may be very different from the ways they would respond under other circumstances. The constraints of the fMRI mean that it is not possible to use procedures to explore cognitive and emotional processes connected with or underlying the "answers" given to the tasks. Even more important, more than one cognitive process, or emotional state, or behavioral response are associated with a particular brain region (Miller, 2008). According to a number of neuroscientists, the same brain regions are involved in a variety of mental states. The identification of a brain region with performance on a particular task is a correlation and not an explanation for cognitive, emotional, and brain processes.

Empiricist approaches that ignore the formulation of definitions to guide the choices of tasks in research and that rely mainly on the researchers' intuitions about what is an appropriate task for a domain are likely to make the types of mistakes I have maintained apply to stimuli like the trolley car dilemmas. For example, attending to the types of issues attributed by Nussbaum (1999) and others to a moral standpoint would result in different empirical approaches from those used in the neuroscience research. Nussbaum, it will be recalled, referred to a moral standpoint as including human beings' dignity and worth, the power of moral choice, and the ability to plan a life in relation to one's own evaluation of ends. Taking these considerations into account, it becomes evident that the trolley car scenarios are hard cases and complex situations that pose myriad considerations for people to sort out. At the least, it is necessary to identify and explore the myriad considerations (e.g., dignity, worth, moral choice) that people might bring to bear and the decision-making process in drawing their conclusions.

As already noted, complex situations are not good starting points for research on morality. Definitions of morality that include its fundamental aspects like welfare, justice, and rights, and which have been supported by a great deal of prior research, would result in a different set of starting points for analyses aimed at productive integrations of biology, brain processes, individual–environment interactions, thought, and emotions. Building on well-established findings in these ways could produce exciting new directions in the study of moral development.

REFERENCES

Abu-Lughod, L. (1993). *Writing women's worlds: Bedouin stories.* Berkeley: University of California Press.

Asch, S. E. (1952). *Social psychology.* Englewood Cliffs, NJ: Prentice-Hall.

Chomsky, N. (1979). *Language and responsibility.* New York: Pantheon Books.

Cushman, F., Young, L., & Hauser, M. (2006). The role of conscious reasoning and intuition in moral judgment: Testing three principles of harm. *Psychological Science, 17,* 1082–1089.

Damasio, A. R. (1994). *Descartes' error: Emotion, reason, and the human brain.* New York: Harper Collins.

Damasio, A. R. (2003). *Looking for Spinoza: Joy, sorrow, and the feeling brain.* New York: Harcourt, Inc.

Dworkin, R. (1977). *Taking rights seriously.* Cambridge, MA: Harvard University Press.

Eisenberg, N., & Fabes, R. A. (1991). Prosocial behavior and empathy: A multimethod, developmental perspective. In P. Clark (Ed.), *Review of personality and social psychology* (Vol.12, pp. 34–61). Newbury Park, CA: Sage.

Festinger, L. (1957). *A theory of cognitive dissonance.* Stanford, CA: Stanford University Press.

Gewirth, A. (1982). *Human rights: Essays on justification and applications.* Chicago: University of Chicago Press.

Greene, J. (2007, October 9). The biology of morality: Neuroscientists respond to Killen and Smetana (1) [Online exclusive]. *Human Development—Letters to the Editor.*

Greene, J. D., Sommerville, R. B., Nystrom, L. E., Darley, J. M., & Cohen, J. D. (2001). An fMRI investigation of emotional engagement in moral judgment. *Science, 293,* 2105–2108.

Habermas, J. (2007). The language of responsible agency and the problem of free will: How can epistemic dualism be reconciled with ontological monism. *Philosophical Explanations, 10,* 13–50.

Haidt, J. (2001). The emotional dog and its rational tail: A social intuitionist approach to moral judgment. *Psychological Review, 108,* 814–834.

Helwig, C. C. (1995). Adolescents' and young adults' conceptions of civil liberties: Freedom of speech and religion. *Child Development, 66,* 152–166.
Hoffman, M. L. (2000). *Empathy and moral development: Implications for caring and justice.* Cambridge, England: Cambridge University Press.
Kant, I. (1964). *Groundwork of the metaphysic of morals.* New York: Harper & Row. (Original work published 1785.)
Killen, M., & Smetana, J. (2007). The biology of morality: Human development and moral neuroscience. *Human Development, 50,* 241–243.
Koenigs, M., Young, L. Adolphs, R., Tranel, D., Cushman, F., Hauser, M., & Damsio, A. (2007). Damage to the prefrontal cortex increases utilitarian moral judgments. *Nature, 446,* 908–911.
Kohlberg, L. (1969). Stage and sequence: The cognitive–developmental approach to socialization. In D. Goslin (Ed.), *Handbook of socialization theory and research* (pp. 347–480). Chicago: Rand McNally.
Kohlberg, L. (1971). From is to ought: How to commit the naturalistic fallacy and get away with it in the study of moral development. In T. Mischel (Ed.), *Psychology and genetic epistemology* (pp. 151–235). New York: Academic Press.
McClosky, M., & Brill, A. (1983). *Dimensions of tolerance: What Americans believe about civil liberties.* New York: Russell Sage.
McDougall, W. (1926). *An introduction to social psychology.* Boston: John W. Luce.
Mill, J. S. (2001). *Utilitarianism.* Indianapolis, IN: Hackett. (Original work published 1863.)
Miller, G. (2008). Growing pains for fMRI. *Science, 320,* 1412–1414.
Neff, K. D. (2001). Judgments of personal autonomy and interpersonal responsibility in the context of Indian spousal relationships: An examination of young people's reasoning in Mysore, India. *British Journal of Developmental Psychology, 19,* 233–257.
Nucci, L. P., & Turiel, E. (1978). Social interactions and the development of social concepts in preschool children. *Child Development, 49,* 400–407.
Nucci, L. P., Turiel, E., & Encarcion-Gawrych, G. (1983). Children's social interactions and social concepts: Analyses of morality and convention in the Virgin Islands. *Journal of Cross-Cultural Psychology, 14,* 469–487.
Nucci, L. P., & Weber, E. (1995). Social interactions in the home and the development of young children's conceptions of the personal. *Child Development, 66,* 1438–145
Nussbaum, M. C. (1999). *Sex and social justice.* New York: Oxford University Press.
Nussbaum, M. C. (2000). *Upheavals of thought: The intelligence of emotions.* Cambridge England: Cambridge University Press.
Perkins, S. A., & Turiel, E. (2007). To lie or not to lie: To whom and under what circumstances. *Child Development, 78,* 609–621.
Piaget, J. (1932). *The moral judgment of the child.* London: Routledge and Kegan Paul.
Piaget, J. (1970). *Structuralism.* New York: Basic Books.

Rawls, J. (1971). *A theory of justice*. Cambridge, MA: Harvard University Press.
Sen, A. (1999). *Development as freedom*. New York: Alfred A. Knopf.
Sen, A. (2006). *Identity and violence: The illusion of destiny*. New York: Norton.
Skinner, B. F. (1971). *Beyond freedom and dignity*. New York: Knopf.
Smetana, J. G. (2006). Social domain theory: Consistencies and variations in children's moral and social judgments. In M. Killen & J. G. Smetana (Eds.), *Handbook of moral development* (pp. 119–153). Mahwah, NJ: Erlbaum.
Turiel, E. (1983). Domains and categories in social–cognitive development. In W. Overton (Ed.), *The relationship between social and cognitive development* (pp. 53–89). Hillsdale, NJ: Erlbaum Associates.
Turiel, E. (1998). The development of morality. In W. Damon (Ed.), *Handbook of child psychology, 5th Edition, Volume 3*: N. Eisenberg (Ed.), *Social, emotional, and personality development* (pp. 863–932). New York: Wiley.
Turiel, E. (2002). *The culture of morality: Social development, context, and conflict*. Cambridge, England: Cambridge University Press.
Turiel, E. (2003). Resistance and subversion in everyday life. *Journal of Moral Education, 32*, 115–130.
Turiel, E. (2006). Thought, emotions, and social interactional processes in moral development. In M. Killen & J. G. Smetana (Eds.), *Handbook of moral development* (pp. 7–35). Mahwah, NJ: Erlbaum.
Turiel, E. (2008a). The development of children's orientations toward moral, social, and personal orders. *Human Development: 50th Anniversary Special Issue; Celebrating a Legacy of Theory with New Directions for Research on Human Development, 51*, 21–39.
Turiel, E. (2008b). Social decisions, social interactions, and the coordination of diverse judgments. In U. Mueller, J. I. Carpendale, N. Budwig, & B. Sokol (Eds.), *Social life, social knowledge: Toward a process account of development* (pp. 255–276). Mahwah, NJ: Erlbaum.
Turiel, E., & Perkins, S. A. (2004). Flexibilities of mind: Conflict and culture. *Human Development, 47*, 158–178.
Wainryb, C., & Turiel, E. (1994). Dominance, subordination, and concepts of personal entitlements in cultural contexts. *Child Development, 65*, 1701–1722.
Waldmann, M. R., & Dieterich, J. H. (2007). Throwing a bomb on a person versus throwing a person on a bomb: Intervention myopia in moral intuitions. *Psychological Science, 18*, 247–253.
Walzer, M. (2007). *Thinking politically: Essays in political theory*. New Haven: Yale University Press.
Watson, J. B. (1924). *Behaviorism*. New York: The People's Institute.
Wegner, D. M. (2002). *The illusion of conscious will*. Cambridge, MA: MIT Press.
Werner, H. (1957). *Comparative psychology of mental development*. New York: International Universities Press.
Wikan, U. (1996). *Tomorrow, God willing: Self-made destinies in Cairo*. Chicago: University of Chicago Press.
Wilson, E. O. (1975). *Sociobiology: The new synthesis*. Cambridge, MA: Harvard University Press.

AUTHOR INDEX

A

Abu-Lughod, L., 316, 329
Ackerly, S. S., 256, 266
Adalbjarnardottir, S., 237, 246
Adams, C. M., 215, 223
Adamson, L., 53, 60
Addessi, E., 27, 36
Adleman, N. E., 213, 221
Adolphs, R., 83, 97, 100, 102, 105, 119, 121, 123, 194, 195, 196, 204, 205, 218, 221, 273, 274, 275, 281, 285, 320, 322, 330
Aggleton, J. P., 172, 184
Agnetta, B., 27, 34
Aharon-Peretz, J., 261, 267
Aichhorn, M., 153, 158
Ainslie, E., 210, 224
Aitken, M., 167, 185
Akbudak, E., 147, 160
Aklin, W. M., 167, 187
Aksan, N., 295, 309
Alessandri, S., 90, 96
Alessandri, S. M., 90, 94
Alexander, M. P., 253, 267
Allison, T., 102, 104, 105, 106, 107, 108, 109, 111, 115, 119, 123
Altschuler, E., 31, 32, 38
Amaral, D. G., 102, 119
Amso, D., 212, 222
Amsterdam, B., 57, 60
Andersen, S. L., 169, 171, 184, 186
Anderson, J. R., 27, 39
Anderson, L. C., 212, 222
Anderson, S. W., 292, 306
Andreiuolo, P. A., 272, 273, 280, 286
Andrews, C., 272, 287
Aniskiewicz, A. S., 280, 283
Ansel, S., 167, 188, 213, 224
Apperly, I. A., 153, 158, 163
Apperly, i. A., 63, 77
Arbreton, A. J. A., 143, 144, 151, 163
Armony, J. L., 83, 94
Arnett, J., 210, 211, 218, 221
Arsenio, W. F., 276, 283
Asch, S. E., 318, 325, 329
Aschersleben, G., 17, 37, 47, 49, 60
Asgari, M., 102, 105, 106, 123
Asperger, H., 133, 135
Assaiante, C., 30, 40
Ayduk, O., 210, 213, 222
Aylward, E., 30, 36
Ayotte, V., 143, 161
Azari, N. P., 183, 188
Aziz-Zadeh, L., 19, 34

B

Badre, D., 84, 94
Baillargeon, R., 14, 39

Baird, A. A., 91, 94, 191–207, 196, 204, 234, 245
Baird, G., 126, 138
Baird, J. A., 24, 34
Bakeman, R., 53, 60
Baker, G. P., 300, 306
Baldwin, D. A., 24, 34, 38, 62, 144, 159
Baldwin, J. M., 144, 159
Balkan, J., 210, 224
Balteau, E., 147, 153, 159
Banaji, M. R., 84, 95, 103, 122, 148, 153, 162
Barbas, H., 252, 266
Barch, D. M., 84, 95, 148, 156, 160
Bard, K., 25, 26, 40
Barker, R. G., 202, 204
Barnes, J. L., 157, 161
Baron-Cohen, S., 23, 30, 32, 34, 63, 77, 102, 106, 111, 115, 119, 120, 125–138, 126, 127, 128, 129, 131, 132, 133, 134, 135, 136, 137, 138, 157, 161
Barr, D. J., 227, 236, 242, 245, 246
Barresi, J., 43–62, 44, 45, 55, 58, 60
Barry, C. M., 149, 159
Barson, A. J., 195, 206
Bartels, M., 147, 161
Barthélemy, C., 30, 40
Bartlett–Williams, M., 20, 34
Bartsch, K., 70, 72, 80
Bassareo, V., 172, 185
Bates, E., 301, 306
Batson, P., 302, 309
Batth, R., 196, 198, 206
Baumeister, R. F., 233, 235, 245, 295, 307
Baxter, L. C., 147, 161
Baxter, M. G., 172, 184
Bechara, A., 183, 184, 252, 253, 267, 292, 306
Beckman, M., 234, 245
Beer, J. S., 83, 94, 147, 153, 162
Beger, B. D., 261, 267
Behne, T., 25, 26, 40, 54, 55, 62
Behniea, H., 102, 119
Bekkering, H., 20, 25, 26, 34, 37
Belleville, S., 131, 137
Bellmore, A. D., 149, 159
Belmonte, M. K., 126, 134, 135
Benes, F. M., 169, 185

Bennett, S. M., 178, 182, 183, 184, 216, 217, 221
Benson, D. F., 86, 97
Benson, J. E., 63–80
Bentin, S., 102, 105, 106, 123
Benton, A. L., 256, 257, 266, 292, 307
Benuzzi, F., 19, 29, 35
Bergman, T., 114, 121
Bermpohl, F., 103, 122
Berndt, T. J., 150, 159
Berns, G. S., 84, 96, 215, 223
Berntson, G. G., 307
Berridge, K. C., 183, 184
Bertenthal, B., 57, 60
Beyth-Marom, R., 210, 222
Bhattacharyya, S., 169, 185
Bialystok, E., 70, 77
Biddle, K. R., 252, 266
Bies, B., 167, 188, 213, 224
Bigbee, M. A., 242, 245
Binkofski, F., 19, 29, 35, 36
Birbaumer, N., 280, 284
Birò, S., 23, 27, 35, 37
Bisarya, D., 131, 136
Bjork, J. M., 176, 178, 180, 181, 182, 183, 184, 216, 217, 221
Bjorklund, D. F., 4, 8, 191, 204
Blair, C., 253, 266
Blair, J., 292, 296, 307
Blair, J. R., 214, 222
Blair, K. S., 271, 272, 273, 274, 275, 276, 277, 278, 279, 280, 281, 282, 283, 284, 292, 307
Blair, R., 272, 273, 276, 280, 282, 283, 285, 286
Blair, R. J., 176, 179, 180, 181, 186, 269–288
Blair, R. J. R., 83, 94, 296, 307
Blake, R., 113, 114, 120
Blakemore, S. J., 155, 160, 171, 184
Blasey, C. M., 213, 221
Blason, L., 21, 41
Blumenfeld, P. C., 143, 144, 151, 163
Blumenthal, J., 87, 96, 169, 170, 186, 196, 198, 205, 206, 211, 212, 223
Boas, D., 91, 94
Boddaert, N., 114, 124
Bonda, E., 104, 106, 120
Bonini, L., 19, 27, 36
Bookheimer, S. Y., 32, 35

Borelli, E., 27, 39
Borg, J. S., 280, 284
Boria, S., 30, 31, 32, 35
Borofsky, L., 154, 162
Botvinick, M. M., 84, 95, 213, 222, 273, 284
Bouchey, H. A., 143, 144, 159, 160
Bourgeois, J. P., 198, 204
Bouthillier, A., 22, 36
Bowles, R., 102, 121
Boyatzis, R. E., 239, 245
Boyer, T. W., 166, 184, 210, 222
Bracceschi, R., 32, 35
Bracken, B., 143, 149, 159
Bramati, I. E., 272, 273, 280, 286
Brammer, M., 87, 92, 97, 147, 161, 272, 287
Brammer, M. J., 111, 120
Brass, M., 20, 37
Braten, S., 22, 34
Braun, C. M. J., 276, 285
Braver, T. S., 84, 95, 103, 120
Breitbart, M. A., 173, 174, 183, 186
Bresnick, S., 143, 160
Breton, C., 64, 70, 77
Brill, A., 316, 330
Brinthaupt, T. M., 152, 159
Brockbank, M., 23, 27, 35
Brooks, R., 22, 38
Brooks-Gunn, J., 57, 60, 198, 206
Brothers, L., 99, 101, 120
Broughton, J. M., 197, 204
Brown, B., 151, 159
Brown, B. B., 202, 204
Brown, E., 111, 120
Brown, S. A., 167, 185
Brown, S. M., 166, 186
Brownell, C. A., 58, 60, 61
Bruckbauer, T., 147, 160
Brunelle, F., 114, 124
Bruner, J., 14, 34, 126, 137
Brunswick, N., 261, 267
Bryant, P., 127, 135, 137
Buccino, G., 19, 20, 21, 29, 35, 36, 37, 41
Buckner, R. L., 103, 124
Bucy, P. C., 102, 121
Budhani, S., 270, 284
Bulgheroni, M., 21, 41
Bullmore, E., 147, 161
Bullmore, E. T., 111, 120

Bunge, S. A., 84, 85, 87, 89, 94, 96, 176, 177, 180, 181, 183, 188, 213, 214, 224, 229, 245
Burack, J. A., 67, 80, 131, 137
Burgess, N., 75, 79
Burgess, P. W., 155, 160, 273, 276, 287
Burgy, L., 150, 159
Burke, K. A., 166, 188
Bush, G., 84, 94, 174, 185
Buxbaum, L. J., 20, 34

C

Cacioppo, J. T., 307
Caggiano, D. M., 178, 182, 183, 184, 216, 217, 221
Caglar, S., 103, 122, 147, 161
Cahill, L., 102, 120
Calder, A. J., 102, 120, 272, 287
Calder, A.J., 100, 102, 104, 106, 109, 111, 122
Calhoun, V. D., 229, 247
Call, J., 25, 26, 35, 37, 40, 54, 55, 62
Callaghan, T., 63, 77
Calu, D. J., 166, 188
Calvo-Merino, B., 19, 29, 35
Calzavara, R., 174, 186
Camak, L., 25, 26, 40
Camarda, R., 28, 39
Campbell, R., 111, 119
Canessa, N., 19, 20, 29, 35, 41
Cantlon, J. F., 27, 40
Caparelli-Daquer, E. M., 272, 286
Carey, S., 197, 204, 261, 267
Carlezon, W. A., Jr., 174, 188
Carlson, S. M., 64, 65, 66, 70, 71, 75, 77, 79, 80, 87, 89, 90, 91, 95, 98
Carmant, L., 22, 36
Carmichael, S. T., 102, 119
Carpendale, J., 289–311
Carpendale, J. I. M., 70, 78, 304, 307
Carpenter, M., 25, 26, 35, 40, 48, 53, 54, 55, 60, 62
Carruthers, P., 194, 204
Carter, C. S., 84, 95, 96, 172, 174, 185, 213, 222, 273, 284
Carter, E. J., 106, 107, 114, 120, 123
Caruana, F., 28, 29, 41
Carver, C. S., 86, 95
Casebeer, W. D., 291, 293, 294, 307

Casey, B. J., 68, 78, 202, 205, 212, 217, 229, 245
Cash, M., 303, 305, 307
Cashon, C. H., 46, 52, 60
Caspi, A., 275, 285
Cassidy, K. W., 71, 78
Castellanos, F. X., 169, 170, 186, 198, 205, 211, 212, 223
Castelli, F., 103, 111, 120
Castiello, U., 21, 41
Cattaneo, L., 30, 31, 35
Cavanna, A. E., 104, 120, 149, 159
Cavell, T. A., 280, 284, 286
Chabane, N., 114, 124
Chandler, M., xiii–xiv, 3–9
Chaput, H. H., 46, 52, 60
Charman, T., 126, 138
Chawarska, K., 113, 114, 121
Chersi, F., 17, 18, 27, 28, 29, 36
Chiavarino, C., 63, 77, 153, 158, 163
Chiron, C., 21, 35
Chomsky, N., 193, 205, 320, 321, 329
Chopra, S., 84, 97
Christoff, K., 155, 159
Chugani, H., 87, 95
Chun, M. M., 102, 103, 121
Church, J. A., 148, 156, 160
Churchland, P. S., 291, 293, 294, 307
Cillessen, A. N., 149, 159
Clark, F., 280, 284
Clark, M. A., 24, 34
Claus, E. D., 86, 95
Claux, M. L., 63, 77
Claxton, L. J., 64, 79
Clements, W., 76, 80
Clubley, E., 125, 136
Cohen, A. L., 148, 156, 160
Cohen, D. J., 113, 114, 121
Cohen, G. L., 168, 172, 185
Cohen, J., 298, 308
Cohen, J. D., 84, 95, 96, 103, 120, 213, 222, 272, 273, 274, 275, 280, 282, 284, 285, 286, 292, 308, 324, 329
Cohen, L. B., 46, 52, 60
Cohn, L. D., 211, 221, 222, 225
Colledge, E., 276, 280, 284, 286
Collette, F., 147, 153, 159
Collins, D. L., 196, 198, 206
Colvin, M. K., 191, 196, 204

Condry, J., 202, 205
Conklin, H. M., 167, 187, 210, 213, 223
Connors, C. M., 147, 163
Cooley, C. H., 144, 159
Cooper, J. C., 84, 97, 147, 153, 162
Copioli, C., 32, 35
Corcoran, R., 128, 136
Corkum, V., 44, 46, 53, 61
Cosmides, L., 151, 161, 300, 301, 310
Cossu, G., 30, 31, 32, 35
Cox, A., 126, 138
Craighero, L., 19, 29, 36, 39
Craik, F. I. M., 147, 159
Crain, R. M., 143, 159
Craven, R., 143, 144, 161
Crawley, A. P., 173, 187
Crick, N. R., 242, 245
Critchley, H. D., 83, 95, 203, 205
Crone, E. A., xiii–xiv, 3–9, 67, 68, 80, 84, 89, 97, 167, 176, 177, 180, 181, 183, 185, 188, 209–225, 210, 211, 213, 214, 216, 218, 222, 224
Cross, D., 70, 72, 80
Cruz, D., 169, 187
Csibra, G., 23, 27, 35, 37, 117, 124
Cunningham, M. G., 169, 185
Cunningham, W., 82, 87, 90, 98
Cunningham, W. A., 81–98, 82, 84, 86, 87, 92, 93, 95
Cushman, F., 273, 274, 275, 281, 285, 300, 308, 320, 322, 329, 330

D

Dabholkar, A. S., 195, 198, 205
Dadds, M. R., 280, 284
Dadlani, M. B., 210, 213, 222
Daffner, K. R., 292, 310
Dager, S., 30, 36
Dahl, R. E., 166, 167, 176, 177, 180, 185, 187, 211, 222, 233, 246
Dalla Vecchia, A., 32, 35
Damasio, A., 195, 205, 291, 292, 306, 307
Damasio, A. R., 83, 95, 217, 222, 253, 257, 266, 276, 284, 292, 306, 307, 322, 329
Damasio, H., 105, 121, 195, 205, 292, 306
Damon, W., 143, 144, 160, 201, 205
Damsio, A., 320, 322, 330
D'Angelo, S. L., 200, 205

Danube, C. L., 176, 178, 180, 181, 184
Dapretto, L., 32, 35
Dapretto, M., 141–163, 148, 151, 154, 156, 162
D'Argembeau, A., 147, 153, 159
Darley, J. M., 272, 273, 274, 275, 280, 282, 285, 292, 308, 324, 329
Dasser, V., 25, 35
David, A. S., 147, 161
Davidson, M. C., 212, 222
Davidson, R. J., 169, 172, 185, 186
Davies, M. S., 32, 35
Davis, M., 83, 87, 95, 97, 102, 120, 172, 185
Davis, M. H., 128, 129, 136
Davis-Dasilva, M., 25, 26, 40
Dawson, D., 111, 120
Dawson, G., 30, 36
de Greck, M., 103, 122
de Oliveira-Souza, R., 258, 261, 267, 271, 272, 273, 280, 286, 294, 299, 309
de Wit, H., 166, 186
Deakin, J., 167, 185
Debus, R., 143, 144, 161
Decety, J., 153, 159, 162
Degueldre, C., 147, 153, 159
Del Fiore, G., 147, 153, 159
Delgado, M. R., 176, 177, 180, 185, 187
Demos, K. E., 147, 160
Denburg, N., 252, 253, 267
Denckla, M. B., 30, 38
Dennis, M., 69, 78, 87, 97
Denny, B. T., 147, 160
D'Entremont, B., 54, 60
D'Esposito, M., 84, 94
Dettmers, C., 29, 36
Dewey, J., 289, 294, 307
Deyoung, C. G., 92, 95
Di Chiara, G., 172, 185
Diamond, A., 30, 36, 68, 69, 78, 87, 95, 118, 120, 198, 205, 212, 222
Dickinson, A., 273, 274, 287
Dickstein, D. P., 271, 286
DiCorcia, J. A., 277, 285
Dieterich, J. H., 292, 311, 327, 331
Dijkstra, M., 210, 224
Dimitrov, M., 253, 266
Dissanayake, C., 57, 61
Dobrowolny, H., 103, 122

Dodge, K. A., 128, 136
Döhnel, K., 73, 80
Dolan, C. V., 212, 223
Dolan, R. J., 83, 94, 95, 100, 102, 104, 106, 109, 111, 122, 203, 205
Donnelly, K., 91, 96
Donnelly, R. E., 69, 78
Dosenbach, N. U., 148, 156, 160
D'Ottavio, G., 21, 41
Downing, P., 106, 122
Downing, P. E., 102, 106, 121
Dray, A. J., 232, 246
Drew, A., 126, 138
Dri, G., 29, 39
Driver, J., 127, 131, 132, 134, 137
D'Souza, S. W., 195, 206
Dumontheil, I., 155, 160
Dupoux, E., 307
Durston, S., 68, 78, 212, 217, 222
Dutton, K., 102, 121
Dworkin, R., 314, 329

E

Eccles, J., 143, 144, 150, 151, 160, 163
Eccles, J. S., 202, 206
Eckstrand, K., 155, 163
Eddy, T. J., 25, 26, 39
Eggan, S., 169, 187
Eickhoff, S. B., 20, 35, 41
Eigsti, I., 210, 213, 222
Eisenberg, N., 279, 286, 315, 329
Eisenberger, N. I., 234, 235, 246
Eisensmith, R. C., 69, 80
Ekman, P., 169, 172, 186
Elkind, D., 200, 205
Ellis, S. C., 210, 217, 218, 224
Emery, N. J., 114, 121
Encarcion-Gawrych, G., 315, 330
Engell, A. D., 274, 280, 282, 285, 292, 308
Enns, J. T., 131, 137
Erickson, S., 169, 187
Ernst, M., 165–189, 166, 167, 168, 171, 172, 173, 174, 176, 179, 180, 181, 182, 183, 185, 186, 187, 188, 214, 217, 222
Ertelt, D., 29, 36
Escola, L., 28, 29, 41
Eshel, N., 176, 179, 180, 181, 186, 214, 222

Eslinger, P. J., 251–267, 252, 253, 257, 258, 261, 266, 267, 272, 273, 280, 286, 292, 307
Espinet, S. D., 92, 95
Etcoff, N. L., 102, 120
Evans, A., 104, 106, 120
Evans, A. C., 198, 205, 206
Exline, J. J., 295, 307
Eysel, U. T., 272, 287
Ezzyat, Y., 153, 162

F

Fabbi–Destro, M., 30, 31, 35
Fabes, R. A., 315, 329
Facchin, P., 29, 39
Fadiga, L., 15, 16, 17, 19, 27, 28, 29, 35, 36, 39, 40, 41, 43, 58, 60, 62
Fair, D. A., 148, 156, 160
Falck-Ytter, T., 24, 36
Farah, M. J., 183, 186
Fecteau, S., 22, 36
Feigenberg, L. F., 227–247, 242, 246
Fellows, L. K., 183, 186
Ferrari, P. F., 17, 18, 19, 27, 28, 29, 36, 39
Ferrugia, J. A., 26, 40
Ferstl, E., 147, 163
Festinger, L., 318, 329
Fiez, J. A., 176, 177, 180, 185, 187
Finch, D.M., 202, 207
Finger, E., 270, 284, 292, 307
Finger, E. C., 270, 286
Fink, G. R., 19, 20, 35, 41, 147, 160
Fischer, K., 57, 60
Fissell, C., 180, 185
Flach, R., 17, 37
Flaherty-Craig, C., 257, 266
Flaherty-Craig, C. V., 292, 307
Flanagan, J. R., 24, 36
Flanagan, O., 307
Flavell, E. R., 57, 60
Flavell, J. H., 57, 60
Fleiss, K., 276, 283
Fletcher, P. C., 261, 267
Flinn, M. V., 194, 205
Flombaum, J. L., 26, 36
Flor, H., 280, 284
Flory, J. D., 166, 186
Flynn, E., 65, 75, 78

Fogassi, L., 5, 9, 15, 16, 17, 18, 19, 27, 28, 29, 35, 36, 38, 39, 41, 43, 58, 60, 62
Fonagy, P., 128, 136
Fong, G. W., 178, 182, 183, 184, 215, 216, 217, 221, 223
Forst, R., 295, 296, 306, 307
Fowler, J. S., 183, 189
Fox, E., 102, 121
Fox, N. A., 167, 188
Frackowiak, R. S., 107, 124
Frackowiak, R. S. J., 75, 79
Frank, M. J., 86, 95
Frassrand, K., 167, 188, 213, 224
Fregni, F., 31, 40
Freud, A., 232, 233, 246
Freund, H. J., 20, 35, 41
Frick, P. J., 280, 288
Friedman, O., 71, 73, 78, 79
Friedman, S., 30, 36
Friston, K. J., 107, 124
Frith, C., 103, 111, 120, 128, 136
Frith, C. D., 68, 73, 78, 99, 100, 102, 103, 104, 106, 109, 111, 115, 121, 122, 153, 160, 261, 267, 270, 277, 281, 284
Frith, U., 23, 30, 32, 34, 99, 103, 111, 115, 120, 121, 123, 126, 127, 131, 135, 136, 137, 153, 160, 261, 267, 270, 277, 281, 284
Frye, D., 67, 69, 78, 80, 88, 90. 112
Fudge, J. L., 173, 174, 183, 185, 186
Fuhrman, R. W., 151, 161
Fuligni, A. J., 154, 162
Funder, D. C., 233, 235, 245
Furby, L., 210, 222
Fusi, S., 29, 39
Fuster, J. M., 212, 223

G

Gabrieli, J. D., 147, 153, 155, 159, 162
Gabrieli, J. D. E., 84, 87, 96, 97
Gallagher, H. L., 68, 73, 78, 261, 267
Gallagher, M., 172, 187
Gallese, V., 5, 9, 13–41, 15, 16, 17, 18, 19, 21, 27, 28, 29, 32, 35, 36, 37, 38, 39, 40, 41, 43, 58, 60, 62
Gallup, G. G., Jr., 253, 267

Galvan, A., 177, 182, 186, 212, 215, 217, 222, 223, 229, 245
Gardner, M., 144, 160, 167, 183, 186, 219, 223
Garmezy, N., 211, 223
Garon, N., 58, 61
Gatenby, J. C., 83, 84, 87, 95, 97
Gattis, M., 25, 34
Gaudette, T., 91, 94
Gazzaniga, M. S., 191, 196, 204
Geddes, L. P. T., 155, 159
Gentilucci M., 28, 39
Gergely, G., 23, 25, 26, 27, 35, 37, 47, 49, 60
German, T. P., 71, 73, 79
Gesierich, B., 17, 18, 27, 28, 29, 36
Gest, S. D., 211, 223
Getz, S., 212, 217, 229, 245
Gewirth, A., 314, 329
Giampietro, V., 272, 287
Giedd, J. N., 87, 96, 169, 170, 186, 196, 198, 202, 205, 206, 211, 212, 217, 223, 252, 267
Gilboa, A., 147, 160
Gilchrist, A. C., 195, 206
Glaser, B. G., 238, 246
Glaser, D. E., 19, 29, 35
Glimcher, P. W., 166, 186
Glover, G., 103, 122, 177, 182, 186, 215, 217, 223
Glover, G. H., 213, 221
Gogtay, N., 87, 96, 155, 163, 211, 223
Golan, O., 134, 136
Goldberg, B., 300, 307
Goldenfeld, N., 133, 136
Goldman, B. D., 114, 123
Goldman-Rakic, P. S., 87, 95, 198, 204
Goldsher, D., 261, 267
Goldstein, J., 104, 106, 107, 108, 115, 123
Good, C. D., 75, 79
Goodhart, F., 127, 135
Gopnik, A., 44, 56, 61, 62, 276, 284
Gore, J. C., 83, 84, 87, 95, 97, 102, 105, 106, 123
Gottlieb, G., 299, 307
Gould, S. J., 191, 205
Gould J., 110, 124
Grabowski, T., 195, 205
Grafman, J., 253, 266, 271, 280, 286, 294, 299, 309

Grafton, S., 280, 284
Grammont, F., 28, 29, 41
Grandin, T., 128, 136
Grant, J., 111, 119
Grattan, L. M., 257, 266
Gray, J. A., 169, 172, 186
Gredeback, G., 24, 36
Greenberg, M. T., 253, 266
Greene, J., 291, 292, 293, 295, 296, 298, 300, 308, 320, 322, 324, 329
Greene, J. D., 272, 273, 274, 275, 280, 282, 283, 285
Greenstein, D., 87, 96, 211, 223
Grezes, J., 19, 29, 35
Griffiths, P. E., 299, 302, 308
Grillon, C., 83, 87, 97
Gross, J., 202, 203, 206
Gross, J. J., 84, 96, 97
Grupe, D. W., 147, 163
Guajardo, J. J., 46, 47, 48, 49, 50, 62
Gurunathan, N., 131, 136
Gusnard, D. A., 147, 160
Guyer, A. E., 168, 186
Guz, G. R., 91, 96

H

Haber, S. N., 174, 186
Habermas, J., 233, 246, 298, 305, 308, 319, 329
Hacker, P. M. S., 300, 306
Haggard, P., 19, 29, 35
Haidt, J., 232, 242, 246, 258, 267, 278, 285, 293, 295, 301, 308, 318, 320, 329
Haines, D. J., 280, 284
Haith, M. M., 114, 121
Hajak, G., 73, 80
Hall, G. S., 210, 223
Halligan, E., 31, 40
Hamann, S., 280, 285
Hamberger, S., 132, 137
Hamburger, S. D., 202, 205
Hanelin, J., 103, 122
Hanjal, J. V., 107, 124
Happe, F., 103, 111, 120, 121, 131, 136, 261, 267, 277, 284
Hardin, M., 165–189, 166, 167, 172, 185
Hare, B., 26, 37
Hare, R. D., 275, 285

Hare, T. A., 177, 182, 186, 215, 217, 223
Harenski, C. L., 280, 285
Hargreaves, D., 150, 160
Hari, R., 19, 37
Hariri, A. R., 166, 186
Harold, R. D., 143, 144, 151, 163
Harris, J. R., 219, 223
Harris, P. L., 91, 96
Hart, D., 143, 144, 160, 201, 205
Harter, S., 141, 142, 143, 144, 149, 151, 159, 160
Hartley, T., 75, 79
Haste, H., 232, 246
Hauser, M., 273, 274, 275, 281, 282, 285, 288, 320, 322, 329, 330
Hauser, M. D., 300, 308
Hauser, S. T., 211, 221, 225
Hawes, D. J., 280, 284
Hayashi, K. M., 87, 96, 211, 223
Head, A. S., 120, 123
Heatherton, T. F., 103, 122, 147, 160, 161
Heberlein, A. S., 105, 121
Hedehus, M., 87, 96
Heekeren, H. R., 272, 273, 280, 285, 292, 308
Heerey, E. H., 83, 94
Heinzel, A., 103, 122
Heiserman, J. E., 147, 161
Heiss, W. D., 147, 160
Helwig, C. C., 271, 272, 275, 276, 278, 287, 316, 330
Henkenius, A. L., 196, 198, 206
Henning, A., 54, 60
Herman, S., 271, 276, 286
Hermann, C., 280, 284
Hershlag, N., 91, 94
Higgins, E. T., 151, 160
Hill, J., 111, 120, 128, 129, 134, 136
Hill, J. J., 134, 136
Hiraki, K., 22, 40
Hix, H. R., 65, 70, 77
Hodges, J. R., 102, 120
Hodgins, S., 276, 285
Hoekstra, R. A., 125, 135
Hoffman, M. L., 202, 205, 315, 330
Holland, P. C., 172, 187
Holloway, R. L., 27, 40
Holmes, C. J., 196, 198, 206
Hommel, B., 17, 37
Hommer, D., 215, 223

Hommer, D. W., 176, 178, 180, 181, 182, 183, 184, 216, 217, 221
Hooper, C. J., 167, 187, 210, 213, 223
Hornak, J., 83, 84, 97
Horner, V., 26, 37
Howard, M. A., 83, 97
Howes, C., 242, 245
Hrabok, M., 304, 308
Hubbard, E. H., 31, 32, 38
Hubbard, J. J., 211, 223
Huber, O., 147, 163
Huettel, S. A., 215, 223
Hughes, C., 64, 65, 75, 78, 276, 277, 280, 285, 286
Hughes, J. N., 280, 284, 286
Huizinga, M., 212, 223
Humphrey, N. K., 25, 37
Humphreys, G. W., 63, 77, 153, 158, 163
Hunnisett, E., 111, 115, 122
Hunyadi, E., 147, 163
Huttenlocher, P. R., 87, 96, 195, 198, 205, 212, 223
Hynes, C., 280, 284

I

Iacoboni, M., 19, 20, 21, 32, 34, 35, 37
Iarocci, G., 131, 137
Ignacio, F. A., 272, 286
Imrisek, S., 65, 79
Inati, S., 103, 122, 147, 161, 191, 196, 204
Inhelder, B., 196, 206
Intskirveli, I., 28, 29, 41
Isomura, Y., 203, 206
Itakura, S., 63, 77

J

Jackson, D. C., 172, 185
Jacob, P., 307
Jacobs, J. E., 150, 160, 202, 206
Jacques, S., 67, 80
James, W., 143, 160
Jarcho, J. M., 147, 149, 150, 156, 161
Järveläinen, J., 19, 37
Jazbec, S., 176, 179, 180, 181, 182, 183, 185, 214, 217, 222
Jeffries, N. O., 87, 96, 169, 170, 186, 198, 205, 211, 212, 223
Jensen, J., 173, 187

Jentsch, J. D., 183, 187
Jernigan, T. L., 196, 198, 206
Jezzini, A., 28, 29, 41
Jiang, Y., 102, 106, 121
Jiang, Y. V., 72, 80
Jin, R. K.-X., 300, 308
Johansson, G., 105, 121
Johansson, R. S., 24, 36
Johnsen, I. R., 84, 86, 95
Johnson, M. H., 106, 117, 121, 124
Johnson, M. K., 82, 84, 87, 95
Johnson, S. C., 47, 60, 147, 161, 163
Johnson, S. H., 20, 34
Jolliffe, T., 131, 134, 135, 136
Jones, J., 113, 114, 121
Jones, L., 280, 284
Jones, R., 127, 134, 135
Jonides, J., 95, 103, 120
Jordan, J. S., 52, 60
Joseph, C. (2004)., 293, 301, 308
Joseph, R. M., 67, 78, 277, 287
Jovanovic, B., 47, 49, 60
Judas, M., 195, 206

K

Kabani, N. J., 155, 163
Kable, J. W., 166, 186
Kagan, J., 91, 94, 96, 196, 197, 199, 206, 234, 235, 236, 246, 279, 285, 293, 308
Kahler, C. W., 167, 187
Kain, W., 68, 78
Kalin, N. H., 172, 185
Kambartel, F., 305, 309
Kan, E., 211, 212, 224
Kanner, K., 132, 136
Kanner, L., 110, 121
Kant, I., 307, 314, 330
Kanwisher, M., 261, 267
Kanwisher, N., 102, 103, 106, 121, 123, 153, 163
Kapur, S., 173, 187
Karmiloff, K., 193, 206
Karmiloff-Smith, A., 111, 119, 193, 206
Kashima, Y., 57, 61
Kastner, S., 273, 285
Kaufman, S., 69, 80
Kawahara-Baccus, T. N., 147, 163
Kawasaki, H., 83, 97

Kelley, W. M., 103, 122, 147, 160, 161
Kelly, J. L., 102, 119
Keltner, D., 83, 94
Kennard, C., 107, 124
Kerns, K. A., 304, 308
Kesek, A., 84, 87, 89, 90, 95, 98
Kessler, J., 147, 160
Keysers, C., 17, 18, 19, 37, 38, 41
Kiehl, K. A., 229, 247
Kihsltrom, J. F., 147, 151, 153, 161, 162
Killen, M., 271, 272, 275, 276, 278, 287, 289, 292, 296, 309, 310, 322, 329, 330
Kim, K. S., 174, 186
Kim, S. Y., 276, 287
King, M. S., 227, 242, 246
Király, I., 25, 26, 37, 47, 49, 60
Kircher, T. T., 147, 161
Klein, S. B., 151, 161
Klin, A., 113, 114, 121
Klingberg, T., 87, 96
Kloo, D., 64, 72, 80
Kluver, H., 102, 121
Knickmeyer, R., 125, 126, 134, 135
Knierim, K., 103, 122
Knight, R. T., 83, 94
Knoblich, G., 17, 37, 52, 60
Knutson, B., 178, 182, 183, 184, 215, 216, 217, 221, 223
Kobayashi, M., 31, 40
Kochanska, G., 295, 309
Koechlin, E., 84, 96
Koenigs, M., 273, 274, 275, 281, 285, 320, 322, 330
Kohlberg, L., 271, 272, 285, 291, 309, 318, 320, 321, 323, 327, 330
Kohler, E., 17, 18, 38, 41
Köhler, W., 25, 37
Kolb, B., 257, 267
Kondo, H., 252, 267
Koob, G. F., 182, 187
Koòs, O., 23, 27, 35
Korsgaard, C. M., 296, 309
Kosson, D. S., 270, 286
Kounelher, F., 84, 96
Kozuch, P., 202, 205
Kramer, R., 271, 272, 285
Kringelbach, M. L., 173, 187
Kronbichler, M., 153, 158
Krueger, F., 271, 286, 294, 299, 309

L

Kruger, A., 25, 26, 40
Kugiumutzakis, G., 22, 38
Kunda, Z., 99, 121
Kwon, H., 213, 223

L

Ladurner, G., 153, 158
Laflamme, L., 21, 35
Lagravinese, G., 19, 29, 35
Laible, D. J., 199, 206
Lamb, S., 279, 285
Lamborn, S. D., 202, 204
Lamm, C., 87, 96
Landry, S. H., 111, 122
Lang, B., 64, 69, 72, 79
Lanza, S., 150, 160
LaPierre, D., 276, 285
Lapsley, D. K., 201, 206
Laureys, S., 147, 153, 159
Laviola, G., 169, 187
Lawson, J., 127, 136
Lazenby, A. L., 69, 78
Le Couteur, A., 112, 122
LeBlanc, C. J., 169, 171, 184
Lecours, A. R., 87, 97
LeDoux, J., 273, 285
LeDoux, J. E., 83, 96, 102, 122, 172, 187
Lee, K., 65, 66, 71, 80
Leekam, S., 127, 131, 137
Leekam, S. R., 111, 115, 122, 123
Legerstee, M., 22, 38
Leibenluft, E., 167, 171, 176, 179, 180, 181, 182, 183, 185, 188, 214, 215, 217, 222, 224
Lejuez, C. W., 167, 187
Lempers, J. D., 57, 60
Lenroot, R., 155, 163
Leonard, A., 276, 280, 283, 284, 286
Leonard, C. M., 211, 212, 224
Leonard, G., 198, 206
Lepage, J. F., 23, 38
LePage, J.-F., 45, 60
Lerch, J. P., 155, 163
Leslie, A. M., 23, 30, 32, 34, 38, 71, 73, 78, 79, 115, 120, 123, 126, 127, 131, 135, 137, 277, 285
Levine, B., 87, 97
Lewis, C., 66, 70, 78, 79, 289, 304, 307
Lewis, D. A., 169, 187

Lewis, M., 57, 60, 87, 90, 91, 92, 93, 94, 96, 97, 198, 206
Lewis, M. D., 87, 96
Lickliter, R., 299, 302, 309
Lieberman, M. D., 4, 9, 141–163, 142, 147, 148, 149, 150, 151, 154, 156, 161, 162, 234, 235, 246
Lillard, A., 63, 77
Linthicum, J., 172, 187
Lipka, R. P., 152, 159
Lipsitt, L. P., 90, 96
Liston, C., 91, 96, 212, 217, 222
Liszkowski, U., 48, 53, 54, 60, 62
Liu, H., 169, 170, 186, 198, 205, 211, 212, 223
Lockhart, P., 30, 38
Lockyer, L., 69, 78
Loeber, R., 275, 285
Loftus, J., 151, 161
Logothetis, N., 293, 309
Lombardo, M. V., 157, 161
Lopez, B., 111, 115, 122
Lord, C., 112, 122
Louie, K., 166, 186
Loveland, K. A., 111, 122
Luciana, M., 167, 187, 210, 213, 223
Ludlow, D. H., 103, 122
Lueck, C. J., 107, 124
Lui, F., 19, 29, 35
Luo, J., 292, 296, 307
Luo, Q., 272, 273, 280, 282, 285
Luo, Y., 47, 60
Luppino G., 28, 39
Lusk, L., 87, 96, 211, 223
Lutchmaya, S., 126, 135
Luu, P., 84, 94, 174, 185
Luxen, A., 159
Lynam, D. R., 275, 285
Lyss, P. J., 169, 171, 184

M

MacArthur, D., 303, 309
MacDonald, A. W., 84, 96
MacIntyre, A., 290, 309
Mack, P. B., 111, 117, 122
MaCrae, C. N., 103, 122
Macrae, C. N., 147, 148, 153, 160, 161, 162
Maguire, E. A., 75, 79
Mailly, P., 174, 186

Maiolini, C., 27, 36
Mallon, R., 277, 285
Mameli, M., 302, 309
Mandell, D. J., 65, 77
Manuck, S. B., 166, 186
Marcovitch, S., 88, 90. 112, 91, 96
Markowitsch, H. J., 147, 160
Markus, H. R., 150, 151, 152, 161, 163
Marsh, A. A., 270, 284, 286, 292, 307
Marsh, H. W., 143, 144, 151, 161
Martin, A., 272, 273, 280, 282, 285
Martin, J., 125, 136
Martin, M. M., 70, 77
Martin-Soelch, C., 172, 187
Martineau, J., 30, 40
Masataka, N., 48, 61
Masten, A. S., 211, 223
Masten, C. L., 154, 162
Masure, M. C., 21, 35
Matelli M., 28, 39
Mathias C. J., 203, 205
Matsuzawa, T., 27, 38
May, J. C., 176, 177, 180, 187
Mazière, B., 21, 35
Mazziotta, J., 19, 21, 37
Mazziotta, J. C., 20, 37
Mazziotta, J. V., 107, 124
McCarthy, G., 102, 104, 105, 106, 107, 108, 109, 110, 111, 112, 115, 117, 119, 122, 123, 162
McClain, C., 173, 174, 183, 186
McCleery, J. P., 31, 32, 38
McClintock, M. K., 307
McClosky, M., 316, 330
McClure, E. B., 168, 171, 176, 179, 180, 181, 182, 183, 185, 188, 214, 215, 217, 222, 224, 271, 286
McClure, S. M., 215, 223
McClure-Tone, E. B., 168, 186
McConnell, S., 195, 206
McDermott, J., 102, 103, 121
McDougall, W., 315, 317, 330
McDowell, J., 299, 303, 309
McFarland, N. R., 174, 186
McGrath, J., 83, 84, 97
McIntosh, A. R., 173, 187
McIntosh, D. N., 32, 38
McKeown, M. J., 104, 106, 107, 108, 115, 123
McNamara, A., 29, 36

Mead, G. H., 144, 162
Mealiea, J., 58, 61
Meinhardt, J., 73, 80
Meltzoff, A. N., 22, 24, 25, 27, 38, 44, 61, 70, 77, 111, 120, 141, 162, 276, 284
Menon, V., 213, 221, 223
Meresse, I., 114, 124
Merz, S., 280, 284
Mesulam, M. M., 292, 310
Metz, U., 47, 49, 57, 62
Miezin, F. M., 148, 156, 160
Mikhail, J., 273, 286, 299, 300, 301, 308, 309
Mikulis, D. J., 173, 187
Mill, J. S., 314, 330
Miller, E. K., 273, 286
Miller, G., 235, 236, 246, 293, 309, 328, 330
Miller, P. A., 279, 286
Mills, B., 210, 217, 224
Milner, A. D., 120, 123
Mineka, S., 83, 97
Mischel, W., 210, 213, 222, 224
Mistlin, A. J., 120, 123
Mitchell, D. G., 276, 280, 284
Mitchell, D. G. V., 270, 276, 280, 284, 286
Mitchell, J. P., 103, 104, 106, 107, 108, 115, 122, 148, 153, 162
Mitchell, T. V., 104, 106, 107, 108, 115, 123
Moffitt, T. E., 275, 285
Mokler, D. J., 102, 122
Moll, F. T., 272, 286
Moll, H., 25, 26, 40, 54, 55, 57, 61, 62
Moll, J., 258, 261, 267, 271, 272, 273, 280, 286, 294, 299, 309
Molnar–Szakacs, I., 19, 21, 37
Monfils, M., 257, 267
Monk, C. S., 176, 179, 180, 181, 182, 183, 185, 214, 217, 222
Montague, P. R., 84, 96, 215, 223
Monti, A., 30, 31, 35
Mooney, R., 18, 39
Moore, C., 43–62, 44, 45, 46, 47, 49, 50, 52, 53, 55, 56, 57, 58, 60, 61, 111, 115, 122
Moore, G. E., 290, 309
Moore, M., 114, 121
Moore, M. K., 22, 27, 38, 141, 162

Morgan, T., 113, 114, 121
Morgane, P. J., 102, 122
Moroz, T. M., 147, 159
Morris, A. S., 144, 151, 163
Morris, J. P., 100, 104, 106, 109, 110, 111, 112, 123, 162
Morris, J. S., 100, 102, 104, 106, 109, 111, 119, 122
Mortimore, C., 134, 135
Morton, J., 276, 280, 283, 284
Mosconi, M. W., 111, 117, 122
Moscovitch, M., 147, 159
Moseley, M. E., 87, 96
Moses, L. J., 64, 65, 66, 70, 71, 74, 75, 77, 79, 80
Moses, P., 255, 267
Mostofsky, S. H., 30, 38
Mottron, L., 131, 137
Mounts, N., 202, 204
Mourao-Miranda, J., 272, 273, 280, 286
Mowrer, S. M., 84, 95
Mrzljak, L., 195, 206
Muesel, L.-A., 92, 93, 97
Müller, U., 65, 79, 88, 90, 112, 289–311, 304, 307
Mundy, P., 111, 122
Murphy, M. N., 201, 206
Murray, E. A., 172, 184
Murray, L., 280, 284
Müsseler, J., 17, 37
Myers, R., 107, 124
Myowa-Yamakoshi, M., 27, 38

N

Nàdasdy, Z., 23, 27, 37
Naito, M., 76, 80
Nakic, M., 272, 273, 280, 282, 285
Nawa, N. E., 183, 188
Needham, A., 24, 40
Neff, K. D., 316, 330
Nelson, E. E., 168, 171, 176, 179, 180, 181, 182, 183, 185, 186, 188, 214, 215, 217, 222, 224
Nestler, E. J., 174, 188
Neville, H. J., 75, 79
Newman, C., 276, 284
Nichols, S., 279, 286
Nickel, J., 183, 188
Nielsen, M., 57, 61

Nieuwenhuis, S., 67, 68, 80, 84, 89, 97
Nissen, A. G., 26, 40
Noll, D., 84, 95
Noll, D. C., 95, 103, 120, 180, 185
Northoff, G., 103, 122
Noseworthy, M. D., 87, 97
Nowicki, S., 18, 39
Nucci, L. P., 271, 275, 276, 282, 286, 296, 309, 315, 330
Nussbaum, M. C., 314, 317, 328, 330
Nystrom, L. E., 95, 103, 120, 180, 185, 272, 273, 274, 275, 280, 282, 285, 292, 308, 324, 329

O

Oberman, L. M., 31, 32, 38, 39
Ochsner, K., 202, 203, 206
Ochsner, K. N., 4, 9, 84, 96, 97, 103, 122, 147, 153, 162
O'Connor, K. J., 83, 87, 97
Odden, H., 63, 77
O'Doherty, J. P., 273, 274, 287
O'Donnell, S., 87, 97
Ody, C., 84, 96
Oh, S., 66, 79
Ohman, A., 83, 95, 97, 203, 205
Ok, S.-J., 47, 60
Olson, C.R., 202, 207
Olson, I. R., 153, 162
Omar, H. A., 200, 205
Onishi, K. H., 14, 39
O'Riordan, M., 127, 131, 132, 134, 135, 137
Osgood, D. W., 150, 160
Osterling, J., 111, 120
Ostry, D., 104, 106, 120
Overman, W. H., 167, 188, 213, 224
Overmeyer, S., 87, 92, 97
Oxford, M., 280, 284, 286
Oya, H., 83, 97
Oyama, S., 299, 309
Ozonoff, S., 67, 68, 79, 132, 137, 276, 277, 287

P

Palfai, T., 67, 69, 78
Panksepp, J., 103, 122
Parkin, L., 76, 80

Parkin, L. J., 71, 76, 79, 80
Parra, C. E., 177, 182, 186, 215, 217, 223
Parrish, J. M., 168, 171, 188
Pascual-Leone, A., 31, 40
Pascucci, T., 169, 187
Passingham, R. E., 19, 29, 35
Patrick, C. J., 280, 284
Patteri, I., 19, 29, 35
Paukner, A., 27, 29, 36, 39
Paul, B., 255, 267
Paul, G., 114, 123
Paul, R., 113, 114, 121
Paulus, 166, 180
Paus, T., 196, 198, 205, 206
Pearlson, G. D., 229, 247
Pedulla, C. M., 167, 187
Peelen, M., 106, 122
Pellegrini, A. D., 4, 8
Pellicano, E., 67, 79
Pelphrey, K. A., 99–124, 100, 101, 104, 106, 107, 108, 109, 110, 111, 112, 114, 115, 117, 119, 120, 122, 123, 153, 162
Penn, J., 177, 182, 186, 215, 217, 223
Pennington, 67
Pennington, 68, 79
Pennington, B., 132, 137
Pennington, B. F., 276, 277, 287
Perez-Edgar, K. E., 167, 188
Perkins, S. A., 316, 330, 331
Perlman, S. B., 99–124, 101
Perner, J., 23, 41, 64, 68, 69, 71, 72, 73, 76, 78, 79, 80, 115, 123, 127, 131, 137, 138, 153, 158
Perrett, D. I., 31, 32, 41, 100, 102, 104, 106, 109, 111, 120, 122, 123
Perry, D. G., 279, 284, 287
Perry, L. C., 279, 284, 287
Perry, Y., 280, 284
Peters, S., 18, 39
Peterson, B. S., 196, 198, 206
Petrides, M., 104, 106, 120
Pezzetta, E., 21, 41
Pfeifer, J. H., 32, 35, 141–163, 148, 151, 154, 156, 162
Phelps, E. A., 83, 87, 97
Phelps, M., 87, 95
Phillips, M. L., 272, 287
Phipps, M., 253, 266

Piaget, J., 24, 39, 47, 50, 58, 61, 196, 197, 199, 206, 269, 271, 272, 282, 287, 295, 297, 303, 304, 309, 315, 320, 330
Pieraccini, C., 30, 31, 35
Pieretti, S., 169, 187
Pike, B., 198, 206
Pine, D. S., 166, 167, 168, 171, 172, 176, 179, 180, 181, 183, 185, 186, 188, 214, 215, 217, 222, 224
Pineda, J. A., 31, 32, 38
Pinker, S., 193, 206
Pipe, J. G., 147, 161
Pitkanen, A., 102, 119
Pitt-Watson, R., 91, 96
Piven, J., 114, 123
Plaisted, K., 127, 131, 132, 134, 135, 137
Plotzker, A., 153, 162
Plumb, I., 111, 120, 128, 129, 134, 136
Polizzi, P., 71, 73, 79
Porro, C. A., 29, 39
Porro, C. A., 19, 29, 35
Posner, M. I., 84, 94, 174, 185
Potter, D. D., 120, 123
Povinelli, D. J., 25, 26, 39, 53, 58, 61
Powell, L. J., 103, 123
Pozdol, S. L., 113, 114, 120
Prather, J. F., 18, 39
Pratt, C., 127, 135, 137
Prehn, K., 272, 273, 280, 285
Premack, D., 23, 39, 99, 123
Price, B. H., 292, 310
Price, J. L., 102, 119, 252, 267
Prigatano, G. P., 147, 161
Prinstein, M. J., 168, 172, 185
Prinz, J. J., 303, 310
Prinz, W., 17, 36, 37, 39, 47, 49, 60
Przuntek, H., 272, 287
Puce, A., 102, 105, 106, 109, 119, 123

Q

Quartz, S. R., 75, 80

R

Racine, T. P., 289, 304, 307
Raichle, M. E., 147, 160
Raine, A., 292, 294, 295, 310
Rakic, P., 198, 204

Ramachandran, T., 103, 122
Ramachandran, V. S., 31, 32, 38, 39
Ramani, G. B., 58, 60
Rameson, L., 150, 162
Ramirez, M., 211, 223
Ramsey, S. E., 167, 187
Rapapport, J. L., 198, 205
Rapoport, J. L., 202, 205
Raste, Y., 111, 120, 128, 129, 134, 136
Ratner, H., 25, 26, 40
Rauch, S. L., 173, 188
Rausch, M., 272, 287
Rawls, J., 314, 331
Ray, R. D., 84, 97
Raye, C. L., 84, 95
Raynaud, C., 21, 35
Read, J. P., 167, 187
Ream, J. M., 155, 159
Redmond, A., 167, 188, 213, 224
Rees, L., 126, 138
Reichman-Decker, A., 32, 38
Reid, M. E., 270, 284, 286
Reilly, J., 255, 267
Reinkemeier, M., 147, 160
Reiss, A. L., 30, 38, 213, 223
Remington, G., 173, 187
Repacholi, B. M., 56, 62
Reyna, V. F., 210, 217, 218, 224
Reynolds, B., 210, 224
Reznick, J.S., 114, 123
Rich, B. A., 167, 188
Richards, J. B., 167, 187
Richards, T., 30, 36
Richell, R., 272, 273, 280, 282, 285
Richell, R. A., 276, 284
Richler, J., 131, 136
Riddell, A. C., 280, 284
Ridderinkhof, K. R., 67, 68, 80, 84, 89, 97
Rilling, J. K., 168, 171, 188
Rimland, B., 132, 137
Rinaldi, J., 111, 120
Ring, H. A., 111, 120
Ritzl, A., 20, 35
Rivers, S., 210, 217, 224
Rizzolatti, G., 5, 9, 15, 16, 17, 18, 19, 20, 21, 27, 28, 29, 30, 31, 34, 35, 36, 37, 38, 39, 41, 43, 58, 60, 62
Robbins, T., 167, 185
Robbins, T. W., 273, 276, 277, 280, 285, 287

Robert, M., 22, 36
Roberts, A. C., 273, 287
Robertson, E. R., 147, 153, 162
Robertson, M., 134, 135
Robinson-Long, M., 251–267
Rochat, M., 13–41, 27, 28, 29, 40, 41
Rochat, P., 27, 34, 63, 77
Rodriguez, M., 210, 213, 224
Roesch, M., 273, 274, 287
Roesch, M. R., 166, 188
Roeser, R., 150, 151, 163
Rogers, S., 31, 40, 132, 137
Rogers, S. J., 67, 68, 79
Rolls, E. T., 83, 84, 97
Rombouts, S. A. R. B., 216, 224
Romeo, R. D., 171, 186
Rosenberg, M., 143, 144, 162
Rotshtein, P., 83, 95, 203, 205
Rowland, D., 100, 102, 104, 106, 109, 111, 120, 122
Rozendal, K., 151, 161
Rozzi, S., 17, 18, 19, 27, 28, 29, 36
Rubia, K., 87, 92, 97
Rubin, E., 113, 114, 121
Ruby, P., 147, 153, 159, 162
Ruffman, T., 76, 80
Rumsey, J., 132, 137, 168, 188
Russell, J., 69, 80, 132, 137, 276, 277, 280, 285
Russell, J. A., 86, 97
Russo, R., 102, 121
Rutherford, M., 131, 136
Rutter, M., 112, 122, 126, 137
Ryan, N. D., 176, 177, 180, 187

S

Saarni, C., 91, 97
Sabbagh, M. A., 63–80, 64, 65, 66, 71, 75, 79, 80
Sahakian, B. J., 167, 185
Saleem, K. S., 252, 267
Samson, D., 63, 77, 153, 158, 163
Samson, Y., 114, 124
Sanfey, A. G., 166, 188
Santelli, E., 32, 35
Santos, L. R., 26, 36, 40
Sasson, N. J., 114, 123
Satpute, A. B., 147, 149, 150, 156, 161

Saxe, R., 72, 80, 103, 104, 110, 113, 114, 116, 117, 123, 124, 153, 163, 261, 267, 282, 288
Saylor, M. M., 24, 34
Scabar, A., 21, 41
Scabini, D., 83, 94
Scahill, V., 127, 136
Scaife, M., 126, 137
Scheier, M. F., 86, 95
Schellenberg, G.D., 30, 36
Scheres, A., 166, 188, 210, 224
Schmajuk, M., 167, 188, 271, 286
Schmidt, H., 272, 273, 280, 285, 292, 308
Schmitz, C., 30, 40
Schmitz, T. W., 147, 163
Schoenbaum, G., 166, 188, 273, 274, 287
Schoonheyt, W., 69, 78
Schultz, L. H., 246
Schultz, R. T., 113, 114, 121, 147, 163
Schulz, L. E., 72, 80
Schurmann, M., 19, 37
Schwintowski, H. P., 272, 273, 280, 285
Schwintowski, H.-P., 292, 308
Scott, A. A., 32, 35
Scott, S. K., 272, 287
Scriver, C. R., 69, 80
Seamans, E., 54, 60
Seitz, R. J., 183, 188
Sejnowski, T. J., 75, 80
Selman, R. L., 201, 206, 227–247, 232, 237, 242, 246
Sen, A., 317, 318, 331
Senior, C., 272, 287
Senju, A., 117, 124
Serra, E., 27, 40
Shah, A., 131, 137
Shah, N. J., 20, 35, 41
Shallice, T., 86, 97, 273, 276, 287
Shamay-Tsoory, S. G., 261, 267
Shannon, B. J., 103, 124
Shaw, P., 155, 163
Shelton, K. K., 280, 288
Sheridan, J. F., 307
Sherman, T., 111, 122
Sherren, N., 257, 267
Shiffrin, N. D., 168, 186
Shimada, S., 22, 40
Shin, L. M., 173, 188
Shiverick, S., 65, 66, 80
Shoda, Y., 210, 213, 222, 224

Shulman, G. L., 147, 160
Shuman, M., 102, 106, 121
Siegler, R. S., 151, 163
Sigman, M., 32, 35, 111, 122
Silverberg, S. B., 144, 151, 163
Silverthorn, P., 280, 288
Simmons, A., 87, 92, 97, 111, 120, 147, 161
Simone, L., 19, 27, 36
Simpson, M. D., 195, 206
Sims, C., 270, 284, 286
Singerman, J. D., 109, 111, 123
Singh, S., 63, 77
Singh, T., 31, 32, 41
Sinnott-Armstrong, W., 280, 284
Skinner, B. F., 298, 299, 310, 317, 319, 331
Skinner, R., 125, 136
Slater, P., 195, 206
Small, S., 29, 36
Smetana, J., 289, 292, 296, 309, 310, 322, 329, 330
Smetana, J. G., 271, 275, 276, 282, 287, 314, 322, 331
Smith, A. R., 176, 178, 180, 181, 184
Smith, E. E., 95, 103, 120
Smith, M., 280, 284
Smith, P. A. J., 120, 123
Smoski, M. J., 113, 114, 120
Snow, J., 271, 286
Snowdon, C., 25, 40
Sodian, B., 47, 49, 57, 62, 73, 80, 127, 137
Sokol, B. W., 289–311
Solodkin, A., 29, 36
Sommer, M., 73, 80
Sommerville, J., 48, 50, 62
Sommerville, J. A., 24, 40
Sommerville, R. B., 272, 273, 274, 275, 280, 282, 285, 292, 308, 324, 329
Song, M., 276, 287
Sonuga-Barke, E., 210, 224
Sowell, E. R., 196, 198, 206, 211, 212, 224
Spear, L. P., 167, 181, 188
Spear, L. R., 215, 224
Spiers, H. J., 75, 79
Spong, A., 127, 136
Sprengelmeyer, R., 272, 287
Staffen, W., 153, 158
Stanger, C., 91, 96
Steele, S., 277, 287
Stein, K. F., 150, 151, 163

Steinberg, L., 144, 151, 159, 160, 163, 166, 167, 183, 186, 188, 202, 204, 210, 215, 217, 218, 219, 223, 224, 229, 233, 246
Stekeler-Weithofer, P., 304, 305, 310
Stenger, V. A., 84, 96, 176, 177, 180, 187
Stern, D. N., 22, 40
Stevens, M. C., 229, 247
Stieglitz, S., 64, 79
Stiles, J., 255, 267
Stone, V., 127, 131, 134, 135, 136
Stone, W., 113, 114, 120
Stortz, K., 299, 302, 308
Stouthamer-Loeber, M., 275, 285
Stowe, R. M., 292, 310
Strauss, A. L., 238, 246
Striano, T., 54, 60
Stuart, G. L., 167, 187
Stuss, D. T., 86, 97, 147, 159, 253, 267
Subiaul, F., 27, 40
Suddendorf, T., 31, 32, 41
Sullivan, M. W., 90, 91, 94, 96
Swain, I. U., 90, 91, 97
Swettenham, J., 126, 138

T

Tager-Flusberg, H., 31, 40, 67, 78, 277, 287
Takada M., 203, 206
Tanaka, M., 27, 38
Tanney, J., 299, 310
Tapanya, S., 63, 77
Tapert, S. F., 167, 185
Taylor, E., 87, 92, 97
Taylor, J. R., 183, 187
Tellegen, A., 211, 223
Terrace H. S., 27, 40
Thagard, P., 233, 247
Theoret, H., 31, 40
Théoret, H., 22, 23, 36, 38, 45, 60
Thoermer, C., 47, 49, 57, 62, 73, 80
Thomas, K. M., 68, 78, 212, 217, 222
Thompson, P. M., 196, 198, 206, 211, 212, 224
Thompson, R. A., 199, 206
Thorne, J. F., 168, 171, 188
Tiggeman, M., 150, 160
Todd, R., 92, 93, 97
Toga, A. W., 196, 198, 206, 211, 212, 224

Tomasello, M., 25, 26, 27, 35, 37, 40, 41, 48, 53, 54, 55, 57, 60, 61, 62, 300, 305, 310
Tomer, R., 261, 267
Tomonaga, M., 27, 38
Tooby, J., 300, 301, 310
Tottenham, N., 212, 217
Towbin, K. E., 271, 286
Trafton, J. G., 151, 161
Trainor, R., 202, 205
Trambley, C., 22, 36
Tranel, D., 102, 105, 119, 121, 195, 205, 252, 253, 266, 267, 273, 274, 275, 281, 285, 292, 306, 320, 322, 330
Trawalter, S., 167, 188, 213, 224
Treland, J. E., 271, 286
Trevarthen, C., 22, 40
Trimble, M. R., 149, 159
Trimble, R., 104, 120
Troje, N., 106, 124
Tronick, E., 22, 41
Tulving, E., 147, 159
Turiel, E., 271, 272, 275, 276, 278, 287, 292, 295, 296, 298, 310, 313–331, 314, 315, 316, 318, 320, 321, 322, 327, 330, 331
Turner, L. M., 113, 114, 120

U

Ugolotti, F., 19, 27, 36
Ullsperger, M., 67, 68, 80, 84, 89, 97
Ulug, A. M., 68, 78
Umiltà, M. A., 17, 18, 27, 28, 29, 37, 38, 41
Underwood, M. K., 242, 247
Ungerer, J., 111, 122
Ungerleider, L. G., 273, 285
Uylings, H. B., 195, 206

V

Vaidya, C. J., 87, 96
Vaituzis, A. C., 87, 96, 211, 223
Vaituzis, C. K., 202, 205
Valentin, V. V., 273, 274, 287
Van Bavel, J. J., 86, 93, 95
Van Der Linden, M., 183, 184
Van der Linden, M., 147, 153, 159

van der Molen, M. W., 167, 176, 185, 212, 222, 223
van Eden, C. G., 195, 206
Van horn, J., 280, 284
Van Leijenhorst, L., 167, 176, 177, 180, 181, 183, 188, 209–225, 211, 213, 214, 216, 217, 224
Van Meel, C. S., 216, 224
van Veen, V., 172, 174, 185
Vander Wyk, B. C., 99–124
Vanderwal, T., 147, 163
Vanhorn, J. D., 191, 196, 204
Vartanian, L. R., 141, 163, 200, 206
Vauss, Y., 202, 205
Vernon, M. K., 202, 206
Villringer, A., 272, 273, 280, 285, 292, 308
Viola, R. J., 123, 162
Visalberghi, E., 26, 27, 36, 39, 41
Vogt, B. A., 202, 207
Vogt, S., 20, 35, 41
Vohs, K. D., 233, 235, 245
Volkmar, F. R., 113, 114, 121
Volkow, N. D., 183, 189
von Cramon, D. Y., 147, 163
von Hofsten, C., 24, 36
Von Wright, G. H., 297, 298, 311
Voss, H., 177, 182, 186, 215, 217, 223

W

Waddington, C. H., 191, 207
Wade, D., 83, 84, 97
Waggoner, A., 84, 95
Wainryb, C., 316, 331
Waldmann, M. R., 292, 311, 327, 331
Walker, J., 111, 119
Walz, K., 91, 94
Walzer, M., 326, 327, 331
Wang, T. S., 87, 89, 91, 95
Warertenburger, I., 292, 308
Warsofsky, I. S., 213, 221
Wartenburger, I., 272, 273, 280, 285
Watson, J., 70, 72, 80
Watson, J. B., 317, 331
Watson, J. D., 107, 124
Webb, S., 30, 36
Weber, E., 315, 330
Wegner, D. M., 319, 331
Weiskrantz, L., 273, 287

Weiss, M., 91, 96
Weiss, M. J., 90, 91, 97
Welcome, S. E., 196, 198, 206, 211, 212, 224
Wellman, H. M., 70, 72, 80
Werner, H., 320, 331
Westenberg, P. M., 167, 188, 211, 216, 221, 222, 224, 225
Westhoff, C., 106, 124
Wexler, A., 103, 123, 153, 163
Whalen, P. J., 83, 97, 102, 120
Wheatley, T., 272, 273, 280, 282, 285
Wheelwright, S., 111, 120, 125, 126, 127, 128, 129, 131, 133, 134, 135, 136, 138
Wheelwright, S. J., 157, 161
White, C. D., 213, 221
Whiten, A., 26, 31, 32, 37, 41
Whitesell, N. R., 143, 160
Wiens, S., 83, 95
Wiens S., 203, 205
Wiesner, L., 113, 114, 121
Wigfield, A., 143, 144, 149, 150, 151, 159, 160, 163
Wiggett, A., 106, 122
Wikan. U., 316, 331
Wilbarger, J., 32, 38
Wilder, L. S., 147, 161
Wilkinson, M., 69, 78
Williams, J., 31, 40
Williams, J. H., 31, 32, 41
Williams, K. D., 234, 235, 246
Williams, L., 65, 77
Williams, S. C. R., 87, 92, 97
Williamson, D. E., 166, 186
Wilson, E. O., 321, 331
Wilson, S. M., 19, 34
Wimmer, H., 23, 41, 127, 138
Wing, L., 110, 124, 132, 138
Wingert, L., 299, 311
Winkielman, P., 32, 38
Winocur, G., 147, 159
Wise, R. A., 172, 189
Wittgenstein, L., 300, 311
Wohlschläger, A., 25, 34
Wohlschläger, A. M., 20, 35, 41
Wolff, P. H., 90, 98
Woo, L. C., 69, 80
Woodruff, G., 23, 39, 99, 123
Woods, R. P., 20, 37, 107, 124

Woodward, A. L., 24, 41, 46, 47, 48, 49, 50, 57, 62
Woodward A., 24, 40
Wootton, J. M., 280, 288
Worsley, K., 196, 206
Wright, C. I., 173, 188
Wright, D., 296, 297, 298, 299, 311
Wright, H. F., 202, 204
Wyland, C. L., 103, 122, 147, 160, 161

X

Xu, F., 65, 66, 71, 80

Y

Yakovlev, P. I., 87, 97
Yang, Y., 68, 78, 292, 294, 295, 310
Yarger, R. S., 167, 187, 210, 213, 223
Yoon, K. S., 143, 144, 151, 163
Young, A. W., 100, 102, 104, 106, 109, 111, 120, 122, 272, 287
Young, L., 104, 110, 113, 114, 116, 117, 124, 273, 274, 275, 281, 282, 285, 288, 300, 308, 320, 322, 329, 330

Z

Zahn, R., 271, 286, 294, 299, 309
Zahn, T. P., 253, 266
Zanolie, K., 216, 224
Zayas, V., 210, 213, 222
Zeki, S., 107, 124
Zelazo, P. D., xiii–xiv, 3–9, 65, 67, 69, 73, 78, 79, 80, 81–98, 82, 84, 85, 86, 87, 88, 89, 90, 91, 92, 93, 94, 95, 96, 97, 98, 253, 266
Zelazo, P. R., 90, 91, 97
Zerwas, S., 58, 60
Ziegler, S., 280, 284
Zijdenbos, A., 169, 170, 186, 196, 198, 205, 206, 211, 212, 223
Zilbovicius, M., 21, 35, 114, 124
Zilles, K., 20, 35, 41
Zimmerman, R. D., 68, 78
Zoia, S., 21, 41
Zuckerman, M., 210, 225
Zvolensky, M. J., 167, 187
Zysset, S., 147, 163

SUBJECT INDEX

A

Amygdala, 83–4, 90, 101–102, 134, 146, 150–152, 155–157, 168–169, 171–176, 181, 184, 217, 219, 236, 261–261, 264–265, 270, 272–274, 276, 280, 292
Anterior cingulate cortex, 73, 84, 89, 134, 171, 173–174, 202–203, 213, 261, 333
Asperger syndrome, 125–129, 130, 131–132, 133–134
  extreme male brain theory and, 133–134
Autism, 6, 13, 30–32, 63, 66–67, 100, 106, 110–116, 119, 125–134, 142, 157–158, 277, 281
  empathizing-systemizing theory and, 6, 128–134
Autism spectrum disorder, 110, 142, 157, 125–126, 129, 131–134

C

Consciousness, 6, 86, 297, 317, 319, 320, *See* Self-awareness; Self-consciousness

D

DLPFC, *See* Dorsolateral prefrontal cortex
DMPFC, *See* Dorsal medial prefrontal cortex
Dopamine, 68–70, 169, 173
Dorsal medial prefrontal cortex, 145, 145–146, 148, 150, 153, 154–155
Dorsolateral prefrontal cortex, 146, 150, 152, 155

E

ERP, *See* Event-related potential
Event-related potential, 87–88, 92
Executive function, 6, 7, 64–77, 82, 87–88, 90–91, 116, 172, 228, 243, 253, 255, 258, 273, 275–277, 283

F

False belief, 14, 23, 63, 65, 66, 71–72, 127
fMRI, *See* Functional magnetic resonance imaging

Functional magnetic resonance imaging, 20, 29, 32, 92, 105–107, 110, 113, 115, 157, 176, 180–182, 212–213, 215–216, 231, 233–236, 251, 258–260, 263–264

G

Goal, 15, 17–29, 31, 33, 43, 45–46, 50–51, 55, 82–84, 86, 88, 106, 109–111, 113, 147, 153, 193–194, 217, 238, 240, 252–253, 257–258, 266, 276, 315–316, 322, 327

I

Imitation, 20, 22, 24, 26–27, 31–32, 44, 53, 128, 196
Intention, 6, 13, 14, 19, 21, 25, 28–30, 33, 43–56, 58–59, 64–65, 100–101, 104–105, 108–118, 142, 153, 199, 232–234, 257, 264, 277, 298, 305
Intentional action, 6, 21, 26, 30–31, 45–49, 51–53, 56, 59, 298
Intentional agent, 6, 26, 43–46, 49, 51–52, 54, 58–59
Intentional understanding, 15, 23, 25, 27, 45, 50–52, 54, 257, 305
Iterative reprocessing model, 6, 81–82, 84, 86

J

Joint attention, 44, 53–54, 111, 114, 126, *See also* Shared attention

M

Medial posterior parietal cortex, 101, 117–118, 145–146, 147–150, 152–157
Medial prefrontal cortex, 68–70, 73, 101, 102–104, 117, 145–146, 148–150, 153–154, 156–157, 171, 180, 181, 253, 255
Mirror neuron, 5–6, 14–23, 25, 27, 29, 31–33, 43–45, 52, 58, 59, 107

Morality, 7, 8, 242, 269, 271, 277, 279, 280–281, 282–283, 289–294, 295–296, 299, 300–301, 303–306
Moral judgment, 104, 314–315
Moral reasoning, 4, 7, 8, 99, 242, 270, 271–272, 279, 280–283, 290, 292–293, 294, 303
Motor System, 15, 18, 25, 28–29, 31
MPFC, *See* Medial prefrontal cortex
MPPC, *See* Medial posterior parietal cortex

O

Orbitofrontal cortex, 83–85, 88–89, 91, 101, 172–173, 180–182, 213–216, 219, 261

P

Perspective taking, 57, 145, 149, 153, 156, 200–202, 257–258, 261, 264, 266
PFC, *See* Prefrontal cortex
Prefrontal cortex, 212, 215, 218, 251, 254–258, 262, 265–266
Problem-solving, 231, 254
Psychopathy, 271, 275–276, 279–281, 292

R

Risk-seeking behavior, 182, 215
Risk-taking behavior, 166–168, 180, 200, 209–212, 214, 215, 217, 220

S

Self-awareness, 6, 57, 91, 142, 193, 198–200, 204, 232, 233, 258, 266
Self-consciousness, 157, 200, 201, 243
Shared attention, 23, 48, *See also* Joint attention; Triadic interaction
Superior temporal sulcus, 16, 100–101, 104–118, 108, 145–146, 153–154, 173, 196, 263–264

T

Theory of Mind, 4, 6, 23, 30, 43–77, 91, 99, 103–104, 105, 114–115, 119, 126, 128, 201, 253, 255, 256, 257, 258, 261, 264, 266, 277, 281, 282

ToM, *See* Theory of Mind
Triadic interaction, 44, 47, 52–55, 58–59
Triadic model, 7, 165, 170–176, 180–181, 183
Trolley problem, 273–275, 280, 282, 292, 323–325, 327–328